Fractal Physiology

THE AMERICAN PHYSIOLOGICAL SOCIETY
METHODS IN PHYSIOLOGY SERIES

White, *Membrane Protein Structure: Experimental Approaches,* 1994

Bassingthwaighte, Liebovitch, West, *Fractal Physiology,* 1994

Fractal Physiology

JAMES B. BASSINGTHWAIGHTE
Center for Bioengineering
University of Washington

LARRY S. LIEBOVITCH
Center for Complex Systems
Florida Atlantic University

BRUCE J. WEST
Physics Department
University of North Texas

NEW YORK   OXFORD
Published for the American Physiological Society by
Oxford University Press
1994

Oxford University Press

Oxford New York Toronto
Delhi Bombay Calcutta Madras Karachi
Kuala Lumpur Singapore Hong Kong Tokyo
Nairobi Dar es Salaam Cape Town
Melbourne Auckland Madrid

and associated companies in
Berlin Ibadan

Published for the American Physiological Society by
Oxford University Press, Inc.,
200 Madison Avenue, New York, New York 10016

Oxford is a registered trademark of Oxford University Press

Library of Congress Cataloging-in-Publication Data
Bassingthwaighte, James B.
Fractal physiology / James B. Bassingthwaighte,
Larry S. Liebovitch, Bruce J. West.
p. cm. — (Methods in physiology series; 2)
Includes bibliographical references and index.
ISBN 0-19-508013-0
1. Chaotic behavior in systems. 2. Fractals.
3. Physiology—Mathematical models.
4. Biological systems—Mathematical models.
I. Liebovitch, Larry S. II. West, Bruce J.
III. Title. IV. Series.
QP33.6.C48B37 1994 599'.01'0151474—dc20 94-13939

9 8 7 6 5 4 3 2 1
Printed in the United States of America
on acid-free paper

Preface

I know that most men, including those at ease with the problems of the greatest complexity, can seldom accept even the simplest and most obvious truth if it be such as would oblige them to admit the falsity of conclusions which they have delighted in explaining to colleagues, which they have proudly taught to others, and which they have woven, thread by thread, into the fabric of their lives.

Joseph Ford quoting Tolstoy (Gleick, 1987)

We are used to thinking that natural objects have a certain form and that this form is determined by a characteristic scale. If we magnify the object beyond this scale, no new features are revealed. To correctly measure the properties of the object, such as length, area, or volume, we measure it at a resolution finer than the characteristic scale of the object. We expect that the value we measure has a unique value for the object. This simple idea is the basis of the calculus, Euclidean geometry, and the theory of measurement.

However, Mandelbrot (1977, 1983) brought to the world's attention that many natural objects simply do not have this preconceived form. Many of the structures in space and processes in time of living things have a very different form. Living things have structures in space and fluctuations in time that cannot be characterized by one spatial or temporal scale. They extend over many spatial or temporal scales. As we magnify these structures, new and ever finer features are continuously revealed. When we measure a property, such as length, area, or volume, the value we find depends on how many of these ever finer features are included in our measurement. Thus, the values we measure depend on the spatial or temporal ruler we use to make our measurement.

To denote objects or processes with multiple-scale properties Mandelbrot (1977) chose the word "fractal." He coined "fractal" from Latin *fractus*, the past participle of *frangere*, "to break," in order to describe the ever finer irregular fragments that appear as natural objects and processes are viewed at ever higher magnification.

An understanding of fractals brings to us the tools we need to describe, measure, model, and understand many objects and processes in living things. Without these tools we simply cannot properly interpret certain types of experimental data or understand the processes that produced those data. In the past, our ability to understand many physiological systems was hampered by our failure to appreciate their fractal properties and understand how to analyze and interpret scale-free structures. Even now, we cannot foresee which systems are fractal or chaotic, or predict that the new tools can be applied, but when one *tries* to find the fractal or chaotic features we *almost always* learn something new about the system.

The values of the measured properties of many physiological systems look random. We are used to thinking that random looking fluctuations must be the result of mechanisms driven by chance.

However, we now know that not everything that looks random is actually random. In a deterministic system, the values of the variables depend on their previous values. There are simple deterministic systems where the fluctuations of the values of the variables are so complex that they mimic random behavior. This property is now called "chaos." The word chaos was chosen to describe the deterministic random fluctuations produced by these systems. Perhaps chaos is a poor word to describe these systems because in normal usage chaos means disordered. Here it means just the opposite, namely, a deterministic and often simple system, with output so complex that it mimics random behavior.

There are now methods to analyze seemingly random experimental data, such as a time series of experimental values, to determine if the data could have been generated by a deterministic process. If that is the case, the analysis is able to reconstruct the mathematical form of the deterministic relationship. These methods are based on the mathematics of nonlinear, dynamic systems. They use many of the properties and ideas of fractals.

Chaos brings us important new insights into our understanding of physiological systems. It gives us the tools to analyze experimental data to determine if the mechanisms that generated the data are based on chance or necessity. Examining the potential for nonlinear chaos in dynamic events often allows us to uncover the mathematical form of physiological mechanisms, and sometimes to predict how their functioning depends on the relevant parameters, and even, least commonly, at this stage of the development of the science, to determine the values and timing of control inputs that can be used to control the physiological system in a predetermined way.

Why fractals and chaos?

The great virtue of introducing the concepts of fractals and chaos is to change our mind set. These concepts serve to nag us into thinking of alternative approaches, and perhaps thereby to fulfill Platt's (1964) admonition to develop the alternative hypothesis in order to arrive at a "strong inference," the title of his article. Platt's idea, in pursuit of enhancing the "scientific method" used by most of us, was that we should not only have a clear hypothesis, expressible in quantitative terms, but also that we should have an alternative hypothesis. This forces the issue: the experimental design should be such that the results discriminate between the two hypotheses, and so long as both were realistic possibilities and one wasn't simply a straw man, science would be forced to advance by eliminating one of the hypotheses, maybe both. The idea is that pressure to think in a different style may lead to the *aha!* of new discovery. Fractals and chaos are alternative hypothetical approaches, vehicles for stimulating the mind. They augment the standard techniques that we are used to, for example making sure that everything adds up, that all the loops connect, and so on. The mere act of trying to devise explanations for natural phenomena in terms of fractals or chaos stimulates us to devise testable explanations that may be of a less traditional sort. This is sufficient to alleviate the worry Tolstoy (1930) expressed.

Both fractals and chaos are *mathematical* ideas. They augment the set of mathematical tools we grew up with. In addition to Euclid's geometry, there is another geometry, fractal, which seems closer to nature, is more complex, and has beauties of its own that challenge the stylish simplicity of Euclid. Fractals use much of mathematics but escape one constraint of Newtonian calculus: the derivatives need not be continuous and sometimes do not exist. The advantage is that fractals may apply at scales so fine that traditional approaches fail, just as quantum mechanics operates at scales unapproachable using continuum mechanics. Chaos too, while built upon linear systems analysis and control theory, gives recognition to failures of exact predictability in purely deterministic systems, despite their being exactly describable by a set of differential equations. Chaos can create fractal "structures," by which we mean maps or loci of the ever-changing state of a system. Though time-dependent signals may be fractal, for the most part we think of fractals as structural. Developmentally the two fields were separate, but in looking at the biology we try to take a traditional view, a fractal view, and a chaotic view, all at once. Students who have had some calculus and the beginnings of differential equations will find this book manageable.

Fractal physiology

In Part I we describe and illustrate the properties of fractals and chaos. Chapter 1 illustrates the wide importance of fractals in physiology by presenting a menagerie of the fractal forms found in living things. Chapter 2 is an introduction, at an elementary mathematical level, to self-similarity, scaling, and dimension. We show that these properties have important consequences for the statistical properties of fractal distributions that must be understood in order to analyze experimental data from fractal structures or processes. In the next four chapters we present a more detailed mathematical description of fractals and how to use them to analyze and interpret experimental data. In Chapter 3 we provide more details about self-similarity and scaling. In Chapter 4 we show how the statistical properties of fractals can be used to analyze correlations. These methods include dispersional analysis (from variances at different levels of resolution), rescaled range analysis, autocorrelation, and spectral analysis. In Chapter 5 we illustrate methods for generating fractals. Understanding the relationships that produce different types of patterns helps us focus in on the mechanisms responsible for different observed physiological patterns. In Chapter 6 we describe fluctuating signals and "chaotic" signals, asking if the fluctuations are different from random ones, and we describe the properties of chaos. In Chapter 7 we show how to use the methods of chaotic dynamics to analyze time series.

In Part II we present physiological applications of fractals and chaos and describe the new information these methods have given us about the properties and functioning of physiological systems. In Chapter 8 we show how the fractal analysis of the currents recorded through individual ion channels has provided new information about the structure and motions within ion channel proteins in the cell membrane. In Chapter 9 we show how fractals have been used to better understand the spread of excitation in nerve and muscle. In Chapter 10 we show how fractals have been used to clarify the flow and distribution of the blood in the heart and lungs. In Chapter 11 we show how an understanding of the rules that generate fractal patterns has helped uncover mechanisms responsible for the growth of organs and the vascular system. In Chapter 12 we provide a summary of mechanisms that generate fractals, to provide a shopping list of causes to consider when fractal patterns are observed. In Chapter 13 we show how methods developed for analyzing chaos in physical phenomena have been used to reveal the deterministic forms arising from seemingly random variations in the rate of beating of heart cells, and in fluctuations in ATP concentration generated by glycolysis.

Acknowledgments

We think of fractals and chaos as a new field. What is remarkable is how many people have been able to guide us in gaining knowledge of the field. These include some of the pioneers, including Benoit Mandelbrot, who phoned up to correct an error, Leon Glass, Danny Kaplan, Paul Rapp, James Theiler, Ary Goldberger, and Steve Pincus. Particularly helpful to the writing and rewriting have been some who reviewed the book or taught a course from it: Bernard Hoop at Harvard, Knowles A. Overholser, Sorel Bosan and Thomas R. Harris at Vanderbilt, Peter Tonelleto at Marquette, Alison A. Carter, Alvin Essig, Wolf Krebs, Leo Levine, Mark Musolino, Malvin C. Teich, Steven B. Lowen, and Tibor I. Toth. At the University of Washington, Spiros Kuruklis provided valued criticism, as did Gary Raymond and Richard King. Eric Lawson provided editing and typesetting skills that made it possible for us to give Oxford University Press photo-ready copy. Angela Kaake obtained, organized, and checked references. We thank the American Physiological Society for the stimulus to write under its auspices. Finally, we thank the authors and publishers of the many fine illustrations we have reproduced from the work of our predecessors, who are acknowledged in the figure legends.

Seattle, Washington	J.B.B.
Boca Raton, Florida	L.S.L.
Denton, Texas	B.J.W.

Contents

Part I

Overview

1

Introduction: Fractals Really Are Everywhere

> Clouds are not spheres, mountains are not cones, coastlines are not circles, and bark is not smooth, nor does lightning travel in a straight line.
>
> Mandelbrot (1983)

1.1 Fractals Are Everywhere

Fractals and chaos bring to us new sets of ideas and thus new ways of looking at nature. With this fresh viewpoint come new tools to analyze experimental data and interpret the results.

The title of Barnsley's 1988 book is *Fractals Everywhere*. It is true that fractals really *are* at least almost everywhere. The essential characteristic of fractals is that as finer details are revealed at higher magnifications the form of the details is similar to the whole: there is self-similarity. This introduction briefly presents a few examples of fractals from physiological systems, just to give a feel for the breadth of systems with fractal properties and their importance to living organisms. This brief list is by no means comprehensive. It is meant only to suggest the ubiquity of fractals, and thus the importance of understanding their properties in the study of physiology. The details of some of these examples and many others appear in Part II.

1.2 Structures in Space

Humans have but 100,000 genes made up from about 10^9 base pairs in DNA. There are about 250 different cell types in the human body, and each has a multitude of enzymes and structural proteins. The number of cells in the body is beyond counting. The number of structural elements in a small organ far exceeds the number of genes: the heart, for example, has about ten million capillary-tissue units, each composed of endothelial cells, myocytes, fibroblasts and neurons. The lung has even more. Consequently, the genes, which form the instruction set, must command the growth of cells and structures most parsimoniously and yet end up with functioning structures that last for decades. They must even contain the instructions for their own repair!

Genes do not specify these anatomical structures directly, but do specify the rules that generate these structures. Repeated application of these rules may then generate structures with ever finer detail. Almost all biological structures display such hierarchical form. Virtually all the vascular systems for organs are constructed like this. A diagram of a kidney's arterial supply is shown in Fig. 1.1. The branching is more or less recursive over a few generations. The branching is not strictly dichotomous, and the simplicity is marred in the kidney by arcades, the arcuate arteries. The glandular structures, the individual nephrons, discharge their effluents into a succession of coalescing ducts leading to the ureter. The vascular structures of glandular organs such as the salivary glands are arrayed to serve the secreting lobular arrangements of epithelial cells, and so are enslaved to the branching arrangement of the organ established by the apparatus for excreting selected solutes from those cells.

If one includes self-similarity in style rather than in a particular structure, then one can look at the alimentary system from the perspective of the sequence of structures from the largest to the smallest. An example of the sequence of ever finer functional structuring is seen in Fig. 1.2, ranging from the tortuosities of the small and large intestine down to the glycocalyx making up the brush borders of the secreting epithelial cells. At each successive level of resolution the observable surface area increases, so that the "fractal" may be surface area versus measuring stick length. In general, biological structures appear to be designed to provide huge surface areas for exchange and for reactions. The lung alveolar surface equals a tennis court (at the scale of alveolar diameters) but the lung holds only five liters of air, a very large ratio of surface to volume.

To uncover the mechanism that generates these fractal patterns, we compare the physiological patterns to the patterns generated by different recursion rules. One simple recursion rule is illustrated in Fig. 1.3. The terminal segments of an asymmetric Y-branch are replaced with each successive generation. In gross form the structure looks the same at each level of resolution, even though it is increasingly complex with more generations. The end-terminal branch is exactly the same as the original simple Y, the root template for the recursion. So we can say it shows self-similarity in two different ways: in the similarity of the overall structure to the initial template defining the recursion rule, and in the reproduction of exactly similar terminae after many generations.

1.3   Processes in Time

The values measured for many physiological processes fluctuate in time. Very often there are ever smaller amplitude fluctuations at ever shorter time scales. This is illustrated in Fig. 1.4 by the time series of values of the volumes of breaths recorded by Hoop et al. (1993). The figure shows tidal volumes for a series of breaths in a rat. The series show moderate fluctuations in the volumes for most of the run, and also

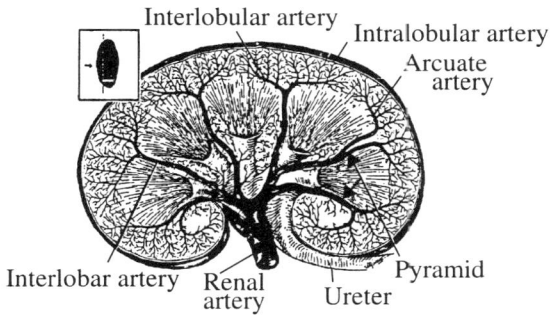

Figure 1.1. Arterial branching system in the kidney in a simplified diagrammatic form. (From Ham, 1957, with permission.)

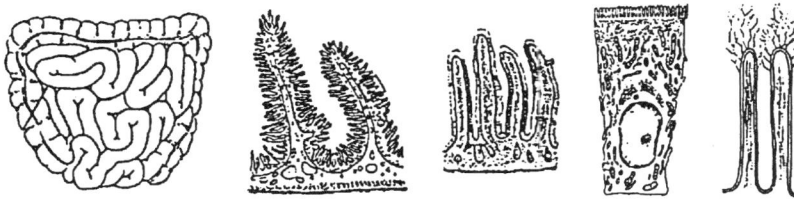

Figure 1.2. The fractal alimentary tract. This "fractal" shows similarity in the scaling, but in actuality each level of scaling can be seen to serve a different aspect of the overall function of the gut. In a strict sense this is not fractal, since no simple recursion rule will serve to define the whole structure, but there is an overall functional self-similarity in the succession of structures and phenomena down to the fundamental functions of secretion and absorption by epithelial cells. (From Bloom and Fawcett, 1975, with permission.)

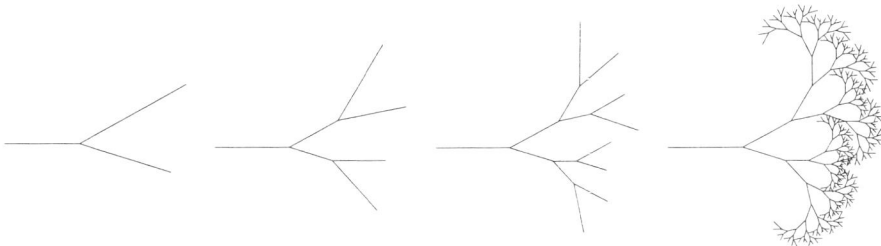

Figure 1.3. A self-similar or fractal recursion. One, two, three, and eight generations are shown. At each generation, the pairs of terminal segments are replaced by a stem and two branches, so this is a pure dichotomous branching fractal, a binary tree, with 2^N terminae after N generations.

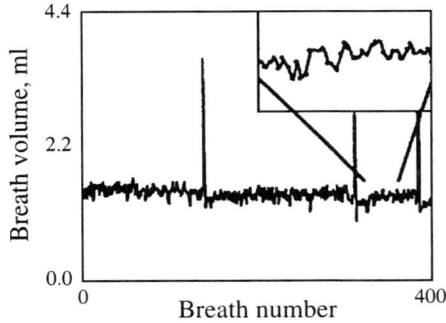

Figure 1.4. Fluctuations in respiratory tidal volume in a 300 gram rat. The inset shows an expanded segment of the volume versus time signal illustrating the similarity in fluctuation at a different scale. (From Hoop et al., 1993, with permission.)

three very large "sighs" of greater volume. An idea of self-similarity is conveyed by the inset graph over a short period, for it shows the same kind of variability.

The electrical state of a cell depends on a complex interaction of the extracellular environment, the ion pumps, exchangers, and channels in the cell membrane, and how their function is modified by the end products of the biochemical pathways triggered by the activation of receptors in the cell membrane. Thus, measurements of the voltage and current across the cell membrane provide important information on the state of the cell. In order to understand the changes that occur when T lymphocytes are stimulated, Churilla et al. (1993) first needed to characterize the electrical properties of inactivated T lymphocytes. They used the whole cell patch clamp technique to measure the voltage fluctuations in inactivated cells. As shown in Fig. 1.5, there are ever smaller fluctuations in the voltage at ever finer time scales.

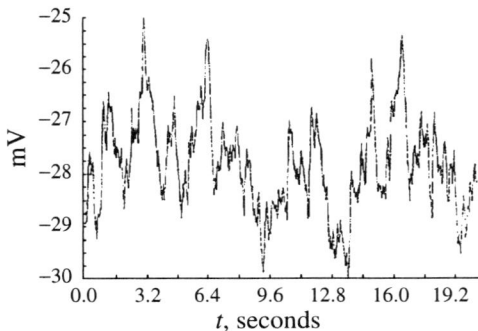

Figure 1.5. Voltage fluctuations shown were recorded using the whole cell patch technique across the cell membrane of inactivated T lymphocytes. Ever smaller fluctuations at ever shorter time scales appear. (Figure from Churilla et al., 1993, with permission.)

Time series are the set of values of a measurement as a function of time. A list of the values of a measurement as a function of one other independent variable can also be thought of and analyzed as a time series. In general all one-dimensional signals can be analyzed using the same sets of tools: for example, the properties of DNA base pairs have been analyzed in this way as a function of length along the DNA. Peng et al. (1992) determined the value x as a function of the distance along a gene, as well as much longer segments of DNA. At the location of each base pair, a variable, x, was increased by one if the base was a pyrimidine, and decreased by one if it was a purine. The "time series" of x values generated in this way is shown in Fig. 1.6. Once again, the fractal form of ever finer fluctuations at ever finer scales is revealed. Voss (1992) also found a similar result analyzing the correlations in all four types of base pairs. These studies revealed for the first time that there are long-term correlations in the base pairs along the length of DNA at scales much larger than the size of genes. These results show that we do not yet understand all the ways information is encoded in DNA.

1.4   The Meaning of Fractals

In all these fractals, structures at different spatial scales or processes at different time scales are related to each other. Fractal analysis can describe and quantify these correlations and suggest mechanisms that would produce them. In Part I we describe in detail the properties that characterize these fractal forms and how those properties can be determined from experimental data. In Part II we give detailed examples of how fractal analysis has led to an understanding of physiological systems that would not otherwise have been possible.

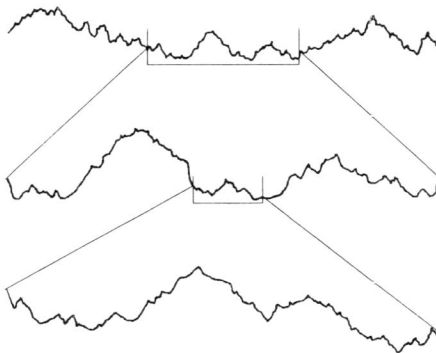

Figure 1.6. DNA sequences mapped at three levels of resolution. The abscissa is the sequence number of nucleic acid bases; the ordinate is a cumulative value starting at zero, increasing by one if the base is a pyrimidine, decreasing by one if it is a purine. (Figure drawn from Peng et al., 1992, by Amato, 1992, reproduced from *Science* with permission.)

Part II

Properties of Fractals and Chaos

2

Properties of Fractal Phenomena in Space and Time

> If the only tool you have is a hammer, you tend to treat everything as if it were a nail.
>
> Abraham Maslow (1966)

2.1   Introduction

As the epigraph suggests, the ability to interpret the full meaning of experimental data is limited by the types of tools that can be used in the analysis. A new analytical tool enlarges the analytical window into the world. The features of the world, previously unseen, that can be made visible by the new instrument of fractals are beautifully surprising. These features give us new insight into anatomical structures and physiological processes.

Consider the work of the physicist L. F. Richardson. He was a Quaker who served as an ambulance driver during World War I. He wondered if the causes of wars would be clearer if we could find the measurable properties of nations or peoples that correlated with the occurrence of such "deadly quarrels." For example, are there more wars between nations that share longer common borders? To answer this question required that he measure the length of national borders, which he did, and which he described in an article that was published seven years after his death (Richardson, 1961).

Although one might think measuring such borders dull work, Richardson found that these borders had fascinating properties that had not been previously appreciated. For example, he studied the west coast of Great Britain on page 15 of the *Times Atlas* of 1900. At coarse resolution the coast is jagged. One might think that as the coastline is examined at finer resolution these jagged segments would be resolved, and thus appear smooth. But that does not happen. As we look closer and closer, we see more and more detail, and smaller and smaller bays and peninsulas are resolved. *At all spatial scales, the coastline looks equally jagged. An object whose magnified pieces are similar to the whole in this way is self-similar.*

The self-similar nature of the coast has quite an interesting effect on the measurement of the length of the coastline. To measure the length of the west coast of Great Britain, Richardson used a divider, which is like a compass used to draw circles, except that it has sharp points at the end of both arms. He kept a fixed distance between the two ends of the divider. He placed the first end on the coastline

11

and swung the second end around until it touched the closest point on the coastline. Then, keeping the second end fixed where it had landed, he swung the first end around until it touched the closest point on the coastline. He continued this procedure all around the perimeter of the coast. The total length of the coast was then given by the number of steps multiplied by the distance between the ends of the divider. When he reduced the distance between the ends of the dividers and repeated the entire measurement of the coastline, the newly resolved small bays and peninsulas increased the measured length of the coastline. Thus, there is no correct answer to the question "How long is the west coast of Great Britain?" Rather, *the length one measures depends on the size of the ruler used to do the measurement.* That is, the length *scales* with the resolution of the instrument used to measure it.

These two properties of *self-similarity* and *scaling* are characteristic of objects in space or processes in time that are called *fractals.* The word "fractal" was coined by Mandelbrot (1977). He assembled, discovered, and popularized a wide collection of such objects, including some whose properties have been studied over the last 300 years. In this chapter we summarize the properties of fractals, and illustrate them with physiological examples. These properties are 1) *self-similarity,* which means that the parts of an object resemble the whole object, 2) *scaling,* which means that measured properties depend on the scale at which they are measured, 3) *fractal dimension,* which provides a quantitative measure of the self-similarity and scaling, and 4) the surprising *statistical properties* of certain fractals and their implications for the design of experiments and the interpretation of experimental results. We describe these properties both in a qualitative way and in a quantitative way using simple algebraic arguments. (More technical mathematical details will be presented in "Asides" which the reader may want to read or avoid.)

2.2   Self-Similarity: Parts That Look Like the Whole

Geometric Self-Similarity

We can construct *geometrically self-similar* objects *whose pieces are smaller, exact duplications of the whole object* (Mandelbrot, 1983). Such geometrically self-similar objects are illustrated in Fig. 2.1. The self-similarity of the geometric form of these objects is not prescribed by an algebraic function, but rather it is specified by means of an algorithm that instructs us how to construct the object. In Fig. 2.1, the original objects are shown on the left. Then each object is shown after one iteration, and then after two iterations of the generating algorithm. For example, at the top, we start with a line segment and remove the middle third of the line segment, and then repeatedly remove the middle third of each remaining piece. The result when this procedure is carried on forever is called the middle third Cantor set. The iterated algorithm for the Koch curve shown in the middle of the figure is to repeatedly add to each edge an equilateral triangle whose new sides are one third the

Figure 2.1. Fractals that are geometrically self-similar. Their form is specified by an iterative algorithm that instructs us how to construct the object. Starting from an initial object, the first three levels of iteration are shown in the construction of three geometrically self-similar fractal objects. *Top:* The iterative algorithm to generate the middle third Cantor set is to repeatedly remove the middle third of each line segment. *Middle:* The iterative algorithm to generate the Koch curve is to repeatedly add to each edge an equilateral triangle whose sides are one third the length of each edge. *Bottom:* The iterative algorithm to generate the Sierpinski triangle is to repeatedly remove triangles that are one quarter the area of each remaining triangle.

length of each old edge. The length of the perimeter of the Koch curve increases by four thirds at each stage of the iteration. The iterated algorithm for the Sierpinski triangle shown at the bottom is to repeatedly remove triangles that are one quarter the area of each remaining triangle.

Statistical Self-Similarity

The pieces of biological objects are rarely exact reduced copies of the whole object. Usually, *the pieces are "kind of like" the whole.* Rather than being geometrically self-similar, they are *statistically self-similar.* That is, the *statistical properties of the pieces are proportional to the statistical properties of the whole.* For example, the average rate at which new vessels branch off from their parent vessels in a physiological structure can be the same for large and small vessels. This is illustrated in Fig. 2.2 for the arteries in the lung (Glenny, unpublished, 1990). Measurements recorded from physiological systems over time can also be fractal. As shown in Fig. 2.3, the current recorded through one ion channel by Gillis, Falke, and Misler shows that there are statistically self-similar bursts within bursts of the opening and closing of these ion channels (Liebovitch and Tóth, 1990).

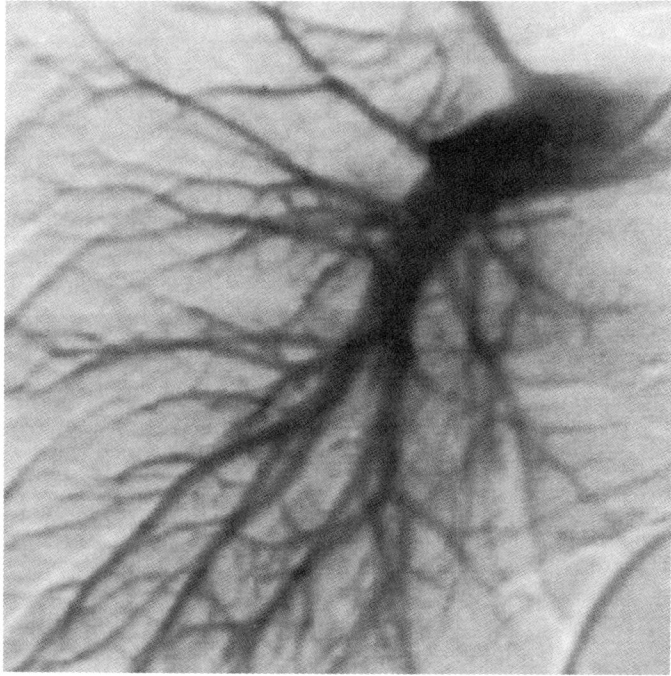

Figure 2.2. Statistical self-similarity in space. In the lung arterial tree, the branching patterns are similar for vessels of different sizes (From R. Glenny, unpublished, with permission.)

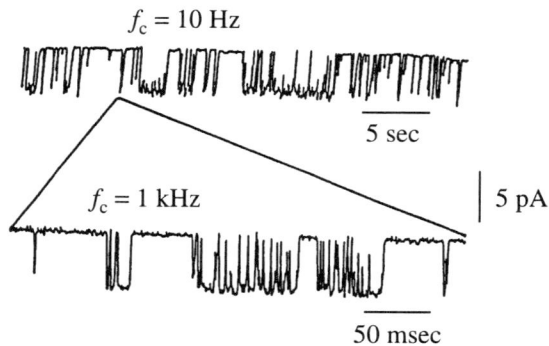

$f_c = 10$ Hz

5 sec

$f_c = 1$ kHz

5 pA

50 msec

Figure 2.3. Statistical self-similarity in time. The recordings made by Gillis, Falke, and Misler of the currents through an ion channel show that there are bursts within bursts of ion channel openings and closings revealed when the current through an individual ion channel is examined at higher temporal resolution. (This figure by Gillis, Falke, and Misler was published in Liebovitch and Tóth, 1990. Reprinted here with permission.)

Examples of Self-Similarity

Many physiological objects and processes are statistically self-similar. Some examples include:

1. Structures with *ever finer invaginations at ever smaller scales* to increase the surface area available for transport. Such structures include the linings of the intestine and the placenta (Goldberger et al., 1990).
2. Systems where the *branching pattern* is similar at different spatial scales. These can be found in the dendrites in neurons (Smith, Jr., et al., 1988; Caserta et al., 1990), the airways in the lung (West et al., 1986), the ducts in the liver (Goldberger et al., 1990), the blood vessels in the circulatory system (Kassab et al., 1993), and the distribution of flow through them (Bassingthwaighte and van Beek, 1988).
3. Processes that occur in a *hierarchical structure*. For example, proteins have different stable shapes called conformational states. These conformational states are separated by energy barriers. Small energy barriers separate shapes that differ in small ways and larger energy barriers separate shapes that differ in larger ways. Thus, there is a hierarchical series of ever larger energy barriers separating ever more different shapes (Ansari et al., 1985; Keirsted and Huberman, 1987), which may be the reason that the kinetics of some proteins, the time course of the changes from one shape to another, have fractal properties (Liebovitch et al., 1987).
4. Processes where local interactions between neighboring pieces produce a global statistical self-similar pattern are called "self-organizing." For example, this can happen in a chemical reaction if the time delays due to the diffusion of substrate are comparable to the time required for an enzymatic reaction. Such patterns can be produced at the molecular level, as in the binding of ligands to enzymes (Kopelman, 1988; Li et al., 1990), at the cellular level, as in the differentiation of the embryo (Turing, 1952), and at the organism level, as in slime mold aggregation (Edelstein-Keshet, 1988).

Mathematical Description of Self-Similarity

Statistical self-similarity means that a property measured on a piece of an object at high resolution is proportional to the same property measured over the entire object at coarser resolution. Hence, we compare the value of a property $L(r)$ when it is measured at resolution r, to the value $L(ar)$ when it is measured at finer resolution ar, where $a < 1$. Statistical self-similarity means that $L(r)$ is proportional to $L(ar)$, namely,

$$L(ar) = k L(r), \tag{2.1}$$

where k is a constant of proportionality that may depend on a.

[Aside]

Statistical self-similarity means the distribution of an object's piece sizes is the same when measured at different scales. Statistical self-similarity is more formally defined in terms of the distribution of values of a measured variable, its probability density function, *pdf*, that gives the number of pieces of a given size of an object. An object is statistically self-similar if the distribution determined from the object measured at scale r has the same form as that determined from the object measured at scale ar, namely, if $pdf\,[L(ar)] = pdf\,[kL(r)]$, so that one *pdf* is scaled to give the same shape as the other. Note that the probability density is given by the derivative of the probability distribution function.

2.3   Scaling: The Measure Depends on the Resolution

What You Measure Depends on the Resolution at Which You Look

We have seen that the lengths of the individual features of a coastline, and thus the total length of the coastline, depend on the measurement resolution. There is no single value for the length of the coast: length scales with measurement resolution. The value measured for any property of an object depends on the characteristics of the object. When these characteristics depend on the measurement resolution, then *the value measured depends on the measurement resolution. There is no one "true" value for a measurement.* How the value measured depends on the measurement resolution is called the *scaling* relationship. Self-similarity specifies how the characteristics of an object depend on the resolution and hence it determines how the value measured for a property depends on the resolution. Thus, *self-similarity determines the scaling relationship.*

Self-similarity Can Lead to a Power Law Scaling

The self-similarity relationship of Eq. 2.1 implies that there is a scaling relationship that describes how the measured value of a property $L(r)$ depends on the scale r at which it is measured. As shown in the following Aside, the simplest scaling relationship determined by self-similarity has the power law form

$$L(r) \;=\; A\,r^{\alpha},\qquad\qquad(2.2)$$

where A and α are constant for any particular fractal object or process.

  Taking the logarithms of both sides of Eq. 2.2 yields

$$\log L(r) \;=\; \alpha\log(r) + b,\qquad\qquad(2.3)$$

where $b = \log A$. Thus, *power law scalings* are revealed as *straight lines when the logarithm of the measurement is plotted against the logarithm of the scale at which it is measured.* This is shown in Fig. 2.4. On the left, the areas measured of mitochondrial membranes are shown to be a power law of the micrograph magnification (Paumgartner et al., 1981). Thus, as the spatial resolution is increased, finer infoldings are revealed that increase the total membrane area. On the right, now considering events in time rather than in space, the effective kinetic rate constant, the probability per second for an ion channel, such as a voltage-dependent potassium channel from cultured hippocampal neurons, to switch from closed to open states, is shown to be an inverse power law of the temporal resolution (Liebovitch and Sullivan, 1987), i.e., $\alpha < 0$ in Eq. 2.2. Thus, the faster one can observe, the faster one can see the channel open and close.

In the analysis of experimental data, a scaling can only extend over a finite range from a minimum to a maximum resolution size. In some cases, these limits are set by the constraints of the measurement technique. In other cases, they are set by the limits of the physiological object or process that produced the data.

The existence of *scaling has important implications for interpretation of experimental measurements from self-similar physiological systems:* 1) There is *no single "true" value* of certain measurements. The value measured depends on the

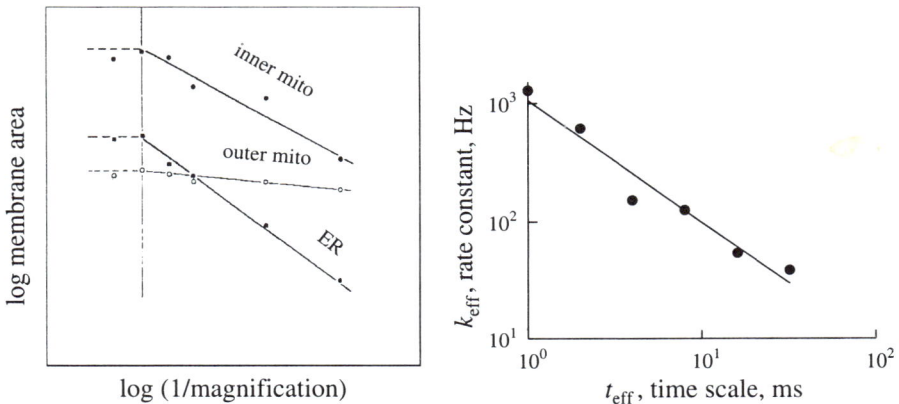

Figure 2.4. Power law scaling relationships in space and time are one form of scaling produced by self-similarity. *Left:* Paumgartner et al. (1981) found that the area measured on electron micrographs for inner and outer mitochondrial membranes and endoplasmic reticulum is a power law of magnification used (from Paumgartner et al., 1981, Fig. 9). As spatial resolution is increased, finer infoldings are recorded that increase the total membrane area measured. *Right:* Liebovitch and Sullivan (1987) measured transmembrane currents and found that the effective kinetic rate constant, which is the probability that an ion channel, such as a voltage-dependent potassium channel from cultured hippocampal neurons, switches from closed to open, is a power law of the time resolution used to analyze the data.

scale used for the measurement. 2) This may mean that *discrepancies* of the values between different investigators measuring the same physiological systems may be due to the *measurements having been done at different resolutions.* Paumgartner et al. (1981) point out this may be one reason the published values of cell membrane areas are markedly different. Thus, it is particularly important that the publication of such values clearly states the resolution at which measurement was made. 3) It may also mean that the *scaling function* that describes how the values change with the resolution at which the measurement is done *tells more about the data than the value of the measurement at any one resolution.*

 Power law scalings that are straight lines on log-log plots are characteristic of fractals. Although not all power law relationships are due to fractals, *the existence of such a relationship should alert the observer to seriously consider if the system is self-similar.* Some scientists derisively joke that everything is a straight line on a log-log plot, and thus such plots should be ignored. In fact, the joke is on them. Such plots are significant *because* they occur so often. They reflect the fact that so many things in nature are fractal.

[Aside]

Scaling relationships determined from self-similarity. Self-similarity   implies the existence of a scaling relationship. In general the scaling parameter k in Eq. 2.1 depends on the resolution scale a. The rule for self-similarity is that there is for some measure a constant ratio of the measure at scale r compared to that at scale ar:

$$L(r)/L(ar) = k \quad \text{for} \quad a < 1. \tag{2.4}$$

Suppose that, as for the Koch curve, there is a power law, such that

$$L(r) = Ar^{\alpha}; \tag{2.5}$$

then, by substitution,

$$kL(ar) = kA(ar)^{\alpha}, \tag{2.6}$$

and Eq. 2.4 can be rewritten:

$$Ar^{\alpha} = kA(ar)^{\alpha}, \tag{2.7}$$

$$k = r^{\alpha}/(ar)^{\alpha} = 1/a^{\alpha} = a^{-\alpha}. \tag{2.8}$$

From Eq. 2.4:

$$L(ra) / L(r) = 1/k = a^{\alpha}.$$ (2.9)

From Eq. 2.6:

$$L(ar) = A(ar)^{\alpha}$$
$$= A a^{\alpha} r^{\alpha}.$$ (2.10)

Because $A r^{\alpha} = L(r)$,

$$L(ar) = L(r) a^{\alpha},$$ (2.11)

and defining $\alpha = 1 - D$,

$$L(ar) = L(r) a^{1-D}.$$ (2.12)

Examples of Power Law Scaling

In addition to those shown in Fig. 2.4, other examples of power law scaling include the diameter of the bronchial passages with subsequent generation of branching in the lung (West et al., 1986); the length of the transport pathway through the junctions between pulmonary endothelial cells (McNamee, 1987, 1990); the time course of chemical reactions if the time delays due to the diffusion of substrate are long compared to the time required for an enzymatic reaction (Kopelman, 1988; Li et al., 1990); the clearance curves which measure the decay with time of the concentration of marker substances in the plasma (Wise and Borsboom, 1989).

A clever "amusing musical paradox" based on the scaling caused by self-similarity was reported by Schroeder (1986). He describes a fractal waveform whose many peaks are produced by adding together many sine waves. If the waveform is played back on a tape recorder at twice its normal speed, all the peaks are shifted to higher frequency. Some of the previously heard notes will be increased in frequency beyond the hearing range and will not be heard, while lower frequencies that were not previously audible will now become audible. Thus, the frequencies heard depend on which frequencies are shifted into the audible range, which depends on the scaling. The scaling was designed so that *when the tape is played* at *higher speed* it produces a sound that seems *lower in frequency*.

Scaling in mammalian physiology, from "mouse to elephant," has been examined fruitfully for metabolism and structure (Thompson, 1942; Schmidt-Nielsen, 1984). An example is scaling of oxygen consumption in mammals ranging in size from a few grams to hundreds of kilograms; the data of Taylor are shown in Fig. 2.5 (McMahon and Bonner, 1983). Such data have been analyzed by Sernetz et al.

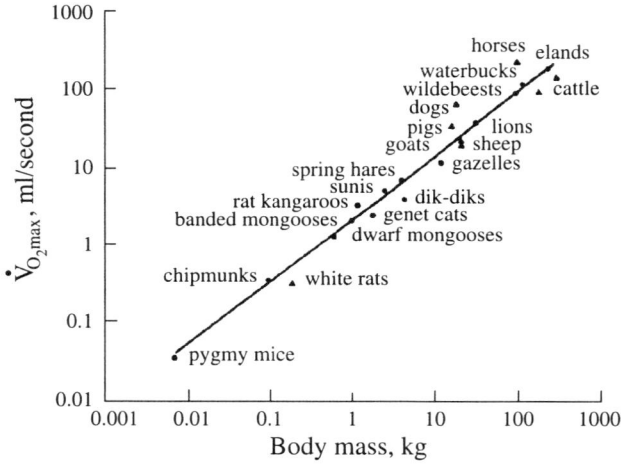

Figure 2.5. Maximal oxygen consumption rates versus body mass of mammals. Data of C. R. Taylor, et al. (1980). *Circles:* Fourteen wild animals; *triangles:* seven lab or domestic species. (Large and small cattle are distinguished.) These maximum rates are ten times basal rates in all animals studied. The slope b (of Eq. 2.14) is 0.809 (thick line) and is not statistically different from the theoretically expected 0.75, but is definitely lower than a fractal excess exponent of 1.0. (Figure provided courtesy of C. R. Taylor et al., 1980.)

(1985) from the viewpoint of the body as a bioreactor in which internal surfaces are the sites of reactions. The generalized proposal is that a metabolic rate increases with body weight, but less in proportion to mass or volume, so that there is "topological excess," meaning that in order to accommodate the metabolic needs the surface areas for metabolism increase at less than length cubed:

$$\text{Rate} = k \cdot M^b$$
$$= k \cdot L^{3b}, \tag{2.13}$$

where Rate is a specific metabolic rate, basal or maximal, k is some constant over many animal sizes, and the exponent b is the topological excess. This implies that the surfaces are fractals with fractal dimension equal to the Euclidean surface dimension plus 1 minus b:

$$D = E + 1 - b. \tag{2.14}$$

With $b = 0.75$, the fractal surface dimension D is 2.25, greater than its Euclidean dimension of 2.

2.4   Fractal Dimension: A Quantitative Measure of Self-Similarity and Scaling

Self-Similarity Dimension

How many new pieces are observed when we look at a geometrically self-similar object at finer resolution? The properties of self-similarity and scaling can be assessed in a quantitative way by using the fractal dimension. There are many different definitions of "fractional or fractal dimension," so called because it has noninteger values between Euclidean dimensions. We will start with definitions that are simpler but limited to specific cases, and then present more sophisticated and more general versions.

When a geometrically self-similar object is examined at finer resolution, additional small replicas of the whole object are resolved. *The self-similarity dimension describes how many new pieces geometrically similar to the whole object are observed as the resolution is made finer.* If we *change the scale by a factor* F, and we find that there are N *pieces similar to the original*, then the *self-similarity dimension* $D_{self\text{-}similarity}$ is given by

$$N = F^{D_{self\text{-}similarity}}. \tag{2.15}$$

We illustrate below why it is reasonable to think of the $D_{self\text{-}similarity}$ given by Eq. 2.15 as a dimension. We can change Eq. 2.15 into another equivalent form by taking the logarithm of both sides and solving for $D_{self\text{-}similarity}$, to find that

$$D_{self\text{-}similarity} = \log N / \log F. \tag{2.16}$$

For example, in the middle illustration of Fig. 2.6, when the scale is reduced by a factor of $F = 3$, then the square is found to have $N = 9$ pieces, each of which is similar to the original square. Hence, from Eq. 2.15 we find that $9 = 3^2$, and thus $D_{self\text{-}similarity} = 2$, or from Eq. 2.16 we find that $D_{self\text{-}similarity} = \log 9 / \log 3 = 2$.

Fig. 2.6 also illustrates why the self-similarity dimension is called a "dimension." Consider objects that are r long on each side. The length of a one-dimensional line segment is equal to r. If we reduce the scale by a factor F, then the little line segments formed are each $1/F$ the length of the original. Hence, F^1 of such pieces are needed to occupy the length of the original line segment, and $D_{self\text{-}similarity} = 1$. The area of a two-dimensional square is equal to r^2. If we reduce the scale by a factor F, then the little squares formed are each $1/F^2$ the area of the original. Hence, F^2 of such little squares are needed to occupy the area of the original square, and $D_{self\text{-}similarity} = 2$. The volume of a three-dimensional cube is equal to r^3. If we reduce the scale by a factor F, then the little cubes formed are each $1/F^3$ the volume of the original cube. Hence, F^3 of such pieces are needed to occupy the volume of the

If N pieces similar to the original are produced, when the length scale is changed by a factor F, *then* $N = F^D$, *and* $D = \log N / \log F$:

$D = 1$ ⟹

3 similar pieces are found when the length scale is reduced by one third:
$N = 3, F = 3, 3 = 3^1$, thus $D = 1$

$D = 2$ ⟹

9 similar pieces are found when the length scale is reduced by one third:
$N = 9, F = 3, 9 = 3^2$, thus $D = 2$

$D = 3$ ⟹

27 similar pieces are found when the length scale is reduced by one third:
$N = 27, F = 3, 27 = 3^3$, thus $D = 3$

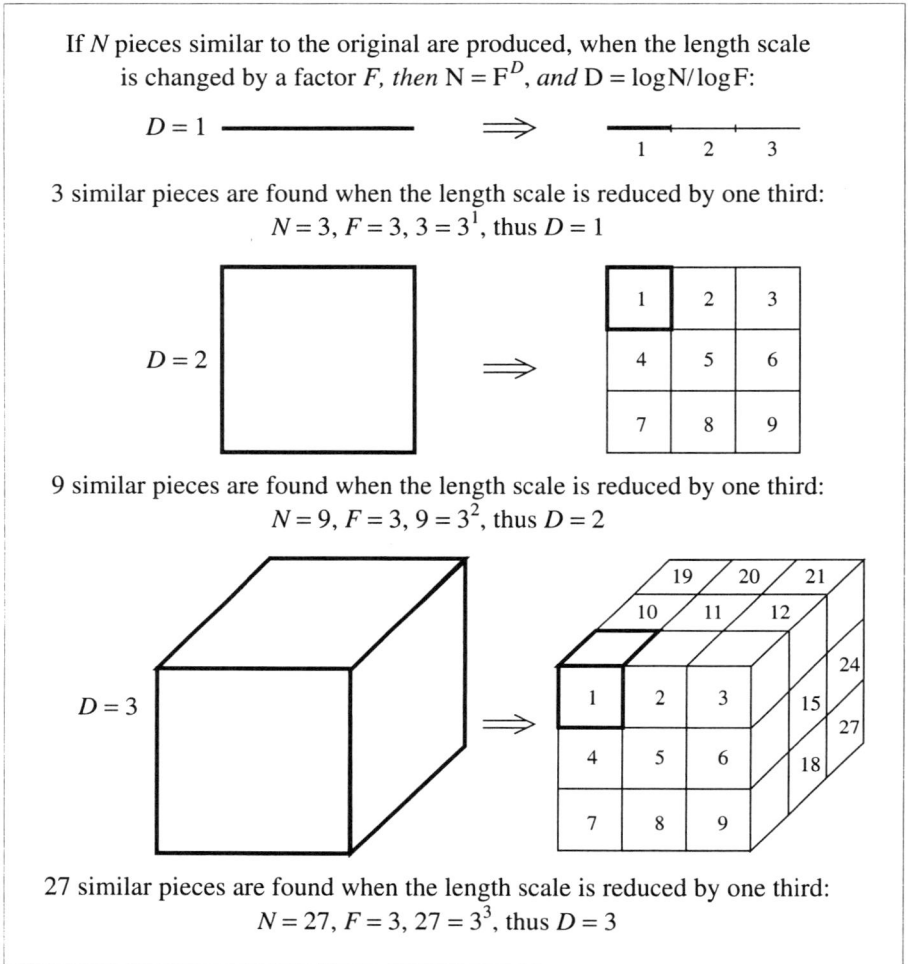

Figure 2.6. The self-similarity dimension tells us how many new pieces similar to the whole object are observed as the resolution is made finer. Objects with self-similarity dimensions of 1, 2, and 3 correspond to our usual ideas that one-dimensional objects occupy lines, two-dimensional objects occupy areas, and three-dimensional objects occupy volumes.

original cube, and $D_{\text{self-similarity}} = 3$. Thus, the self-similarity dimension is consistent with our usual notions of the properties of integer dimensional objects.

For the perimeter of the Koch curve shown in Fig. 2.7, when the scale is reduced by a factor $F = 3$, then $N = 4$ pieces are found. Thus, from Eq. 2.16 we find that $D_{\text{self-similarity}} = \log 4 / \log 3 \approx 1.2619$. That is, the properties of this curve are between those of a one-dimensional line segment and a two-dimensional area. A one-dimensional line segment of finite length occupies only a one-dimensional space and not an area. An infinitely long line segment might be so wiggly that it could

By self-similarity:

Number of pieces: 1 $\Longrightarrow$

$N = 4$ similar pieces are found when the length scale is reduced by a factor $F = 3$.

$$N = F^D \quad \Longrightarrow \quad D = \frac{\log N}{\log F} = \frac{\log 4}{\log 3} = 1.2619$$

By scaling:

scale: r $r/3$

The length is 4/3 its original value when the scale is reduced by a factor 3.

$$L(r) = Ar^{1-D} \qquad L(r/3) = (4/3)\,L(r)$$
$$L(r) = Ar^{1-D}$$
$$A(r/3)^{1-D} = (4/3)Ar^{1-D}$$
$$(1/3)^{-D} = 4$$
$$D = \log 4/\log 3 = 1.2619$$

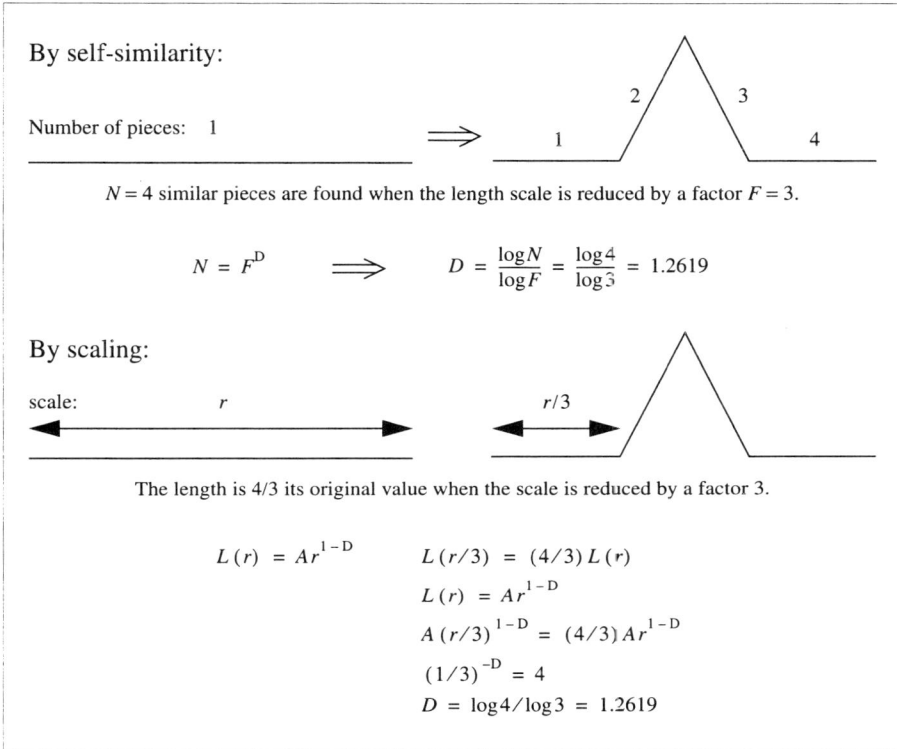

Figure 2.7. The fractal dimension of the perimeter of the Koch curve can be determined from self-similarity (*top*) or from power law scaling (*bottom*).

touch every point in a two-dimensional plane and thus occupy an area. The Koch curve is between these two cases. It has a dimension that is not an integer such as 1 or 2. It has a fractional dimension. The perimeter of the Koch curve is also infinitely long. As we examine it under increasing magnification we find that it has more and more pieces. There are enough such pieces that it occupies more than just a one-dimensional line segment, but not so many pieces that it occupies a two-dimensional area. The word "fractal" was coined to remind us that such objects have many *fragments* and can be characterized by a *fractional dimension* (Mandelbrot, 1983), which usually takes on *noninteger* values and which therefore lies between Euclidean dimensions.

Hausdorff-Besicovitch Dimension and Capacity Dimension

Generalizations of the self-similarity dimension can be used to analyze statistically self-similar objects. The self-similarity dimension can only be used to analyze geometrically self-similar fractals where the pieces are smaller exact copies of the whole. Thus, we need a generalization of the self-similarity dimension such as the *Hausdorff-Besicovitch dimension* or the *capacity dimension* (Falconer, 1985; Barnsley, 1988; Edgar, 1990) to determine the fractional dimension of irregularly shaped objects. These two dimensions are quite similar and the technical differences between them are described in the following Aside. We use the word "fractal" dimension to refer to these definitions of dimension.

The ideas that underlie these dimensions are illustrated in Fig. 2.8. The dimension of the space that contains the object is called the *embedding dimension*. This space will have dimension 1, 2, or 3. A "ball" consists of all the points within a distance r from a center. In one dimension balls are line segments, in two dimensions balls are circles, and in three dimensions balls are spheres. We use balls whose dimension equals the embedding dimension of the space that contains the object. We "cover" the object with balls of radius r. We make sure that every point in the object is enclosed within at least one ball. This may require that some of the balls overlap. We now find $N(r)$, the minimum number of balls of size r needed to cover the object. *The capacity dimension, D_{cap}, tells us how the number of balls needed to cover the object changes as the size of the balls is decreased.* That is,

$$D_{cap} = \lim_{r \to 0} \log N(r) / \log(1/r) . \tag{2.17}$$

The reference situation implied here is that at $r = 1$, one ball covers the object. A clearer definition is

$$D_{cap} = \log [N(r)/N(1)] / \log(1/r) , \text{ or in general,} \tag{2.18}$$

$$D_{cap} = \log [N(ra)/N(a)/\log(a/ra)] , \text{ and} \tag{2.19}$$

$$D_{cap} = -d\log N(r)/d\log r . \tag{2.20}$$

The capacity dimension of Eq. 2.17 is a generalization of the self-similarity dimension of Eq. 2.16 that is based on: 1) counting the number of balls needed to cover all the pieces of the object, rather than the number of exact small replicas of the whole object, 2) using the size r that is equal to the reciprocal of the scale factor F, and 3) taking the limit as the size of the balls becomes very small.

If the capacity dimension is determined by using balls that are contiguous nonoverlapping boxes of a rectangular coordinate grid, then the procedure is called "box counting." That is, we count $N(r)$, the number of boxes that contain at least one point of the object for different box sizes r and then determine the fractal dimension from Eq. 2.17. There are very efficient algorithms available for such box counting (Liebovitch and Tóth, 1989; Block et al., 1990; Hou et al., 1990).

[Aside]

Some mathematics of fractal dimensions. Below are some brief descriptions of different ways of defining and calculating fractional dimensions and references that describe them in greater detail:

D_{cap} = capacity dimension

$N(r)$ = minimum number of balls of radius r needed to cover the set

$$D_{cap} = \lim_{r \to 0} \frac{\log N(r)}{\log(1/r)} = \frac{\log[N(ra)/N(a)]}{\log(a/(ra))}$$

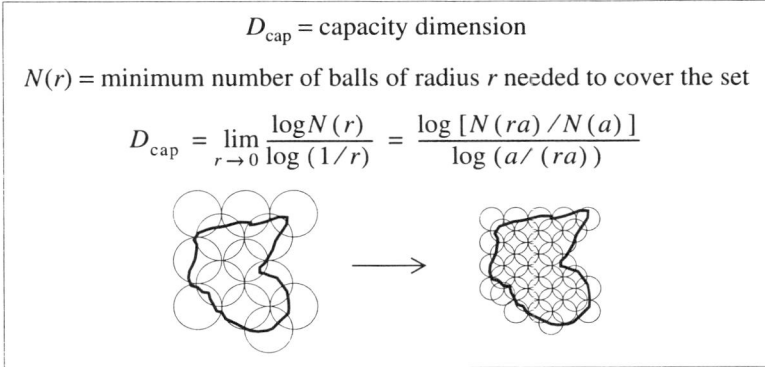

Figure 2.8. Estimating the capacity dimension, D_{cap}, for a two-dimensional object that is statistically self-similar. The minimum number of circles (two-dimensional "balls") of a given size needed to cover each point in the object is determined. The capacity dimension tells us how that number of balls changes as the size of the balls is decreased.

Hausdorff-Besicovitch dimension: A fractal object is a set defined on a metric space. A metric space has a function that defines the distance between any two points in the space (Barnsley, 1988, p. 11; Edgar, 1990, p. 37). The diameter of a set is the distance between the most distant points in the set (Barnsley, 1988, p. 200; Edgar, 1990, p. 41). An outer measure is a non-negative function defined on all the subsets of a set, such that the measure of a (countably infinite) union of subsets is less than or equal to the sum of the measures of those subsets (Halmos, 1950). Cover the object with a union of subsets A_i each of which has diameter less than or equal to r. The s-dimensional Hausdorff outer measure $H(s, r)$ is the minimal value for all such covers of the sum of the diameters of these subsets raised to the power s (Barcellos, 1984, pp. 112–114; Falconer, 1985, p. 7; Barnsley, 1988, p. 220; Edgar, 1990, p. 147), that is

$$H(s, r) = \inf \left(\sum_i (\text{diameter } A_i)^s \right). \tag{2.21}$$

In the limit as the diameter of the covering subsets $r \to 0$, there exists a unique value of $s = D_{H\text{-}B}$, which is defined as the Hausdorff-Besicovitch dimension (Barcellos, 1984, pp. 112–114; Falconer, 1985, p. 7; Barnsley, 1988, p. 201; Edgar, 1990, p. 149), such that

$$\lim_{r \to 0} H(s, r) \to \infty \quad \text{for all} \quad s < D_{H\text{-}B},$$

$$\lim_{r \to 0} H(s, r) \to 0 \quad \text{for all} \quad s > D_{H\text{-}B},$$

$$\lim_{r \to 0} H(s, r) \quad \text{exists and is nonzero for} \quad s = D_{H\text{-}B}. \tag{2.22}$$

Capacity dimension: The capacity dimension is very similar, but not identical to the Hausdorff-Besicovitch dimension. Both these dimensions examine the minima of a function of all covers as the diameters of the covering subsets approach zero. However, each uses a slightly different function. In the capacity dimension we determine $N(r)$, the least number of open balls of radius r required to cover the object. The capacity dimension D is then given by

$$D = \lim_{r \to 0} [\log N(r) / \log (1/r)] . \tag{2.23}$$

The capacity is defined by Grassberger and Procaccia (1983, p. 190) and Eckmann and Ruelle (1985, p. 620), and its use is extensively illustrated through many examples by Barnsley (1988, Ch. 5). The Hausdorff-Besicovitch dimension provides more information than the capacity dimension because fractals with the same D_{H-B} can have different values for the limit $_{r \to 0} H(D_{H-B}, r)$. The capacity dimension can be used to define a potential function that is similar to limit $_{r \to 0} H(D_{H-B}, r)$ but that can differ from it for some fractals of the same dimension (Falconer, 1985, pp. 76–79).

Generalized dimensions: The dimensions described above are based on an equal weighting of each point of an object. Other dimensions are based on weightings that measure the correlations among the points of an object. The most often used is the correlation dimension D_{corr}. This definition is based on the principle that $C(r)$, the number of pairs of points within radius r of each other, is proportional to r^D, namely

$$D_{corr} = \lim_{r \to 0} \log C(r) / \log (r) , \text{ where} \tag{2.24}$$

$$C(r) = \lim_{N \to \infty} (1/N^2) \sum_{j=1}^{N} \sum_{i=j+1}^{N} H(r - |r_i - r_j|) , \tag{2.25}$$

and N is the number of points measured in an object, $H(x)$ is the Heaviside function that is if $x < 0$ then $H(x) = 0$, otherwise $H(x) = 1$, and the norm $|r_i - r_j|$ is the distance between the two points r_i and r_j.

Another important generalized dimension is the information dimension D_{info}, which is based on a point weighting that measures the rate at which the information content called the Kolmogorov entropy changes. It is given by

$$D_{info} = \lim_{r \to 0} S(r) / \log (r) , \text{ where} \tag{2.26}$$

$$S(r) = - \lim_{N \to \infty} \sum_{i=1}^{N} p_i \log p_i , \tag{2.27}$$

and p_i is the probability for a point to lie in ball i. There is an infinite set of generalized dimensions based on different weightings of the correlations between the points of an object. These dimensions are described by Parker and Chua (1989), Grassberger and Procaccia (1983), and Hentschel and Procaccia (1983) and are comprehensively reviewed by Moon (1987, Ch. 6), Schuster (1988, Ch. 5), Gershenfeld (1988, Sec. 6), and Theiler (1990).

Each of these dimensions can yield different values for the fractal dimension of the same object. Eckmann and Ruelle (1985) showed that

$$D_{H\text{-}B} \leq D_{cap},\tag{2.28}$$

and Grassberger and Procaccia (1983) showed that

$$D_{corr} \leq D_{info} \leq D_{cap}.\tag{2.29}$$

In practice the differences are usually much smaller than the statistical error in their estimates (Parker and Chua, 1989).

$$*\quad*\quad*$$

The topological dimension of the space in which the fractal is embedded. The dimensions in the previous section describe the space-filling properties of the fractal set. The *topological dimension* describes the connectedness between the points in the fractal set. The topological dimension is always an integer. For curves, surfaces, or solids, the topological dimension $D_T = 1$, 2, or 3. Below are two brief descriptions of different ways of determining the topological dimension:

Covering dimension: In a minimal covering of a union of overlapping subsets of arbitrarily small size, each point of the space will be contained in no more than $D_T + 1$ subsets (Pontryagin, 1952, pp. 16–17; Mandelbrot, 1983, pp. 411–412; Edgar, 1990, pp. 95–100). For example, if we cover a plane with circles, each point in the plane lies in three circles or fewer, and thus the topological dimension of the plane is two.

Inductive dimension: A point within a three-dimensional volume can be contained by boundaries that are two-dimensional walls. A point in each two-dimensional wall can be contained by boundaries that are one-dimensional lines. A point in each one-dimensional line can be contained by boundaries that are zero-dimensional points. A point in each zero-dimensional point is already contained without any additional boundary. Thus, counting how many times we must take the boundary of the boundary of the space to reach a zero-dimensional point, determines the topological dimension of the space (Mandelbrot, 1983, pp. 410–411; Edgar, 1990, pp. 79–95).

Examples of topology: For a dichotomously branching system one may ask whether the object is one-dimensional or two-dimensional. Clearly, a branching tree requires embedding in three-dimensional space, while a flattened vine on a wall would be embedded in two-dimensional space. The rule is that a one-dimensional line can be covered by two-dimensional circles in a plane, and that the circles intersect only two at a time. This rule holds for the branched line in Fig. 2.9, so we can see that any set of branching lines can be considered as one-dimensional. The idea of the rule is shown in Fig. 2.9, where three different coverings are used: the

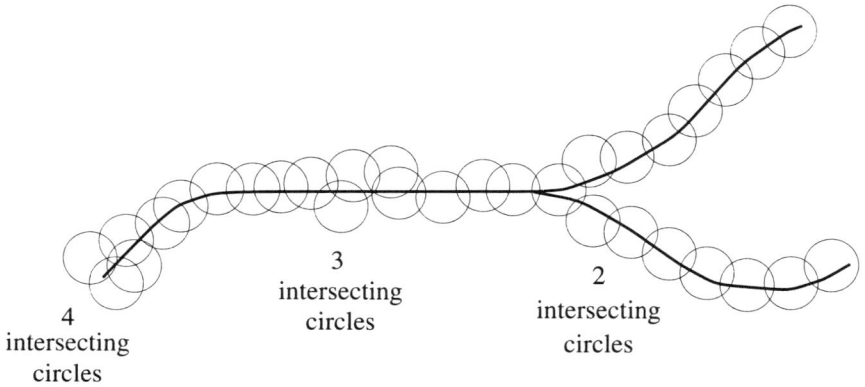

3
intersecting
circles

2
intersecting
circles

4
intersecting
circles

Figure 2.9. Covering a line with intersecting circles. The branched line is one-dimensional because only overlapping pairs of intersecting circles are needed to cover it.

covering four-way intersections at the left end and the three-way intersections provide superfluous coverage—the two-way intersecting circles suffice to cover the line, even through a branch point, no matter how small the circles are made. To cover a two-dimensional curved surface completely, spheres must be used, and *three* must intersect, as in Fig. 2.10. Likewise a point, with topological dimension zero, is covered by a line; removal of the point of a line breaks it into two parts.

The "zeroset" of a line is the intersection of a plane with a line, and the intersection is a point. A line is the zeroset of a surface, the intersection of a plane

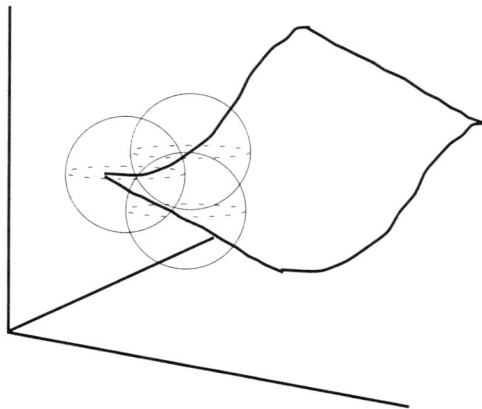

Figure 2.10. Covering a surface requires triplets of intersections of three-dimensional balls, thereby defining the surface, no matter how tortuous, as two-dimensional.

with a surface. A surface is the zeroset of a three-dimensional object, the intersection of a plane with the object. All of these will be of rather complex form if the form of the original object is complex. The zeroset of a three-dimensional apple tree is its intersection with a plane and is a composite of disconnected planes through branches and trunk and of lines through leaves (if we ignore their thickness). A familiar linear zeroset is a coastline, the intersection of a plane at altitude zero with an irregular land mass. Other levels are also zerosets, e.g., the 1000 meter contour level on a topographical map. A planar intersection with a fractal surface is a fractal boundary.

The definition of a "fractal". In the previous section we defined several *fractal or fractional dimensions* D that describe the space-filling properties of an object and the *topological dimension* D_T that describes the connectedness of an object. Mandelbrot (1983, p. 361) defines a "fractal" as a set for which

$$D > D_T. \tag{2.30}$$

When the fractional dimension exceeds the topological dimension then many new pieces keep appearing as we look at finer details. Very loosely, the topological dimension tells us about the type of object the fractal is, and the fractional dimension tells us how wiggly it is. For example, a line segment that has topological dimension $D_T = 1$ could be so long and wiggly that it nearly fills a two-dimensional area, and thus its fractional dimension $D \approx 2$. Since $2 > 1$, it is a "fractal." Mandelbrot (1983, p. 362) also notes that there are some sets that we would very much like to call fractal, where $D = D_T$. For a mathematical discussion see the book by Falconer (1990).

The Fractal Dimension Can also be Determined from the Scaling Properties of a Statistically Self-Similar Object

The power law scaling is a result of self-similarity. The fractal dimension is based on self-similarity. Thus, the power law scaling can be used to determine the fractal dimension. The power law scaling describes how a property $L(r)$ of the system depends on the scale r at which it is measured, namely,

$$L(r) = A \, r^{\alpha}. \tag{2.31}$$

The fractal dimension describes how the number of pieces of a system depend on the scale r, namely, we can approximate Eq. 2.17 as

$$N(r) = B \, r^{-D}, \tag{2.32}$$

where *B* is a constant. *The similarity of Eq. 2.31 and Eq. 2.32 means that we can determine the fractal dimension* D *from the scaling exponent* α, *if we know how the measured property* L(r) *depends on the number of pieces* N(r). For example, for each little square of side r, the surface area is proportional to r^2. Thus, we can determine the fractal dimension of the exterior of a cell membrane or a protein by measuring the scaling relationship of how its surface area depends on the scale r. Examples of this procedure are given in the following. *Since it is often easy to measure a property at different scales of resolution, this scaling method is a very useful technique to determine the fractal dimension.* However, to use this approach we must derive the relationship between the fractal dimension D and the scaling exponent α. This relationship depends on the property that is measured.

Examples of determining the relationship between the fractal dimension D *and the scaling exponent* α. We now give some examples of how to determine the fractal dimension D from the scaling exponent α. The procedure is to derive the function of the dimension $f(D)$, such that the property measured is proportional to $r^{f(D)}$. The experimentally determined scaling of the measured property is proportional to r^α. We equate these powers of the scale r, namely $f(D) = \alpha$, and then solve for the dimension D.

For example, let's say we find that the total length we measure for a line is proportional to r^α, where r is the resolution used to make the measurement. We measure the length of a line segment at scale r by breaking it up into pieces each of which has length r. Eq. 2.32 tells us that the number of such pieces is proportional to r^{-D}. The total length measured is the number of such pieces times the length of each piece. Thus the total length is proportional to r^{-D} multiplied by r, which is proportional to r^{1-D}. Since Eq. 2.31 tells us that the length is proportional to r^α, we find that $1 - D = \alpha$, and so

$$D = 1 - \alpha \quad \text{for lengths.} \tag{2.33}$$

As another example, let's say that the total area we measure is proportional to r^α, where r is the resolution used to make the measurement. We measure the area of an object at scale r by breaking up the object into pieces each of which has area r^2. The number of such pieces is proportional to r^{-D}. The area measured is the number of such pieces times the area of each piece. Thus the area is proportional to r^{2-D} and hence $2 - D = \alpha$, and so

$$D = 2 - \alpha \quad \text{for areas.} \tag{2.34}$$

If the average density of an object that is within a radius R of a point is proportional to R^α, then this scaling relationship can be used to determine the fractal dimension. The object can be considered an aggregate of many identical small subunits. The number of these subunits, and thus their mass, within radius R, is proportional to R^D. For an object in a plane, such as that shown in Fig. 2.11, the total area within a radius R is proportional to R^2. Thus the surface density σ, the mass within radius R divided by the area within radius R, is given by

$$\sigma = \sigma_0 R^{D-2}, \tag{2.35}$$

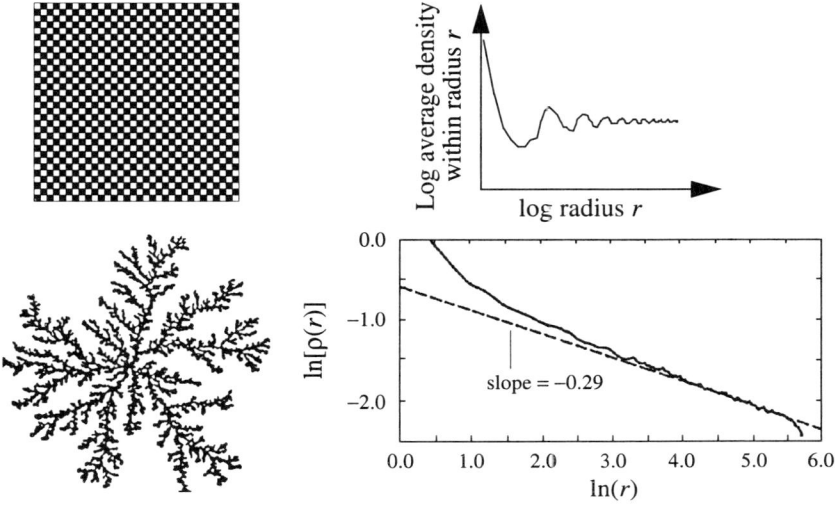

Figure 2.11. Application of the max-radius method to two objects. Estimation of density as a function of radius of enclosing circle. *Top:* The average surface density exists for the nonfractal checkerboard. As shown on the right, as the average density is measured at larger radii, it converges to a finite limit. *Bottom:* The diffusion limited aggregation is a fractal where the average surface density does not exist. It is statistically self-similar, that is, it has small empty spaces between the small branches, and ever larger empty spaces between the ever larger branches. As shown on the right, as the average surface density is measured at larger radii, more of the larger empty regions are included and thus the average surface density decreases. The straight line on this log-log plot indicates that the average surface density has the power law scaling of Eq. 2.31. The slope of this line is equal to the scaling law exponent $\alpha = -0.29$. The scaling of the average surface density with radius given by Eq. 2.35 implies that fractal dimension $D = \alpha + 2 = 1.71$. (Bottom panels from Meakin, 1986b, Figs. 10, 12, with permission.)

where σ_0 is a constant. Hence, the fractal dimension $D = \alpha + 2$. For an object in three-dimensional space, the total volume within a radius R is proportional to R^3. The density, the mass divided by the volume, is given by

$$\rho = \rho_0 R^{D-3}, \tag{2.36}$$

where ρ_0 is a constant. Hence, the fractal dimension $D = \alpha + 3$.

Power law scaling for the Koch curve. We determined the fractal dimension of the Koch curve in Fig. 2.7 by self-similarity. In this section we determine the exponent α of the scaling relationship of the Koch curve, which we use in the next section to determine its fractal dimension.

The perimeter $L(r/3)$ of the Koch curve measured at scale $r/3$ is $(4/3)$ of the value of the perimeter $L(r)$ measured at scale r, namely,

$$L(r/3) = (4/3) L(r) .\tag{2.37}$$

Since the scaling relationship has the form $L(r) = A r^{\alpha}$, then

$$L(r) = A r^{\alpha} \quad \text{and}\tag{2.38}$$

$$L(r/3) = A (r/3)^{\alpha} .\tag{2.39}$$

Substituting Eqs. 2.38 and 2.39 into Eq. 2.37, and simplifying, yields

$$3^{1-\alpha} = 4 ,\tag{2.40}$$

and by taking the logarithms of both sides and solving for α yields

$$\alpha = 1 - \log 4/\log 3 .\tag{2.41}$$

Determining the Fractal Dimension from the Scaling Relationship

We showed that for the measurement of lengths of one-dimensional objects that the fractal dimension D is related to the scaling exponent α by

$$D = 1 - \alpha\tag{2.42}$$

and for E-dimensional objects D and α are related by

$$D = E - \alpha ,\tag{2.43}$$

where E denotes the Euclidean dimension of the space in which an object is embedded. We also showed that the scaling exponent for the length of the perimeter of the Koch curve in Fig. 2.7 is given by

$$\alpha = 1 - \log 4/\log 3 .\tag{2.44}$$

Hence, using Eqs. 2.42 and 2.44 we find that $D = \log 4 / \log 3$, the same value of the fractal dimension as was determined by the self-similarity dimension.

Another example of using the scaling relationship to determine the fractal dimension is the analysis of an object called a diffusion-limited aggregation. The average surface density within a radius R is the number of pixels occupied by the object divided by the total number of pixels within that radius. The plot of the average surface density versus radius is a straight line in the log-log plot. This means that the average surface density measured within radius R is a power law function of R. The slope of that line is equal to the exponent α of the scaling relationship, and was found to be $\alpha = -0.29$. In the Aside we showed that for the measurement of average surface density, $D = \alpha + 2$, and thus the diffusion limited aggregation has fractal dimension $D = 1.71$ (cf. Fig. 2.11).

Other examples where the scaling relationship and the relationship between the fractal dimension D and the scaling exponent α have been used to determine the fractal dimension include:

1. The electrical impedance of electrodes coated with fractal deposits is a power law of the frequency used to make the measurements. Liu (1985) showed that the fractal dimension of the impedance is given by $D = \alpha + 3$.
2. The intensity of light, X rays, or neutrons scattered through small angles in materials such as silica, sandstone, and coal is a power law of the angle. When the scattering is produced within the bulk of a material then $D = -\alpha$, and when the scattering is produced by the surface of the material then $D = \alpha + 6$ (Schmidt, 1989).
3. When the vibrational energy states in a protein have a density distribution $g(f)$ with frequency f where $g(f)$ is proportional to f^{D-1}, then the electron spin relaxation rate is a power law function of the absolute temperature and the fractal dimension $D = (\alpha - 3)/2$ (Elber and Karplus, 1986).
4. When the variance of a time series of measurements increases as a power law of the window used to sample the data, then the time series has fractal dimension $D = 2 - \alpha/2$ (Voss, 1988; Feder, 1988).
5. When the power spectrum of a time series is a power law of frequency f, that is, when it is proportional to f^{β} for $-3 < \beta < -1$, then the time series has fractal dimension $D = 2.5 + \beta/2$ (Mandelbrot, 1977; Voss, 1988).

Multifractals: Objects with distributions of fractal dimensions. In the previous sections we assumed that an object can be completely described by a single fractal dimension. Sometimes no unique fractal dimension can be used to characterize an object because it is a composite of fractals of different dimensions. Such a distribution of fractal dimensions is called a *multifractal*. The distribution of fractal dimensions often has the same form in many different objects. More information about multifractals may be found in the works by Feder (1988), Jensen (1988), and Schuster (1988).

2.5   The Surprising Statistical Properties of Fractals

Processes Where the Mean Does Not Exist

For a fractal process, the moments, such as the arithmetic mean, may not exist. How can that happen? If we measure a quantity in an experiment, and then perform the same experiment a number of times, can't we always punch those measured values into a hand calculator, sum them, divide by the number of experiments, and thus determine their average? Those numbers give us the mean of one *realization* of the process that produced the data but not necessarily the *mean of the process* itself. In statistical theory, this one realization of the experimentally measured values is one *sample* taken from the entire *population* of all the possible experimental values. The mean determined from that sample of data is called the *sample mean* and the mean of the entire population is called the *population mean*. We can always determine the sample mean by averaging our experimentally measured values. We use the sample mean to estimate the value of the population mean. If the population mean exists, then as we collect more data the sample mean converges to a fixed value that we then identify as the best estimate of the population mean. On the other hand, if as we analyze more and more data the value of the sample mean keeps changing, we must conclude that the population mean of the process does not exist. This is exactly what happens for some fractal processes. As more data are analyzed, *rather than converge to a sensible value, the mean continues to increase toward ever larger values or decrease toward ever smaller values*. This is illustrated in Fig. 2.12.

An example of a fractal where the mean does not exist is a diffusion-limited aggregation. This fractal consists of the black pixels shown on the white background in Fig. 2.11. The average surface density is the number of black pixels divided by the total number of pixels within a radius. This structure is self-similar. Thus, there are small empty spaces between the small branches, and ever larger empty spaces between the ever larger branches. Hence, as the average surface density is measured at larger radii, more of the empty spaces are included, and thus the average surface

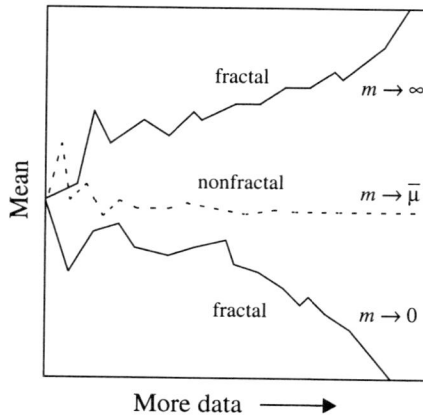

Figure 2.12. An experiment samples the values from the population of possible values. The sample mean is the arithmetic average of the experimentally measured values. If as we collect more data the sample mean converges to a fixed value, then we then identify that value as the population mean. However, for many fractal processes, as more data are analyzed the sample mean continues to increase or decrease, and never converges to a finite nonzero value. Thus, the population mean for these cases is not defined.

density is less. *There is no "true" value of the average surface density.* The value we measure depends on the scale at which the measurement is done. The average surface density approaches zero, rather than a finite nonzero value, as we increase the radii over which it is measured. This is in striking contrast to the statistics that we have become used to, as illustrated by the nonfractal checkerboard pattern in Fig. 2.11, where the average surface density approaches a constant value as we increase the radii over which it is measured.

Gerstein and Mandelbrot (1964) analyzed the intervals between action potentials recorded from different neurons. Some neurons had such a large number of long intervals between action potentials that these few intervals were of primary importance in determining the value of the mean interval between action potentials. As more data were analyzed, even longer intervals were found, and so the mean interval between action potentials increased. The trend was such that if an infinite amount of data could have been recorded, the mean interval between the action potentials would have been infinitely long. In some contexts this phenomenon has been called fractal time (Shlesinger, 1987).

[Aside]

Fractals with infinite means: The St. Petersburg game and the distribution of random mutations. The properties of fractal distributions with infinite means are well illustrated by a problem that is famous in the history of probability theory.

Consider an ordinary coin toss game where the player wins $1 if the coin falls heads and wins nothing if the coin falls tails. One game consists of tossing the coin and determining the winnings. The average winnings per game are the sum over all possible outcomes, of the probability of each outcome multiplied by the winnings of that outcome. Thus, the average winnings per game are $(1/2)(\$1) + (1/2)(\$0) = \$0.50$. As we play more games, the average winnings per game converge to $0.50, as shown in Fig. 2.13.

Now consider the coin toss game proposed by Niklaus Bernoulli and popularized by his cousin Daniel Bernoulli, which is now known as the St. Petersburg game (Feller, 1968; Dutka, 1988). Each game consists of tossing a coin until it falls heads. If it falls heads on the first toss the player wins $2. If it falls tails, heads, the player wins $4. If it falls tails, tails, heads, the player wins $8. If it falls tails, tails, tails, heads, the player wins $16 . . . etc. That is, when the probability of an outcome is 2^{-k}, the player wins $\$2^k$. The average winnings per game are the sum over all possible outcomes, of the probability of each outcome multiplied by the winnings of that outcome. Thus, the average winnings per game are $(2^{-1})(\$2^1) + (2^{-2})(\$2^2) + (2^{-3})(\$2^3) + (2^{-4})(\$2^4) + . . . = \$1 + \$1 + \$1 + \$1 . . . = \$\infty$. As we play more games, the average winnings per game does not converge to a finite value, and continues to increase, as shown in Fig. 2.13. This was also called the "St. Petersburg Paradox" since the bettor and the gambling house could not agree on the initial fair wager, and the house could foresee a possibly infinite loss. Feller (1968) solved the paradox by showing that they should agree on a finite number of trials.

There are physical and biological processes that are similar to the St. Petersburg Game. These processes have been called fractal time processes (Scher et al., 1991).

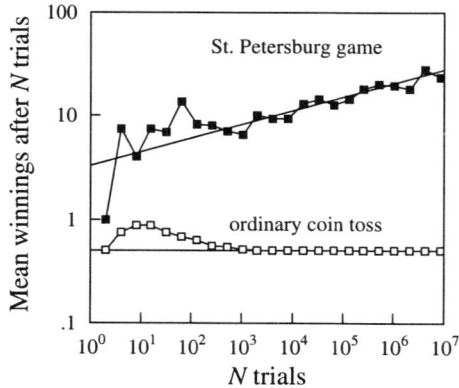

Figure 2.13. Computer simulation of the average winnings per game after N games of an ordinary coin toss and the St. Petersburg game. The mean winnings per game of the ordinary coin toss converges to a finite value. The mean winnings per game of the St. Petersburg game continues to increase as it is averaged over an increasing number of games, that is, it approaches infinity rather than a finite well defined value.

An important biological example is the number of random mutations produced in a clone of cells (Luria and Delbruck, 1943; Lea and Coulson, 1949; Mandelbrot, 1974; Cairns et al., 1988). We start with one cell in culture that will divide many times through many generations. With each cell division there is a small constant rate that a mutational event will occur. A mutational event early in the culture will produce many more daughter cells than if it occurs later in the culture. The final number of mutant cells therefore depends both on the number of mutational events and when they occur. The distribution of mutations tells us if we run the same experiment many times, how often we will find a given number of mutant cells out of the total number of N cells in the culture dish at the end of the experiment. To derive this distribution, first consider what would happen if there was only one mutational event that occurred after the culture had already grown to $N/2$ cells. If the probability of a mutational event per cell is $2r$, then the probability that the mutational event will happen during this time for $N/2$ cells is equal to $2r(N/2) = rN$. Since there will be no time left for the mutant cell to double, there will be only one mutant cell at the end of the experiment. However, if a mutational event occurred when the culture grew from $N/4$ to $N/2$ cells, there will be one doubling time and thus 2^1 mutant cells at the end of the experiment. The probability that the mutational event will happen during this time for $N/4$ cells is equal to $2r(N/4) = rN(2^{-1})$. If a mutational event occurred when the culture grew from $N/8$ to $N/4$ cells, there will be two doubling times and thus 2^2 mutant cells at the end of the experiment. The probability that the mutational event will happen during this time for $N/8$ cells is equal to $2r(N/8) = rN(2^{-2})$. Thus, each mutational event with probability proportional to $rN2^{-k}$ will produce 2^k mutant cells at the end of the experiment. This is the same probability distribution as the St. Petersburg Game and therefore it has the same statistical properties, including the fact that the average number of mutant cells expected in the dish at the end of the experiment is not defined.

Processes Where the Variance Does Not Exist

In the previous section we discovered that the mean may not exist, that is, it may approach zero or infinity, rather than a nonzero finite value, as more data are analyzed. The variance, which equals the square of the standard deviation, has been traditionally used to describe the amount of uncertainty in a measurement. This interpretation is based on the theory of errors developed for distributions that have a Gaussian form called the "normal" error curve (see for example, West, 1985). In this section we show that the variance may also not exist, that is, rather than approach a well-defined value, it may approach infinity as more data are analyzed.

Self-similarity means that the small irregularities at small scales are reproduced as larger irregularities at larger scales. These increasingly larger fluctuations become apparent as additional data are collected. Hence, *as additional data are analyzed from a fractal object or process, these ever larger irregularities increase the measured value of the variance.* In the limit as the amount of data analyzed continues to increase, the variance continues to increase, that is, the variance becomes infinite.

This surprising behavior has been observed in many different systems, including the fluctuations in the prices of cotton (Mandelbrot, 1983) and the profiles of rocks (Brown and Scholz, 1985). This behavior is illustrated for physiological processes in space and in time in Fig. 2.14. Bassingthwaighte and van Beek (1988) measured the density of microspheres deposited in the tissue to determine the volume of blood flow per gram of heart tissue. As the heart is cut into finer pieces, the variance of the flow per unit mass increases as the size of the pieces used to make the measurement decreases. Teich (1989) and Teich et al. (1990) measured the intervals between action potentials recorded from primary auditory nerve fibers. As the intervals were analyzed over longer window lengths, the variance in the mean firing rate increased.

[Aside]

$1/f^\beta$ *power spectra is characteristic of self-similarity and infinite variance.* Self-similarity means that the small irregularities at small scales are reproduced as larger irregularities at large scales. Small scales correspond to high frequencies, and large scales to low frequencies. Hence, the amplitude of the fluctuations is small at high frequencies and large at low frequencies. The power spectrum of such data is usually proportional to $1/f^\beta$, where f is frequency and β is a constant. This power spectrum is divergent at low frequency. That is, the longer the interval of data that is analyzed, the larger the power in the fluctuations. This is equivalent to the increase in the variance as additional data are analyzed. The inverse power law form of this type of power spectra is caused by the self-similar nature of the fluctuations.

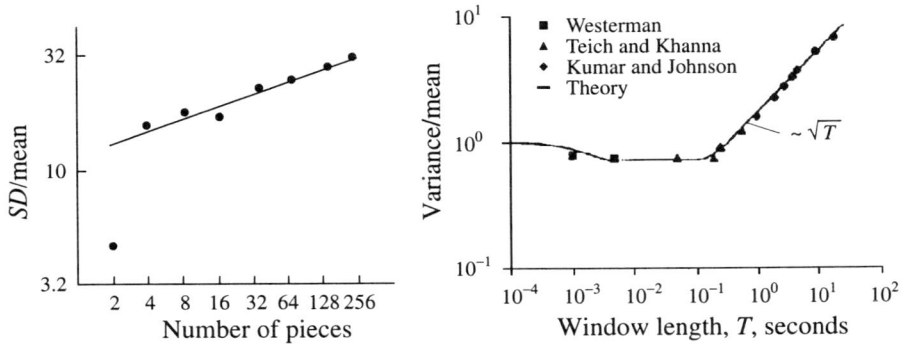

Figure 2.14. Examples where the variance is not defined, that is, it increases in value as more data are analyzed. *Left:* Bassingthwaighte and van Beek (1988) found that the relative dispersion (the standard deviation divided by the mean) in the blood flow in the heart increases as a power law of the number of heart pieces used to make the measurement. *Right:* Observing the rate of firing of action potentials in cat auditory nerve, Teich (1989) found that the Fano factor (the variance in the number of spikes in an interval of duration T divided by the mean number of spikes over intervals of length T) increased as a power law of the window length T over which the data was analyzed, for window lengths longer than 100 ms. (*Left:* Bassingthwaighte and van Beek, 1988, Fig. 4; *right:* Teich, 1989, Fig. 4a, ©1989 IEEE.)

Examples of such $1/f^\beta$ power spectra, in space and in time, are illustrated in Fig. 2.15 and mechanisms leading to their generation are discussed by West and Shlesinger (1989, 1990).

Properties and Problems of Fractal Data Where the Mean and Variance Are Not Defined

Most scientists know about the mathematical properties of processes that have a "normal" or Gaussian distribution, where the moments, such as the mean and variance, have well defined values. However, these processes are only one small subject in the theories of probability and statistics. For example, in the previous sections we showed that the mean may decrease or increase forever, or the variance may increase forever, as more data are analyzed. This behavior is typical of fractals. Although mathematicians have studied such deportment for over 200 years, this knowledge is not familiar to most working scientists. Yet many things in nature have exactly these fractal properties, which has important implications for the design of experiments and the interpretation of experimental results. In the past, many of these properties have been ignored or misunderstood, and thus an understanding of many physiological systems has been hampered. We shall now describe more of these

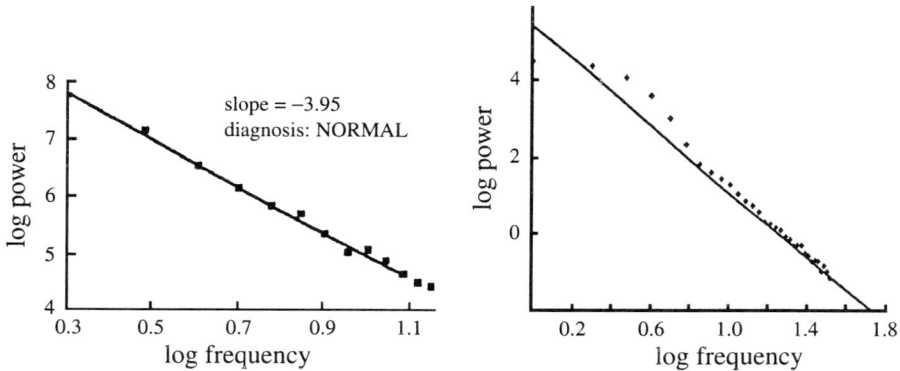

Figure 2.15. The power spectrum describes the amplitude of the fluctuations at different spatial or temporal frequencies. It has an inverse power law form for fractals in space and time. *Left:* Power spectrum of the fluctuations in the spatial distribution of radioisotope in the liver measured by Cargill et al. (1988, their Fig. 2a). *Right:* Power spectrum of the fluctuations in time of the ventricular depolarization in an electrocardiogram measured by Goldberger et al. (1985a, their Fig. 2).

properties and also see that there are gaps in our knowledge of how to apply these ideas to analyze experimental data. This may be frustrating, in that we may not know how to analyze some features in our data, but it is also exhilarating to know that mathematics is not a dead science and it is fun to watch new mathematical and statistical ideas develop and be published during our lifetime.

The first property concerns averages over small intervals such as regions in space for spatial objects or windows in time for temporal processes. We may find *very different values when experimental data are averaged over different intervals.* The self-similarity of a fractal implies that there are dense regions within dense regions within dense regions. . .and also sparse regions within sparse regions within sparse regions. Thus, averages of the experimental values in the densest of the dense regions differ from their averages in the sparsest of the sparse regions. The increase of the variance, as ever larger intervals are analyzed, means that the averages determined between those intervals are substantially different. For example, the surface density averaged around one square centimeter of the diffusion limited aggregation in Fig. 2.11 will be very different if we compare one measurement centered on the pattern to one caught between its branches. Another example is the mean firing rate recorded from a neuron where there are bursts within bursts within bursts of action potentials. The average firing rate is very high within the increasing hierarchy of bursts and very low in the increasing hierarchy of quiet intervals between bursts.

These fractals have *no "true" value for the mean, in the sense that the value measured depends on the scale used for the measurement.* When different groups perform the same measurement using different scales, then their results can differ. Paumgartner et al. (1981) suggest that this may help to explain the different values reported by different groups for the areas of membranes in cells. This means that it is particularly important that the publication of measurements clearly state the spatial or temporal scale at which the measurement was done.

These variations in the mean are a consequence of the fact that the mean does not converge to a well-defined value. If the mean of a process converges to a well-defined value, then we have a natural scale by which to measure events in space or time. If we observe over much longer than this natural scale, it will appear that events occur at a constant rate. However, when the mean is infinite, *no natural scale exists by which to gauge measurements*. We cannot choose a scale large enough to form a well defined finite average of the rate of events. Self-similarity implies that no natural spatial or temporal scale exists, that is, there are events at all scales.

A process that generates data is characterized by a set of *parameters*. A linear system whose parameters are not changing in time is termed "stationary," and its response to a particular stimulus does not change from one time to another. The moments, such as the mean, of processes with "normal" distributions are typically used to determine those parameters. However, when the processes are fractal, and thus those moments are not defined, *it is not known how to determine the parameters of the underlying process*. For example, it is not known how to determine the rate of mutational events from the number of mutant cells in a clone, because the mean is infinite for this distribution. One method to analyze such data has been to determine the rate of mutational events that yields the distribution of the number of mutant cells per experiment that best fits the data (Cairns et al., 1988; Stewart et al., 1990). This approach is not very satisfactory because it is not known what criterion to use to determine the goodness of fit and because this method requires very large amounts of data to define the entire distribution. Furthermore, if it is not known how to determine the value of the underlying parameters, even *less is known about how to tell if an experimental change has produced a statistically significant change in those parameters*. For example, Davis (1989) proposed that the mutational rate increases when DNA is more often exposed because it is being transcribed more often. In repeating those experiments Liebovitch (unpublished) found 8% and 85% more mutant cells in clones where transcription was increased, but we do not know how to determine if those amounts of change are significant.

It is also not known how to determine *if the mechanisms that produce the experimental data are changing in time*, which is equivalent to asking if the parameters of the process are changing. It is commonly believed that 1) if the mean is constant during a time interval, then the process that produces the data has remained constant, and 2) if the mean changes significantly during a time interval, then the process that produces the data has changed. This is *not* true. *The means measured over different periods can be substantially different even if the process is not changing in time*. We cannot determine if a generating process is constant in time by simply monitoring the constancy of its moments, and the system appears nonstationary even when no parameters have changed. For example, consider again the electrical recordings from a neuron where there are bursts within bursts within bursts of action potentials. The average firing rate will change when recordings are made at the onset of a hierarchy of bursts. This does not mean that the underlying state of the neuron has changed. It is due rather to the self-similar nature of the firing pattern. Another example is the currents recorded through individual ion channels, which often show periods of great activity and inactivity. A channel may flicker open and closed many times, and then suddenly remain closed almost indefinitely. This sudden change could be due to a modification of the channel protein. However, processes with infinite mean or infinite variance can also show such sudden changes, without any modification in the channel protein. Thus, it is difficult to determine if a change in the mean number of openings and closings is due to a *change* in the attributes of a process with finite variance, or an *unchanging* process *with infinite mean or infinite variance*.

[Aside]

The properties of the moments of fractals are due to the inverse power law form of fractal distributions. The fact that the moments of fractal processes may not converge to well defined values is due to the inverse power law form of fractal distributions. The number of pieces $N(r)$ of size r of a fractal with dimension D is given by Eq. 2.32, namely $N(r) \propto r^{-D}$. If the experimentally measured property $L(r)$ depends on size r as r^s, then the average measured for the property $<L>$ over all scales will be given by

$$\langle L \rangle \propto \int_0^\infty r^s \, r^{-D} \, dr. \qquad (2.45)$$

The value of this integral will be zero, infinite, or converge to a finite value depending on the values of D and s, and how they change in the limit as r approaches 0 and ∞.

The statistical properties of fractals are described by stable distributions. Many scientists are familiar with Gaussian distributions, which are also called "normal" curves, or distributions that approach normalcy when large amounts of data are collected, which are called asymptotically normal. The mean and variance of these distributions are finite well defined values. However, these distributions are a subset of a much larger class of distributions that are called "stable distributions" (Feller, 1971; Holt and Crow, 1973; Mandelbrot, 1983, pp. 367–370; Montroll and West, 1979, 1987; Cambanis and Miller, 1980, 1981). Stable distributions have the property that the distribution of linear combinations of two independent variables has the same shape as the original distribution. Stable distributions can have moments that are either finite or infinite.

Gaussian distributions are stable distributions with finite mean and finite variance. These properties satisfy the conditions of the Central Limit Theorem, which means that as more experimental data are analyzed, the mean is determined with increased precision.

There are also stable distributions with *infinite variance*, which Mandelbrot and others call *Lévy stable*. This property does *not* satisfy the conditions of the Central Limit Theorem in its usual form, which means that the variance of linearly combined independent variables can be the *same* as that of each distribution. Thus, as more experimental data are analyzed, the precision of the determination of the mean does *not* improve.

2.6   Summary

Fractal objects in space or fractal processes represented by their time series have four important properties: self-similarity, scaling, fractal dimension, and surprising statistical properties. 1) *Self-similarity* means that every small piece of the object resembles the whole object. The fractal is *geometrically self-similar* if each small piece is an exact replica of the whole object. The fractal is *statistically self-similar* if the statistical properties of each small piece, such as the average value of a measurement made on the piece, are proportional to those properties measured on the whole object. 2) Statistical self-similarity implies that the value measured for a property depends on the scale or resolution at which it is measured, and this is called *scaling*. The form of the scaling is very often a *power law*. Thus, we find a straight line when the logarithm of the values measured for a property are plotted versus the logarithm of the resolutions at which the property is measured. Although not all systems manifesting power law relationships are fractal, the existence of such a relationship should alert us to consider seriously that the experimental system under study might have fractal properties. 3) The self-similarity and scaling can be quantitatively measured by the *fractal dimension*. The fractal dimension tells how many new pieces are resolved as we examine the fractal at finer resolution. This dimension is a generalization of the integer dimensions used to categorize positions, lengths, areas, and volumes. Fractals have fractional, noninteger values of dimension, and thus their properties are distinct from those of points, lines, surfaces, and solids that have integer dimension. 4) The *statistical properties* of fractals can also have power law scalings. As more data are analyzed, the mean, that is, the average value measured for a property, can become increasingly small or increasingly large, while the variance can become increasingly large. Thus, both the *mean and variance can be undefined* for fractals.

These properties have important consequences for the measurement and interpretation of data from fractal objects or processes. 1) There is no "true" value for the mean of a property measured in that the value depends on the scale used for the measurement. Thus, it is particularly important to report the spatial or temporal scale at which a measurement is made. Some of the discrepancies in measurements of the same physiological properties reported by different groups may be due to the measurements being performed at different scales. 2) How the mean measured changes with the scale of the measurement, as characterized by the fractal dimension, may tell more about the object or process than the value measured at any one scale. 3) Most scientists are not familiar with the statistical properties of systems, such as fractals, that have infinite mean or infinite variance. In these systems the values measured change with the amount of data analyzed and do not converge to a well defined value. Moreover, the values averaged over two different neighborhoods in space or windows in time may differ significantly, even though the process that generated the data remains constant. This contradicts the widely believed idea that if the local mean has changed, then the process that produced it must have changed. 4) The existence of such fractal properties should serve as an incentive to examine the physiological system at other scales to confirm its fractal nature and also to consider what mechanisms might generate such structures at many different scales.

Future Directions

Different objects or processes can have the same fractal dimension. Thus, the dimension alone does not completely characterize the system. Hence, other tools are needed to characterize other properties that are not differentiated by the fractal dimension alone. The spectral dimension is one such measure (Orbach, 1986; Pfeifer and Obert, 1989), but others are also needed. Many objects or processes cannot be uniquely characterized by a single fractal dimension. There needs to be further development of multifractals (Feder, 1988; Jensen, 1988; Schuster, 1988) and techniques to characterize distributions of fractal dimensions.

The mathematics of systems with infinite mean or infinite variance has been studied for a long time. They are called stable processes (Feller, 1971; Holt and Crow, 1973). However, only recently have there been many physical examples of such systems (see, for example, Montroll and West, 1979, 1987; Montroll and Shlesinger, 1984). There are many statistical methods to analyze experimental data to estimate the parameters of processes that have finite moments. However, very little is known about how to analyze experimental data to determine the parameters for processes that have infinite moments, or how to tell if the parameters are significantly different in different data sets or different periods of time.

Background References

Because of the explosion in both theory and experiment over the past decade there is no one book that is both a clear and a comprehensive introduction to fractals. However, we can suggest several excellent books, each having its own specialization and level of mathematical sophistication.

The existence and properties of fractals became widely known through the books *Fractals: Form, Chance and Dimension* and *The Fractal Geometry of Nature* by Mandelbrot (1977, 1983), who accurately describes each book as a "casebook" of examples and advises the reader to "browse and skip" through them. His examples help one develop a powerful intuition, but often he presents too little, or too much, mathematics and so the details of the concepts in these books are difficult for many readers to understand. Nonetheless, these books show an appreciation of beauty, history, and advocacy that makes the reading worthwhile.

A nonmathematical introduction to fractals can be found in *Chaos: Making a New Science* by Gleick (1987). There are several introductory books for those comfortable with mathematics at the elementary calculus level. *Fractals* by Feder (1988) provides an excellent extended review and derivation of fractal properties with their application to physical problems. A similar, but shorter, review of fractal properties by Pfeifer and Obert, and their application to chemical problems, can be found in the *The Fractal Approach to Heterogeneous Chemistry*, edited by Avnir (1989). *The Science of Fractal Images*, edited by Peitgen and Saupe (1988), has an excellent review of the overall properties of fractals by Voss, and many algorithms for generating fractal images.

Several books provide a more rigorous introduction to the mathematical properties of fractals. *Fractals Everywhere* by Barnsley (1988) is an excellent introduction that is mathematically rigorous but starts at an elementary level and uses many words, pictures, and examples to make the mathematics accessible to those intimidated by other mathematical texts. Another excellent introduction is *Measure, Topology, and Fractal Geometry* by Edgar (1990), which also starts from the elementary concepts but proceeds at a more sophisticated mathematical level. A mathematically sophisticated terser review of fractal properties is given in *The Geometry of Fractal Sets* by Falconer (1985).

Two excellent books cover specialized topics but also include more general review material. *The Beauty of Fractals* by Peitgen and Richter (1986) concentrates on iterated functions and includes their application to the physics of magnetism through renormalization groups. *Fractals Everywhere*, by Barnsley (1988), mentioned above, elucidates links between fractals and iterated function systems.

Other books review a variety of fractal applications in different fields. In biology, *Fractal Physiology and Chaos in Medicine* by West (1990) gives an excellent overview of the mathematics used in biomedical applications as well as examples of the successes these concepts have had in providing new insights in such phenomena. A collection of articles on the application of fractals to biological systems may be found in a special issue of the *Annals of Biomedical Engineering*, edited by McNamee (1989). Reviews of applications in physics and chemistry include: *Fractals in Physics* edited by Pietronero and Tosatti (1986), *On Growth and Form: Fractal and Non-fractal Patterns in Physics* edited by Stanley and Ostrowsky (1986), *Random Fluctuations and Pattern Growth: Experiments and Models* edited by Stanley and Ostrowsky (1988), *The Fractal Approach to Heterogeneous Chemistry* edited by Avnir (1989), and *A Random Walk Through Fractal Dimensions* by Kaye (1989).

Several companies now specialize in books, computer software, slides, and videos about fractals. These include: 1) Amygdala, Box 219, San Cristobal, NM 87564, 2) Media Magic, P.O. Box 2069, Mill Valley, CA 94942, and 3) Art Matrix, P. O. Box 880, Ithaca, NY 14851, one of whose pretty fractal videos is called "Nothing But Zooms."

3

The Fractal Dimension: Self-Similar and Self-Affine Scaling

To see a World in a Grain of Sand
And a Heaven in a Wild Flower,
Hold Infinity in the palm of your hand
And Eternity in an hour.

William Blake, Auguries of Innocence (1800–1803)

3.1   Introduction

In the previous chapter we provided an overview of the utility of fractal concepts in biology, and indeed of their usefulness in science in general. We now describe these concepts in greater detail.

The goal of modeling in biology, as elsewhere in science, is to obtain simple models that capture the essential features of the structure or process being investigated. From physics one recalls such expressions as Ohm's law ($E = IR$), where E is the voltage measured across a resistance R and I is the current flowing through the resistor. This law describes the phenomenology of current flow in conductors and much research has gone into constructing detailed microscopic models that justify the macroscopic *law*. Another example is the "perfect gas law" ($PV = RT$), where P is the pressure of a gas contained in a volume V at a temperature T and R is a known constant. Of course, such simple "laws" are not usually available in biology, but one does often encounter relations of the form $Y = \alpha M^{\beta}$. These so-called allometric equations relate the Y, which is a physiological, morphological or ecological variable, to M, which in most cases is the body mass, and where α and β are constants. The exponent β is usually not a rational number, which is to say that it cannot be represented as the ratio of two integers (see, e.g., Calder, 1984, and MacDonald, 1983). The traditional discussions of these allometric growth laws rely on simple physical arguments resulting in rational noninteger exponents, and the observed deviation of the data from their rational values is often assumed to be the result of experimental limitations. We now put these power laws into a much broader context.

We discussed power law scaling in Chapter 2 and pointed out that such behavior is consistent with a fractal underlying process. It is worthwhile to point out that the label "power law" is used in a number of quite distinct situations. In the first, it refers to the power spectrum for a dynamical process. The correlation function for

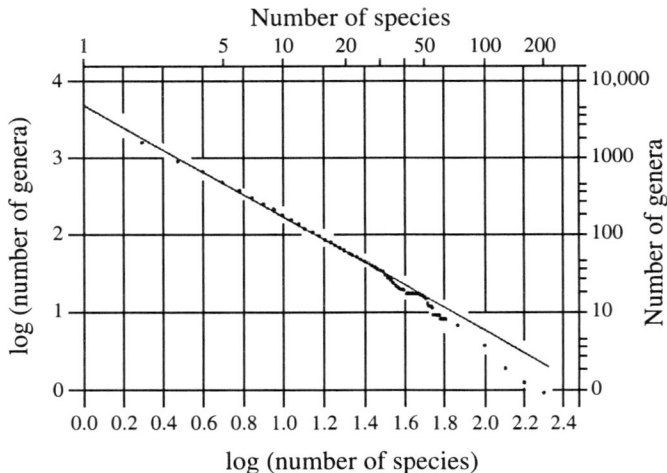

Figure 3.1. Relation between number and size of genera of all flowering plants, plotted logarithmically (Willis,1922).

such a process increases algebraically with time t^α, so that its Fourier transform, the power spectral density, depends inversely on the frequency $1/f^{\alpha+1}$. This is also called an inverse power law spectrum. The second use of the term power law applies to a probability density. It was used in this latter sense in Chapter 2. Probability densities that behave as an inverse power of the variate are also called hyperbolic distributions, and the variates are called hyperbolic random variables. We shall find that the inverse power law spectrum may be related to *chaos* in dynamical processes, and that hyperbolic processes have an intrinsic scaling that can often be related to the fractal properties. Finally, expressions such as the allometric growth law mentioned above have the basic power law form. We shall refer to all these latter cases as hyperbolic relations, since the name allometric growth law carries a connotation that is not always appropriate.

We now give examples of power law scalings. Willis (1922) collected data on various natural families of plants and animals and graphed the number of genera as ordinate and the number of species in each genus as abscissa. A typical result is displayed in Fig. 3.1, where such a plot is given for all flowering plants. On this log-log graph paper a power law graphs as a straight line with a slope determined by the power law index.

A second example is Zipf's law from linguistics, which is concerned with the relative frequency of word usage (Zipf, 1949). In Fig. 3.2 we show the fraction of the total word population in a sample versus the rank order of frequency usage that a given word takes in the word population considered, i.e., the most frequent word is first, the second most frequent is second, and so on. The distribution density is $1/f$, where f is the relative frequency of the word in the sample population. While for English the power of f is unity, other languages also have an inverse power law

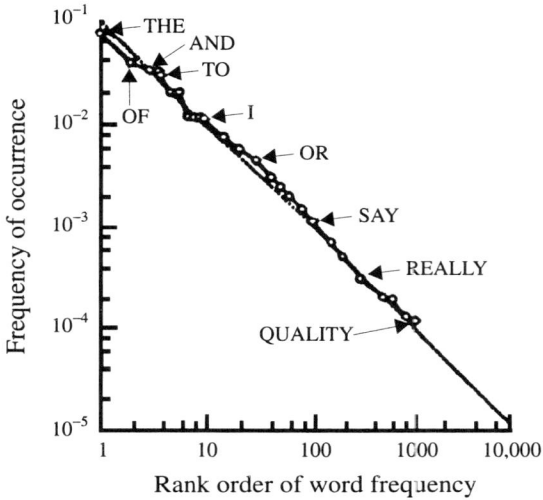

Figure 3.2. Word frequency is graphed as a function of rank order of frequency of usage for the first 8728 words in the English language (from Shannon, 1948). The inverse power law is that of Zipf (1949).

distribution $1/f^{\alpha}$ with an α greater than unity and with a value that varies from language to language. The greater the value of α the steeper the slope of the curve, indicating that words of a given rank are less frequently used relative to the size of the overall vocabulary than others are in languages with a smaller value of α. This is most easily seen by normalizing the distribution to the size of the vocabulary in a given language so that the integrated value of the distribution over word frequency is unity.

[Aside]

These power law distributions are known to have a statistical self-similarity resulting in the distribution density obeying a *functional scaling law* (*renormalization group property*). Consider $P(f) \sim 1/f^{\alpha}$, so that if we multiply f by a parameter λ then $P(\lambda f) = \lambda^{-\alpha} P(f)$, indicating that the new distribution is given by the original times a factor. This particular functional relation $P(f) = \lambda^{\alpha} P(\lambda f)$ is referred to as a renormalization group relation because the connection between functions at scaled values of the independent variable can be shown to give rise to a group. A group is a collection of objects sharing some property in common, say, for example, the group of rational real numbers. The renormalization group relation is the mathematical expression of the underlying process not having a fundamental scale. *This lack of a single scale is similarly manifest in the divergence of the central moments.* See, for example, West and Salk

(1987). Consider the μth moment of f written as $<f^{\mu}>$, where the brackets denote an average over the distribution $P(f) = c/f^{\alpha}$ and c is the normalization on the interval $[\varepsilon, \infty]$, $\varepsilon > 0$. We obtain $<f^{\mu}><\infty$ if and only if $\alpha > (\mu + 1)$ and $<f^{\mu}> = \infty$ if $\alpha \leq (\mu + 1)$. Classical scaling principles are based on the notion that the underlying process is uniform, filling an interval in a smooth, continuous fashion, and thereby giving rise to finite averages and variances. The new principle is one that can generate richly detailed, heterogeneous, but *self-similar structures* at all scales. We shall discuss how these hyperbolic distributions manifest self-similarity later in this chapter. Thus, such things as length, area, and volume are not absolute properties of systems but are a function of the unit of measure.

$$* \quad * \quad *$$

As we discovered in Chapter 2, the properties of many systems are determined by the scale of the measuring instrument. For example, such things as length are a function of the size of the ruler. Consider the case studied by Richardson (1961), where he estimated the length of irregular coastlines to be

$$L(r) = L_0 r^{1-D}. \tag{3.1}$$

Here $L(r)$ is the measured length of the coastline when a measuring rod of length r is used and L_0 is the length measured with a ruler of unit length. The constant D is calculated from the slope of the straight line on a log-log plot:

$$\log L(r) = \log L_0 + (1 - D)\log r. \tag{3.2}$$

For a classical smooth line the dimension D equals 1, and $L(r)$ becomes a constant independent of r, i.e., the coefficient of $\log r$ vanishes when $D = 1$. For an irregular coastline it was found that $D > 1$ as shown in Fig. 3.3, where we see that the data for the total length of coastlines and boundaries fall on straight lines with slopes given by $(D - 1) > 0$. For these data we find that $D \approx 1.3$ for the coast of Britain and $D = 1$ for a circle, as expected. Thus we see that $L(r) \to \infty$ as $r \to 0$ for such an irregular curve since $(1 - D) < 0$. Such a curve was termed "fractal" by Mandelbrot.

Historically, we learned from Michael Woldenberg at SUNY Buffalo that the difficulty in measuring coastlines was defined by Penck (1894), who showed that the apparent length increased as measurements were made on maps of larger ratios of map dimension to true dimension. His data also showed a linear relationship between $\log r$ and $\log L(r)$ with $D = 1.18$ for the Istrian peninsula in the North Adriatic. Perkal (1966) examined the length determined by rolling a circle along the contour of a coastline, showing that the apparent length increases inversely with the radius of the circle. Likewise, he observed that when using an overlay of circles to measure areas enclosed by a crooked boundary the error in the method diminished as the circle diameters were diminished. Even though both Penck and Perkal defined the issues, they did not resolve it as Richardson did by providing a simple relationship between the measured length and the length of the measuring stick.

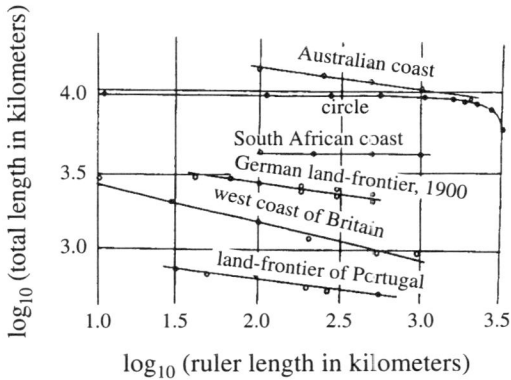

Figure 3.3. Plots of various coastlines in which the apparent length $L(r)$ is graphed versus the measuring unit r. (From Richardson, 1961.)

To fully appreciate the significance of the fractal concept we need to understand the way in which scaling ideas have been traditionally used in biology. It is only from such a perspective that one can reinterpret existing data sets and gain new insight into the mechanisms underlying erratic biomedical processes.

3.2   Branching in the Lung: Power Law Scaling

Perhaps the most compelling feature of all physiological systems is their complexity. *Capturing the richness of physiological structure and function in a single model presents one of the major challenges of modern biology, and fractals offer the hope of doing just this.* On a static (structural) level, the bronchial system of the mammalian lung serves as a useful example of such anatomic complexity. One sees in this treelike network a complicated hierarchy of airways, beginning with the trachea and branching down on an increasingly smaller scale to the level of the smallest bronchioles (see Fig. 3.4).

Any successful model of pulmonary structure must account not only for the small-scale measurements, but also for the global organization of these smaller units. In the larger-scale structures that we observe with the unaided eye, we see the extreme variability of tube lengths and diameters, constrained by a high level of organization. The variability is not only evident from generation to generation as we proceed down the bronchial tree, but is also apparent within a given generation. The term "generation" is used to label the number of splittings the bronchial tree has undergone starting from the trachea. The first generation of tubes is comprised of just two members, the left and right mainstem bronchi. The second generation consists of four tubes, and so forth (cf. Fig. 3.4).

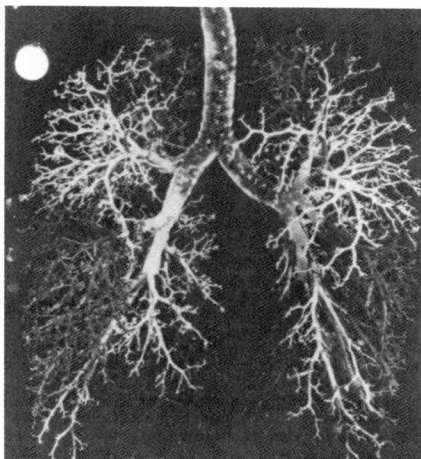

Figure 3.4. The illustration shows a plastic cast of the human bronchial tree, from the trachea to the terminal bronchioles. The mammalian lung has long been a paradigm of natural complexity, challenging scientists to reduce its structure and growth to simple rules. (From West and Goldberger, 1987.)

Over the past several years quantitative models have been developed that suggest a mechanism for the "organized variability" inherent in physiological structure and function. The essential concept underlying this kind of constrained randomness is that of *scaling*.

The human lung has two dominant features, irregularity and richness of structure, along with organization. Both these characteristics are required to perform the gas exchange function for which the lung is designed. As the bronchial tree branches out, its tubes decrease in size. The classical theory of scaling assumes that their diameters should decrease by about the same ratio from one generation to the next. If $d(z)$ is the diameter at generation z then $d(z) = qd(z-1)$, where q is the constant ratio between generations. If this ratio were 1/2, for example, the relative diameters of the tubes would be 1/2, 1/4, 1/8, . . . an exponential decline. Weibel and Gomez (1962) measured the tube diameters for 22 branchings of the bronchial tree in the human lung. On semi-log graph paper this scaling prediction is that the data (the average tube diameter) should lie along a straight line (cf. Fig. 3.5) with the slope given by ln$q$, where q is the scaling parameter.

In Fig. 3.5 we see that the data are consistent with classical scaling with q about 0.6 for ten generations after which a striking deviation from exponential behavior appears. Weibel and Gomez (1962) attributed this deviation to a change in mechanism from convection by airflow in the larger branches of the bronchial tree to that of molecular diffusion in the alveolar ducts and alveoli. West et al. (1986) proposed instead that the observed change in the average diameter can be explained equally well without recourse to such a change in flow properties. Recall that the

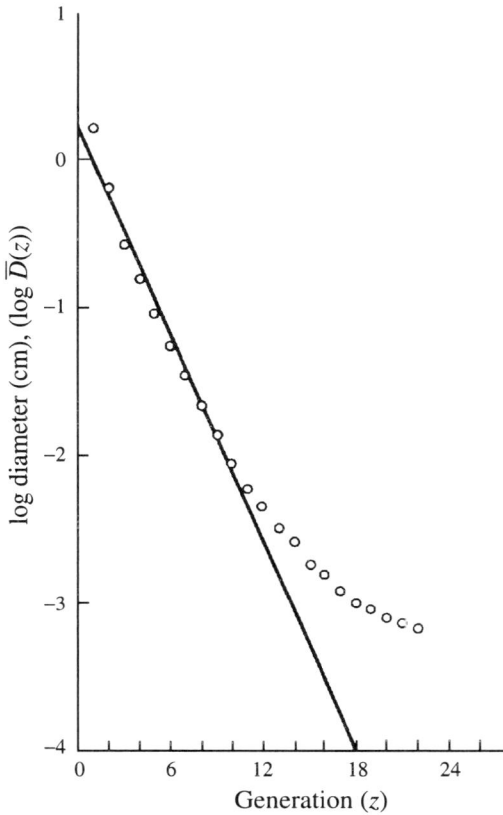

Figure 3.5. The human lung cast data of Weibel and Gomez (1962) for 23 generations are indicated by the circles, and the prediction using the exponential form for the average diameter is given by a straight line. The fit is quite good until $z \approx 10$, after which there is an asymmetric deviation of the anatomic data from the theoretical curve. (From West et al., 1986.)

arguments we have reviewed neglect the variability in the linear sizes at each generation and use only average values for the lengths and diameters. *The distribution of linear sizes at each generation accounts for the deviation in the average diameter from a simple exponential form. Thus, the small-scale architecture of the lung, far from being homogeneous, is asymmetrically structured.*

When Fig. 3.5 is redrawn on a log-log plot, we now see that the data fall closer to a straight line (Fig. 3.6). Such power law scales are similar to the scaling principles of allometry, as elucidated for morphometries by D'Arcy Thompson (1961), and for physiologic and metabolic processes by Schmidt-Nielsen (1984). Within a species or a given organ, a fractal relationship may dominate. In comparing species other necessities intervene. Just as mass influences shape and bone strength, the larger

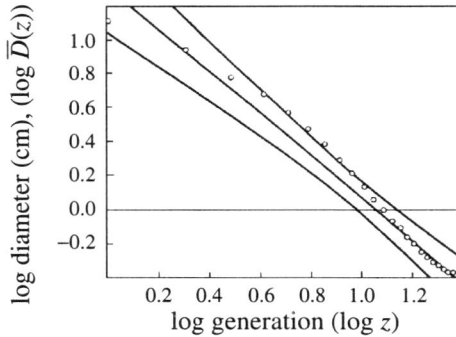

Figure 3.6. Log-log plot of lung bronchial diameters from the data of Weibel and Gomez (1962). The dominant character of the functional dependence of the average bronchial diameter on generation number is closer to an inverse power law rather than an exponential. The two additional lines indicate the 99% confidence interval about the inverse power law behavior. In addition to this inverse power law dependence of the average diameter on z, there is curvature suggesting a periodic variation of the data about this power law behavior. (From West et al., 1986.)

size of an organ necessitates increasing the fraction of the organ blood volume, that is, the volume of large arteries and veins must be higher even when capillary density is constant.

3.3 A More Complex Scaling Relationship: Weierstrass Scaling

Scaling relationships can be more complex than just a power law. One such scaling was developed by Karl Weierstrass. He was a teacher and later a colleague of Cantor, who was keenly interested in the theory of functions. He formulated a particular series representation of a function that is continuous everywhere but is not differentiable. See Deering and West (1992) for some historical perspective. His function was a superposition of harmonic terms: a fundamental with a frequency ω_0 and unit amplitude, a second periodic term of frequency $b\omega_0$ with amplitude $1/a$, a third periodic term of frequency $b^2\omega_0$ with amplitude $1/a^2$, and so on (cf. Fig. 3.7). The resulting function is an infinite series of periodic terms, each term of which has a frequency that is a factor b larger than the succeeding term and an amplitude that is a factor of $1/a$ smaller. These parameters can be related to the Cantor set discussed earlier if we take $a = b^H$ with $0 < H < 1$. Thus, in giving a functional form to Cantor's ideas, Weierstrass was the first scientist to construct a fractal function. Note that for this concept of a fractal function, or fractal set, there is no smallest scale. For $b > 1$ in the limit of infinite N the frequency $b^N\omega_0$ goes to infinity, so there is no limit or highest frequency.

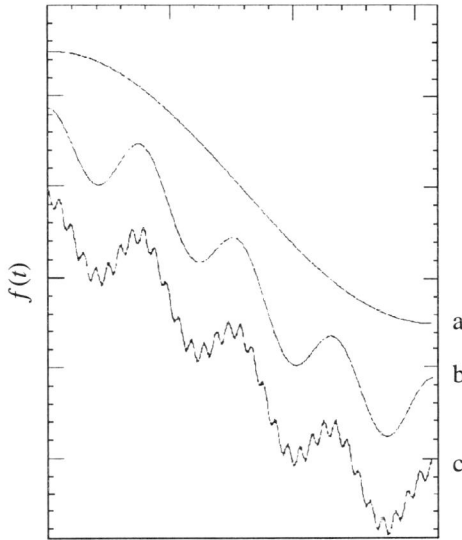

Figure 3.7. Harmonic terms contributing to the Weierstrass function; (a) a fundamental with frequency ω_0 and unit amplitude; (b) a second periodic term of frequency $b\omega_0$ with amplitude $1/a$ and so on until one obtains (c) a superposition of the first 36 terms in the Fourier series expansion of the Weierstrass function. We choose the values $a = 4$ and $b = 8$, so that the fractal dimension is $D = 2 - 2/3 = 4/3$. (From West, 1990, his Fig. 2.2.7, with permission.)

[Aside]

Consider what is implied by the absence of a limit to the frequencies in the Weierstrass function depicted in Fig. 3.8. At first glance the curve seems to be a ragged line with many abrupt changes in direction. If we now magnify a small region of the line, indicated by the box (a), we can see that the enlarged region appears the same as the original curve. If we now magnify a small region of this new line, indicated by box (b), we again obtain a curve indistinguishable from the first two. This procedure can be repeated indefinitely. This equivalence appears to be the same as self-similarity, but it is not quite. To see this difference, let us write out the mathematical expression for a generalized Weierstrass function:

$$X(z) = \sum_{n=-\infty}^{\infty} 1/a^n \left[1 - \cos b^n \omega_0 z \right]. \qquad (3.3)$$

This form was first suggested by Lévy and later used by Mandelbrot (1983). It is interesting to note that if the "time" z on the left-hand side of Eq. 3.3 is scaled by b, then a rearrangement of the index n on the right hand side yields the scaling relation

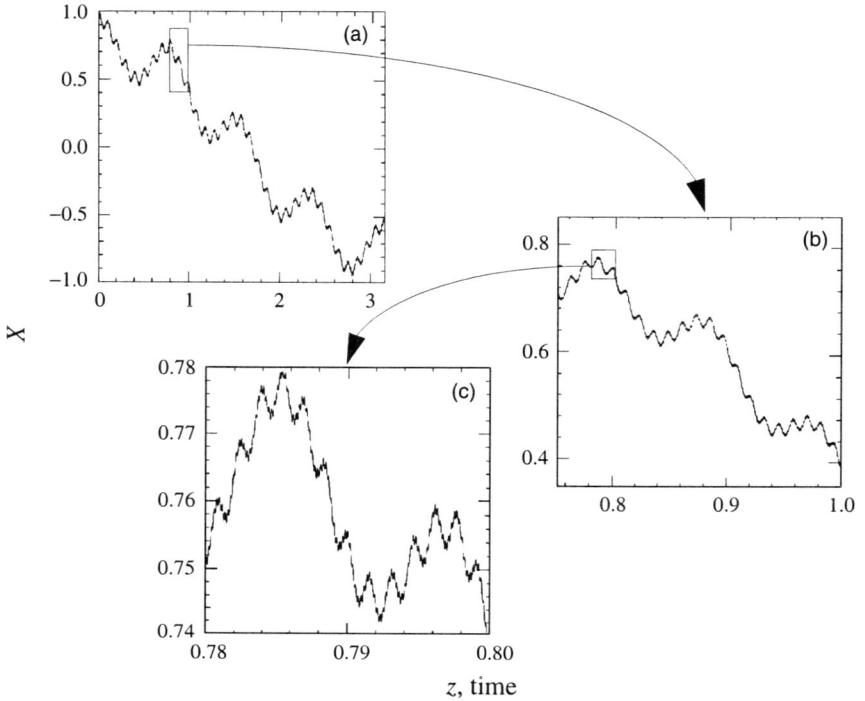

Figure 3.8. Magnifications of the Weierstrass function. Smaller and smaller wiggles are superposed on the Weierstrass curve constructed in Fig. 3.7 so that the curve looks like the irregular line on a map representing a very rugged seacoast. Inset (b) is a magnified picture of the boxed region of inset (a). The curve in (b) appears qualitatively the same as the original. The boxed region in (b) is then magnified to obtain the curve in (c), which is qualitatively indistinguishable from the first two. This procedure can in principle be continued indefinitely. (From West, 1990, his Fig. 2.2.8, with permission.)

$$X(bz) = aX(z) \quad \text{or} \quad X(z) = X(bz)/a. \tag{3.4}$$

Eq. 3.4 expresses the property of $X(z)$ being *self-affine*, which is to say that if a, $b > 1$ then a graph of the function on a scale bz could be scaled down by a factor $1/a$ to obtain the desired function on a scale z. *Whereas self-similarity expresses the fact that the shapes would be identical under magnification, self-affinity expresses the fact that X and z have to be scaled by different amounts for the two views to become identical.*

Solving the renormalization group scaling relation, Eq. 3.4, gives the dominant "time" behavior of the infinite series, Eq. 3.3. Assume a solution of the form

$$X(z) = A(z) z^{H}. \tag{3.5}$$

Insert the assumed form of the solution, Eq. 3.5, into Eq. 3.4 to obtain

$$A(z) z^H = \frac{1}{a} A(bz) (bz)^H .$$

(3.6)

In this expression z^H from both sides of the equation cancel and we obtain

$$b^H = a, \text{ and}$$

(3.7)

$$A(z) = A(bz) .$$

(3.8)

Thus the power law index H is related to the two parameters in the series expansion by $H = \ln a / \ln b$, just as we found earlier. Also the function $A(z)$ is equal to $A(bz)$ so that it is periodic in the logarithm of the variable z with period $\ln b$. The algebraic (power law) increase in $X(z)$ with z is a consequence of the scaling property of the function with z. Thus, we see that if z is increased by a factor r (z goes to rz) then X goes to $r^H X$, i.e., it is magnified by a factor r^H. For $b > a > 1$ we have $0 < H < 1$. For a one-dimensional fractal coastline $H = 2 - D$, where D is the fractal dimension.

3.4   Branching in the Lung: Weierstrass Scaling

The importance of fractals in medicine has been emphasized in a recent editorial in the British medical journal *The Lancet* (1992): ". . . the concept of fractal geometry is likely to prove very productive in areas of medicine where Euclidean geometry fails. With the availability of high-speed computers and high-resolution graphical displays we can now see that the iteration of very simple formulae can produce infinitely complex structures—perhaps the infinitely rich complex forms of nature will be revealed as repetitions of simple formulae after all."

The Fractal Lung

How do the abstract concepts of self-similar and self-affine scaling, renormalization group theory, and fractal dimensionality relate to physiological structures? Here we examine the architecture of the lung as a paradigm for the application of these ideas. The classical model of bronchial diameter scaling, as we saw, predicts an exponential decrease in diameter measurements. However, the data indicate marked divergence of the observed anatomy from the predicted exponential scaling of the average diameter of the bronchial tubes beyond the tenth generation. These early arguments assume the existence of a single characteristic scale governing the decrease in bronchial dimensions across generations. If, however, the lung is a fractal structure, no smallest scale will be present. Instead there should be a distribution of scales contributing to the variability in diameter at each generation. Based on the preceding arguments, the subsequent dominant variation of the average bronchial diameter with generation number would then be an inverse power law, not an exponential (West et al., 1986), as suggested by Figs. 3.5 and 3.6.

The arguments leading to the exponential form of the dependence of the average diameter of the bronchial tube with generation number z neglected the variability in the linear sizes at each generation and use only average values for the tube lengths and diameters. The fractal assumption, on the other hand, focuses on this neglected variability: the observed deviation of the average diameter from a simple exponential dependence on z results from the distribution in fluctuations in the sizes with generation. If we interpret the generalized Weierstrass function $W(z)$ as the diameter of the bronchial tree, after z branches then the series has two distinct contributions. One is the singular behavior of the inverse power law, which is the dependence of the average bronchial diameter W on generation number z, and the other is an analytic, short-scale variation of the measured diameter, which is averaged out in the data. The parameter b is a measure of the interval between scales that contribute to the variation in the diameter, and the parameter a denotes the importance of that scale relative to its adjacent scales. In the case of the lung, in addition to the single scale assumed in the traditional models, the fractal model assumes no single scale is dominant, but instead there is an infinite sequence of scales, each a factor of b smaller than its neighbor, that contribute to the structure. Each such factor b^n is weighted by a coefficient $1/a^n$. This is exactly analogous to the weighting of different frequencies in the generalized Weierstrass function given above. We choose $pb = 1/a$, where p is the probability that a scale size b is included in the lung and satisfies the condition $pb < 1$. Averaging this function over the smaller-scale variations yields the average diameter $D(z)$,

$$D(z) = (1 - pb) [d(z) + pd(z/b) + p^2 d(z/b^2) + \ldots] , \qquad (3.9)$$

where $d(z)$ is characterized by a single scale. Eq. 3.9 is the expanded form of a renormalization group relation,

$$D(z) = pD(z/b) + (1 - pb) d(z) . \qquad (3.10)$$

This scaling law, using the methods discussed in the previous chapter, produces an average diameter of the form

$$D(z) = \frac{A(z)}{z^H} , \qquad (3.11)$$

with $H = \ln(1/p)/\ln b$ and again $A(z)$ is periodic in $\ln z$ with period $\ln b$. The solution to the renormalization group relation, Eq. 3.9, given by Eq. 3.11 is only mathematically valid for large z. Thus, we would have expected that Eq. 3.11 would be a faithful description of the lung asymptotically, if at all, and not for small z. A good fit to the data from four mammalian species, humans, dogs, rats and hamsters, is $A(z) = A_0 + A_1 \cos(2\pi\ln z/\ln b)$. From the quality of the fit to the data it appears that for z beyond the fifth or sixth generation we are in the asymptotic regime.

In order for the series (Eq. 3.9) to converge we must have $p < 1$. We interpret $pd(z/b)$ to be a fraction p of the branching process that has lengths drawn from a distribution having a mean b times larger than that in $d(z)$. Similarly, the term $p^2 d(z/b^2)$ accounts for paths along the bronchial tree which by generation 2 experiences lengths b^2 that have larger mean than that in $d(z)$. This occurs with probability p^2. This argument is repeated for each of the terms in the renormalization series (Eq. 3.9) for $D(z)$.

Finally, for $p \ll 1$ few if any such increased diameters occur early in the branching process of the bronchial tree so that

$$D(z) \simeq d(z) + O(p), \qquad (3.12)$$

and the average diameter is exponential, the form given by classical scaling. Here we indicate that the neglected factors contribute to Eq. 3.12 terms of order p and are therefore negligible for small values of z. By generation ten there are $2^{10} = 1024$ branches so that if $p = 10^{-3}$, there are at most a few special branchings in the first ten generations. However, as we pointed out, the representation Eq. 3.11 is only expected to be valid for large z so we would not expect p to have such small values. From the data depicted in Fig. 3.5 we see that the average diameter of the bronchial tree has the form of a modulated inverse power law distribution. The overall slope of the curve yields the power law index H, and the period of oscillation yields the parameter $\ln b$. In Fig. 3.6 we indicate the inverse power law fit to the data to emphasize the variation of the data about this line, and the lines for the 99% confidence interval about the inverse power law are indicated. This harmonic modulation is unlikely to represent a random variation, since it is apparent in multiple plots, using data from different mammalian species and from two different laboratories. Shlesinger and West (1991) obtain from these data $H = 0.86$ and 0.90 for the dog and hamster, respectively, from the slopes of the average diameter curves and $b \approx 9$ as the period of oscillation of $A(z)$. In a similar way we obtain $H = 1.26$ and 1.05 for the human and rat, respectively, with $b \approx 11$. These parameters yield $p_{\text{human}} \approx 0.05$, $p_{\text{rat}} \approx 0.08$, $p_{\text{dog}} \approx 0.15$ and $p_{\text{hamster}} \approx 0.12$, which are overestimates of the probabilities since they are based on the complete data sets rather than only $z \geq 6$. Said differently, we expect the exponential form from classical scaling to accurately describe the early stages of the bronchial tree where the network may have a fundamental scale. However, at some intermediate z-value there should be a transition from the exponential to the modulated inverse power law (Eq. 3.11) behavior that correctly describes the asymptotic region of the bronchial network. The parameters p and H should therefore be determined by the data from these latter branches. However, the values quoted herein use the full network and are therefore overestimates of the true values of these parameters. From Eq. 3.11 we see that the average diameter is an inverse power law in the generation index z modulated by the slowly oscillating function $A(z)$ just as is observed in the data. In point of fact, we find that the present model provides an excellent fit to the lung data in four distinct species: dogs, rats, hamsters, and humans. The quality of this fit shown in Fig. 3.9

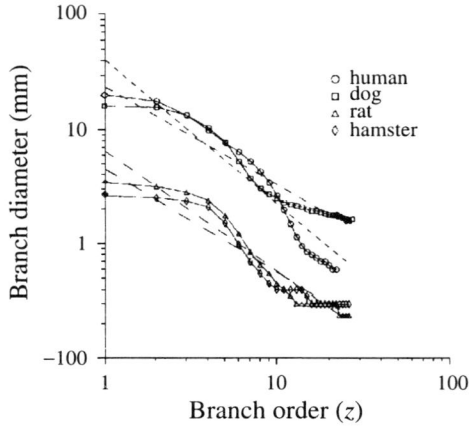

Figure 3.9. The variation in diameter of the bronchial airways is depicted as a function of generation numbers for humans, rats, hamsters, and dogs. The modulated inverse power law observed in the data of Raabe et al. (1976) is readily captured by the function $F(z) = [A_0 + A_1 \cos(2\pi \ln z / \ln b)]/z^H$. (From Nelson et al., 1990.) The dashed lines indicate the inverse power law, and the solid lines are used as an aid to the eye.

strongly suggests that the renormalization group relation captures a fundamental property of the structure of the lung that is distinct from traditional scaling. In Fig. 3.10 we display the point by point fit of the mammalian lung for three of the species, fitted with the full model, which fits the data closely in each case. Furthermore, the data show the same type of scaling for bronchial segment lengths and therefore for airway volumes.

Other Such Structures

On a structural level the notion of self-similarity can also be applied to other complex physiological networks. The vascular system, like the bronchial tree, is a ramifying network of tubes with multiple scale sizes. To describe this network Cohn (1954, 1955) introduced the notion of an "equivalent bifurcation system." The equivalent bifurcation systems were examined to determine the set of rules under which an idealized dichotomous branching system would most completely fill space. The analogy was based on the assumption that the branchings of the arterial system should be guided by some general morphogenetic laws enabling blood to be supplied to the various parts of the body in some optimally efficient manner. The branching rule in the mathematical system is then to be interpreted in the physiological context. This was among the first physiological applications of the self-similarity idea, predating the formal definition of fractals.

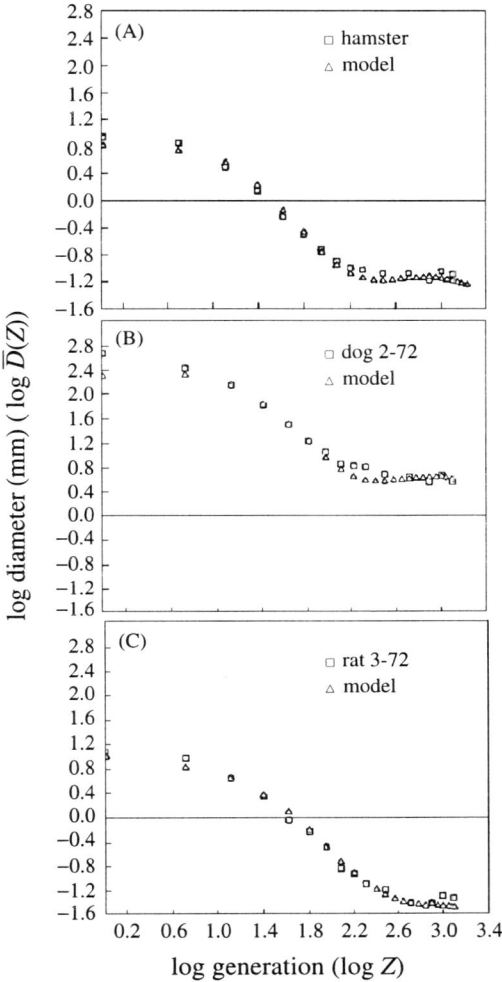

Figure 3.10. The point-by-point fit of the average diameter for (A) hamster, (B) dog and (C) rat using the fractal model equation (3.2) with $A(z) = A_0 + A_1 \cos(2\pi \ln z / \ln b)$. (From West et al., 1986.)

Modern treatments of blood flow often rely on fractal models for vascular networks; the dichotomous branching network of vessels represents the arterial tree and connects to a similar venous network (van Beek et al., 1989; Bassingthwaighte and van Beek, 1988). Regional myocardial blood flow is very heterogeneous. The estimated relative dispersion *RD* (standard deviation divided by the mean) of the distribution of blood flow increases when the sample pieces into which a heart is divided are made smaller. Meinhardt (1982) presented a set of models for growth based on branching to supply domains which required nutrients and which expressed growth factors when inadequately supplied. Likewise, Wang and Bassingthwaighte (1990) provide an algorithm for dichotomous branching to supply cells distributed in domains of various geometries.

3.5   Other Examples

The first example of the fractal concept applied to microbial species was in the analysis of the growth patterns of *Streptomyces griseus* and *Ashbya gossypii*. Here the cell is the smallest aggregating unit and the colony is the aggregate. It was found by Obert et al. (1990) that the global structure of the branched mycelis is describable by a fractal dimension which increases during morphogenesis to the value 1.5. They determined that the mycelial *A. gossypii* changes during growth in such a way that it makes a transition from a *mass fractal* to a *surface fractal*, whereas *S. griseus* does not. A mass fractal is a structure in which the whole mass of the organism is fractal, whereas a surface fractal is one in which only the surface (border) is a fractal. To distinguish between the two, Obert et al. (1990) used two different box-counting methods. The number of boxes used to cover the aggregate is given by $N_{box}(\varepsilon) = C\varepsilon^{-DB}$, where D_B is the fractal dimension determined by box counting. The box mass (*BM*) method was applied to the entire mass of the mycelium, yielding the fractal dimension $D_B = D_{BM}$. With the box surface (*BS*) method, only those boxes intersecting the surface of the mycelium are used, yielding the fractal dimension $D_B = D_{BS}$. Note that for a mass fractal $D_{BM} = D_{BS}$, whereas for a surface fractal the two dimensions are distinct (cf. Fig. 3.11).

Another phenomenon in which the concept of surface fractal is important is in heterogeneous chemistry, in particular for surface reactions on proteins where areas of high fractal dimension may give rise to high receptor selectivity (see Mandelbrot, 1983). Goetze and Brickmann (1992) have numerically examined the self-similarity of 53 protein surfaces to experimentally determine the dimensional structure of these biological macromolecules. They found that the surfaces are self similar within a range of 1.5Å to 15Å, and point out that a local surface dimension may be useful for molecular recognition as well as being important for biomolecular evolution.

Goetze and Brickmann's results showed that protein surfaces have an exponential scaling with respect to the size of the particles, i.e., the surface area $A = A_0\varepsilon^{-D_s}$. They find that the average fractal dimension increases with increasing size of the biomolecule so that large molecules appear to have more surface irregularity than do smaller ones to a potential substrate of molecules of a given size (Fig. 3.12).

It is not only the surface of proteins to which the fractal concept has been applied. Stapleton et al. (1980) examined the physical properties of several proteins and found that they could interpret the twisted overall structure of these macromolecules as a self-avoiding random walk (SAW). A SAW is one in which the walker is not allowed to intersect his own path and therefore has a long-term memory. The self-

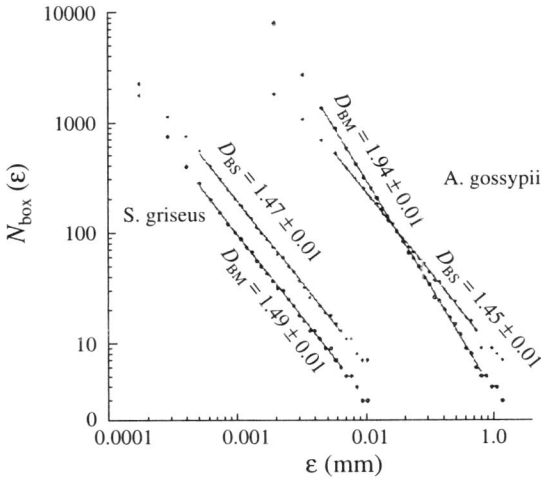

Figure 3.11. Evaluation of the mycelia shown in Obert et al. (1990). Experimental values of number of boxes $N_{box}(\varepsilon)$ versus box length ε (mm), determined by the BS (+) and BM (◊) methods. Scaling behavior was well over one decade, as indicated by the regression lines. (From Obert et al., 1990.)

avoiding character of the walk mimics the sinuous backbone of the protein. These results show that the protein occupies a space of fractal dimension $D = 1.65 \pm 0.04$. This agrees well with the results of SAW that yield a theoretical dimension of $5/3$ to within 1%.

Many other fractal-like structures in physiology are also readily identified by their multiple levels of self-similar or self-affine branchings or foldings—for example, the bile duct system, the urinary collecting tubes in the kidney, the convoluted

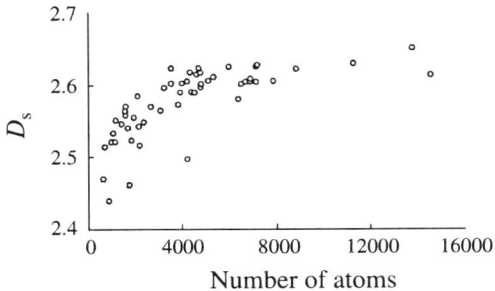

Figure 3.12. Surface fractal dimensions D_s of proteins as functions of the molecular size (measured by the number of atoms in the molecule). (From Goetze and Brickmann, 1992.)

surface of the brain, the lining of the bowel, neural networks, and the placenta (Goldberger and West, 1987b). The fractal nature of the heart is particularly striking. The cardiac surface is traversed and penetrated by a dichotomously branching system of coronary arteries and veins. Within its chambers, branching strands of connective tissue, the chordae tendineae, anchor the mitral and tricuspal valves, and the electrical impulse is conducted by a fractal network, the His-Purkinje system, composed of specialized myocardial cells (with high conductivity through gap junctional connections) embedded within the muscle.

3.6   Summary

In this chapter we have emphasized the importance of inverse power law distributions and how such distributions arise from the self-similarity of the underlying process. The property of self-similarity implies that a scaling relation exists between the structure observed at one scale and that found at subsequent scales. However, this is not the familiar scaling used in biology that assumes processes are continuous and uniformly fill space, but is rather a new kind of scaling that is filled with gaps. Examples presented in this chapter for which the application of these ideas has provided new insight are the branching networks of the bronchial tree and the vascular system, the fingering growth patterns of microbial species, and the irregular structures of proteins.

4

Fractal Measures of Heterogeneity and Correlation

> Science is said to proceed on two legs, one of theory (or, loosely, of deduction) and the other of observation and experiment (or induction). Its progress, however, is less often a commanding stride than a kind of halting stagger—more like the path of the wandering minstrel than the straight-ruled trajectory of a military marching band.
>
> Timothy Ferris (1988)

4.1   Introduction

If anything is characteristic of biological signals it is that they fluctuate. Yet, like the staggering progress of science, there is direction and correlation. We often need to characterize correlations, that is, the degree and extent of similarity of a measured property as it varies in space and time. The use of the normal curve assumes the fluctuations are independent, not correlated. However, in nature these fluctuations are more structured. For example, structured correlations occur in the variation in density of water in a cloud, the variation in regional blood flow, and local receptor densities in an organ.

In these cases the correlation between neighboring regions is closer than that between separated regions. The variation often shows "self-similarity upon scaling," namely, *the correlation between neighbors of 1 cubic centimeter is the same as the correlation between neighbors of 1 cubic millimeter.* Modern statistical textbooks scarcely touch the issue of how to analyze data when the variance depends on the choice of unit. Standard textbooks, for example, those of Rogers (1974) and Snedecor and Cochran (1980), give useful methods for assessing estimates of means, for testing differences between regions, and so on, but in general these approximations do not recognize that the estimates of variance may be dependent upon assuming a particular unit size in space or time. The classic statistics works of Feller (1968) and Cramér (1963) come close to discussing some of the implications of fractal relationships. Shaw and Wheeler (1985) in their textbook on geographical analysis recognize the problem, and state that no solutions are available. The best handling of the predicament is done in the field of mining geostatistics, where a technique initiated by Krige (1951) was found useful in determining the extent of field of mineral deposits. This is well described in a text by Journel and Huijbregts (1978), but Krige's technique is not yet thoroughly developed. Journel and Huijbregts did advance Krige's technique, but the models for spatial variances are purely empirical. The works of Ripley (1977, 1979) provide methods for measuring spatial variation, and give a framework for further work in which scale-independence will play a role.

These ideas were not wholly new; they were foreshadowed in the work of Perrin (1908), as translated by Mandelbrot (1983) and which we have abridged:

> Consider, for instance, the way in which we define the density of air at a given point and at a given moment. We picture a sphere of volume v centered at that point and including the mass m. The quotient m/v is the mean density within the sphere, and by *true* density we denote some limiting value of this quotient. . . . The mean density may be notably different for spheres containing 1,000 cubic meters and 1 cubic centimeter. . . .
>
> Suppose the volume becomes continually smaller. Instead of becoming less and less important, these fluctuations come to increase. For scales at which the Brownian motion shows great activity, the fluctuations may attain 1 part in 100, and they become of the order of 1 part in 5 when the radius of the hypothetical spherule becomes of the order of a hundredth of a micron. . . .
>
> Let our spherule grow steadily smaller. Soon, under exceptional circumstances, it will become empty and remain so henceforth owing to the emptiness of intra-atomic (and *intermolecular*) space; the true density *vanishes* almost everywhere, except at an infinite number of isolated points, where it reaches an infinite value. . . .
>
> Allow us now a hypothesis that is arbitrary but not self-contradictory. One might encounter instances where using a function without a derivative would be simpler than using one that can be differentiated. When this happens, the mathematical study of irregular continua will prove its practical value.

The fractal form of such correlations is not mandated by the mathematics. Thus, when a biological system exhibits such a form one should look for its anatomic, physiologic, dynamic or phylogenetic basis. In this chapter we show how to analyze the heterogeneity, or variation, in a local property by determining how the variance depends on the size of the units used to measure it.

The Critical Concept Is That of Scale

If observers of natural phenomena had each time asked, "What is the scale of the unit at which I should make my measures?", or "Do my estimates change when I change my measuring stick?", then the fractal nature of biological phenomena would have been long since revealed. Once this question is asked, the doors open to seeing the self-similarity that was so evident in the illustrations in Chapter 1. One stimulus to ask the question was our need to understand heterogeneity of tissues and its basis in structure and function.

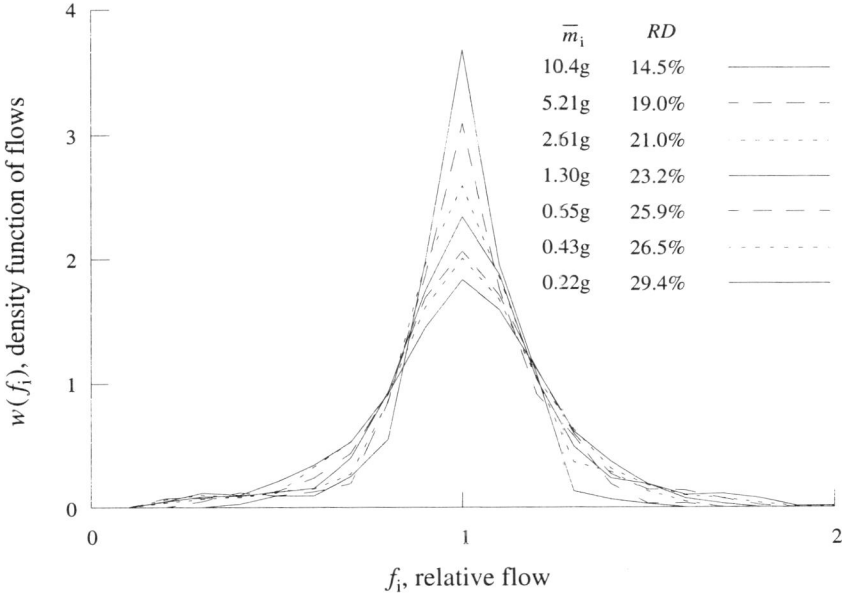

$\overline{m}_i$	RD	
10.4g	14.5%	————
5.21g	19.0%	– – – –
2.61g	21.0%	· · · · · ·
1.30g	23.2%	————
0.55g	25.9%	– – – –
0.43g	26.5%	· · · · · ·
0.22g	29.4%	————

Figure 4.1. Probability density functions of myocardial blood flows at seven levels of resolution. Composite data from 11 sheep using a "molecular microsphere," iododesmethylimipramine, which deposits at submillimeter resolution in proportion to flow. The relative dispersion, RD, is given for each *pdf* at a resolution defined by the average mass $\overline{m}$. (Figure reproduced from Bassingthwaighte et al., 1989, with the permission of the American Heart Association.)

For example, the regional blood flows within an organ have this kind of variability. Flows can be estimated on a local level using tracers that are deposited in proportion to the flow when they are completely extracted from the blood passing through capillaries. Fig. 4.1 illustrates that when the resolution is high, the probability density function of the flows around their mean is broad; using tissue pieces of 0.22 g the coefficient of variation of the flows (the standard deviation divided by the mean) was 29%. However, when the pieces of tissue were larger, the apparent overall variation was less; with 10.4 g pieces the coefficient of variation was only 14.5%. Taking only large pieces gives only the average flow within each piece, thereby smoothing the data, not revealing the variation within the piece. No matter how small the pieces are, there is further variation within each, as can be demonstrated using autoradiography.

The scale dependence of the measure of heterogeneity shown in Fig. 4.1 illustrates the problem of how to define the variation. The relationship between RD and $\overline{m}$ cannot *a priori* be expected to be fractal; on the contrary, a fractal relationship is only one of infinite possibilities. If the relationship *is* fractal, then insight is gained *and* a mathematically simple relationship exists.

Measures of Heterogeneity

In this chapter we cover three basic scale-independent methods of assessing spatial and temporal heterogeneity. The methods are

1. Dispersional analysis using the Standard Deviation (*SD*) divided by the Mean, which is the relative dispersion, *RD*, at each of a succession of scales of resolution. (*RD* is identical to the coefficient of variation.)
2. Hurst rescaled range analysis using the range of cumulative deviations from the mean divided by the *SD* at each of a succession of interval lengths.
3. The autocorrelation function, r_n, where r_n is the correlation coefficient for regions of any specified size separated by n regions.

Our illustrations are mainly of one-dimensional data, functions of time, or functions of spatial distance (assuming isotropy in two-space or three-space). Ideally, all three methods should give the same information on the variation and the degree of correlation between parts of the data set. However, they differ in accuracy and reproducibility.

The variance around the mean or its square root, the standard deviation, *SD*, serves as a measure of dispersion. We make use of the relative dispersion, *RD*, which is identical to the coefficient of variation, *CV*. The reason for preferring the term "relative dispersion" to "coefficient of variation" is that "relative dispersion" is directly descriptive, whereas "coefficient of variation" is used most often in a different context, for example, as the normalized sum of squares of differences between two functions. The *RD* or *CV* is dimensionless:

$$RD = SD/(\text{mean of a distribution}) . \tag{4.1}$$

Many distributions are close enough to being symmetrical, even if not really Gaussian, that their spread is adequately defined by the *RD*. Although the heterogeneity of tissue properties has often been ignored in the past, modern techniques in tissue staining at the microscopic level, and imaging by positron emission tomography or nuclear magnetic resonance at the macroscopic level, have demonstrated that virtually every tissue is heterogeneous in one or several features. The fractal description, when applicable, offers a simple relationship between observations at the microscopic level and those at the macroscopic level.

[Aside]

The general correlation coefficient, r, is defined as covariance divided by variance:

$$r = \text{Cov}\,(Y_i, Y_j)\,/\text{Var}\,(Y) , \tag{4.2}$$

where Y_i and Y_j represent subsets of the observations Y. To be more explicit for nearest-neighbor pieces, $n = 1$, or at a separation n the autocorrelation function r_n is

$$r_n = \frac{\left(\dfrac{N}{N-n}\right) \sum\limits_{i=1}^{i=N-n} Y_i Y_{i+n} - \left(\sum\limits_{i=1}^{i=N} Y_i\right)^2 / N}{\sum\limits_{i=1}^{i=N} Y_i^2 - \left(\sum\limits_{i=1}^{i=N} Y_i\right)^2 / N}. \tag{4.3}$$

For a signal occurring in 1-dimension (the Euclidean dimension, E, is 1), the fractal dimension lies between 1 and 2. In general the fractal dimension lies between E and $E + 1$, and is therefore a measure of how far the roughness extends toward the next-higher dimension.

When nearest-neighbor pieces are positively correlated the fractal dimension D lies between 1.0 and 1.5. With $r_n = 0$ for all n, the fractal D is 1.5, and the components are unrelated, purely random. Values above 1.5 indicate negative correlation between nearest neighbors. As D approaches 2.0, the correlation coefficient approaches -1.0, which one can imagine for a crystalline matrix of uniformly spaced ones and zeroes. This does not work exactly but fits with the general perspective that smooth signals have D near 1, and do not fill much of the plane, whereas high Ds, approaching 2.0, fill much of the plane. Consider the one-dimensional function, $Y(t)$. To fill the Y,t plane as fully as possible with one Y value at each t requires that Y values at neighboring t values be negatively correlated. Using D to characterize variation has an additional virtue: D shows the degree of spatial correlation, and its spread with distance. This gives D great potential for the analysis of regional properties and functions.

In this chapter we make use of the Hurst coefficient H, which is simply related to D for one-dimensional data:

$$H = 2 - D, \tag{4.4}$$

and in general $H = E + 1 - D$, where E is the Euclidean dimension, 1 in this case.

4.2   Dispersional Analysis

Relative Dispersion of One-Dimensional Signals

Fractal relationships can often be used to describe the apparent heterogeneity over some fairly large range of unit sizes. This allows for a concise description of the heterogeneity of the property. The general relationship was observed empirically by Bassingthwaighte (1988) to exhibit a linear relationship between the log of the variance, or the RD, and the log of the size of the observed unit:

$$RD(m)/RD(m_0) = (m/m_0)^{1-D}, \tag{4.5}$$

and the fractal dimension D is

$$D = 1 - \frac{\log\,[RD\,(m)\,/RD\,(m_0)\,]}{\log\,(m/m_0)}, \tag{4.6}$$

where m is the element size used to calculate RD and m_0 is the arbitrarily chosen reference size, where "size" is the number of points aggregated together in a one-dimensional series, or a length over which an average is obtained, or the mass of a tissue sample in which a concentration is measured. One reason for calling this a fractal relationship rather than just a power law relationship is that the possible slopes of the relationship are bounded by limits, and the fractal dimension, D, gives insight into the nature of the data. In this particular case, a value for D of 1.5 is found when the coefficient of variation of the densities (or the relative dispersion, RD) represents random uncorrelated noise, and the lower limit of $D = 1.0$ represents uniformity of the property over all length scales.

A second reason for calling this a fractal relationship is that the fractal dimension D gives a measure of the spatial correlation between regions of defined size or separation distance. The correlation between nearest neighbors is derived from this in Section 4.4:

$$r_1 = 2^{3\,-\,2D} - 1, \tag{4.7}$$

where r_1 is the correlation coefficient (van Beek et al., 1989). If a fractal relationship is a reasonably good approximation, even if only over a decade or so, then it will prove useful in descriptions of spatial functions and should stimulate searches for the underlying bases for correlation; for myocardial flow heterogeneities the basis appears to be a consequence of the fractal nature of the branching vascular network.

The goal of this section is to explore fractal variances and to illustrate the ideas expressed above, using numerical experimentation. The tests are posed by sets of data in two dimensions, where the characteristics of the data are known in advance and the degrees of correlation are either known or calculable. We begin with an exploration of random noise without correlation, then examine spatial functions that have correlation. Finally, the topic is illustrated via an application to studies of myocardial blood flow.

Estimating D from the RDs

The basic idea is to form groups from the data each consisting of m consecutive data points. First, the mean of each group is determined and then the relative dispersion (standard deviation divided by the mean) of these group means. We then find how the mean of each group and the relative dispersion of these means changes as the groups are enlarged to contain more and more data points. This process is shown for the signal in Fig. 4.2. The top shows the signal, ungrouped. The second row is the means of groups of two, the third row, groups of four, etc. The apparent variation diminishes as the number of data points in each binning period, τ, enlarges.

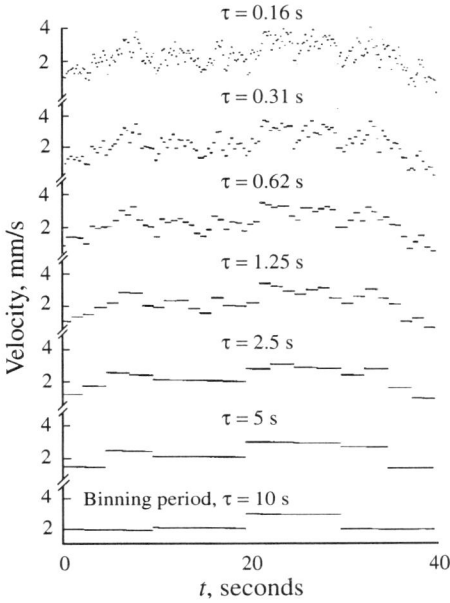

Figure 4.2. Successive averaging of pairs of elements of a one-dimensional correlated signal. The group sizes enlarge by a factor of two for each row; the group means are plotted over the range included in the group. Fractal $D = 1.38$, $H = 0.62$. (From Bassingthwaighte et al., 1988b, their Fig. 3, with permission.)

[Aside]

Estimating the fractal dimension from the variation in a one-dimensional signal.

1. Define the signal. Consider the simple case of a signal measured at even intervals of the independent variable, for example, as a function of position along a line: f_i at $t = t_1, t_2, \ldots t_n$. The chosen constant time interval between samples, Δt, may be dictated by the resolution of our measuring apparatus, or may be defined arbitrarily by taking a subset of a larger data set. In our example, there are 256 data points obtained at even time intervals, shown in Fig. 4.2. This is like having groups of size m, with $m = 1$.

2. Calculate the mean $\overline{f}$ of the whole set of observations, f:

$$\overline{f} = \Sigma f_i / n. \tag{4.8}$$

3. Calculate the standard deviation of the set of n observations:

$$SD = \frac{1}{n} \sqrt{n \Sigma f_i^2 - (\Sigma f_i)^2}, \tag{4.9}$$

where the denominator is n, using the sample SD rather than the population SD where $n - 1$ would be used. Calculate the RD for this group size:

$$RD(m) = SD(m)/\bar{f}. \qquad (4.10)$$

4. For this grouping, each group consists of one datum. Thus, $m = 1$, and the mean of each group equals the value of the one datum in that group. Thus, in Eq. 4.9, the f_i are the individual data values and $n = 256$.

5. For the second grouping aggregate adjacent samples into groups of two data points and calculate the mean for each group, the SD for the group means and the RD for the group means. For "groups" with two members the result is in the second row of Table 4.1. For this $m = 2$, $n = 128$, and $m \times n = 256$.

6. Repeat step 5 with increasing group sizes until the number of groups is small. The results in Table 4.1 show that as the number of data points in each group increases the RD decreases. For each $m \times n = 256$.

7. Plot the relationship between the number of data points in each group and the observed RDs. A plot of the logarithm of $RD(m)$ versus the logarithm of the number of data points in each group gives a straight line if the signal is a simple fractal, as is seen in Fig. 4.3.

8. Determine the slope and intercept for the logarithmic relationship. This can be done by taking the logarithms of m and RD in Table 4.1 and estimating the linear regression by the standard Y on X regression, which assumes no error in m. (One can also use nonlinear regression to fit a power law equation of the form of Eq. 4.5, but the results will usually be similar.) Any of the group sizes might be the chosen reference size, m_0. For example, choosing 8 as the reference group size, then we have

$$RD(m) = RD(8) \cdot (m/8)^{1-D} = 0.29 \cdot (m/8)^{-0.38}, \qquad (4.11)$$

and if we choose $m_0 = 1$, then the equation is that given in Fig. 4.3. To use Eq. 4.6 directly with a single value of m would be like finding the straight line in Fig. 4.3 by drawing a line between two of the points instead of using all the points, so using the slope of the regression line is better.

9. Calculate the fractal D. The power law slope is $1 - D$, so in this case $D = 1.38$. The standard linear regression calculation provides a measure of the confidence limits on the slope. Nonlinear optimization routines may be used to provide estimates of confidence limits on the estimate of the slope from the covariance matrix.

Estimating Variation in Two-Dimensional Spatial Densities

To illustrate the problem in estimating variation and the approach to its solution, we begin with a unit square surface on which points are distributed with equal probability in all locations, using a random number generator to choose values for X

Table 4.1. Relative dispersions, *RD*, of subgroups of a
256-point correlated signal

Grouping	m, number of points in each group	n, no. of groups	$RD(m)$ of the group means
1	1	256	0.62
2	2	128	0.48
3	4	64	0.37
4	8	32	0.29
5	16	16	0.22
6	32	8	0.17
7	64	4	0.13

and *Y* in the interval 0 to 1. An example of the result is shown in Fig. 4.4. The points are not positioned uniformly, despite the fact that if enough were calculated they would approach statistical uniformity. The *problem* is to estimate the variation in the density of dots over the area and determine any correlations or patterns that exist; or to determine unambiguously that none exist.

The value of the mean is known, 8192 points per unit square. To measure the variation in local point density we might overlay grids with differently sized pixels, such as are shown in Fig. 4.5. The *RD* for the number of points for the four-pixel grid was 2.2%, for the sixteen-pixel grid it was 4.6%, and for the sixty-four-pixel grid it was 8.9%. This was extended to finer grids. This illustrates the problem, that is, which estimate is the "correct" variation, and how can it be chosen?

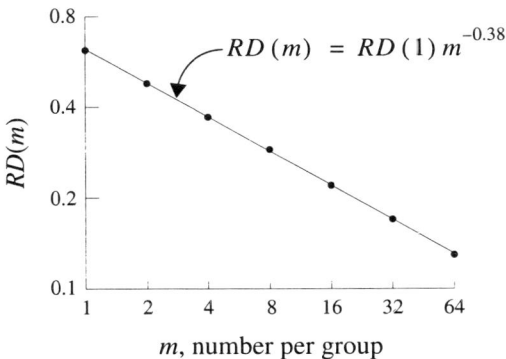

$$RD(m) = RD(1) \, m^{-0.38}$$

Figure 4.3. Fitting of the data of Fig. 4.2 and Table 4.1. The log of the relative dispersion (*RD*) of the mean of each group is plotted versus the log of the number of data values *m* in each group. The log-log regression is a good fit. Fractal $D = 1.38$; $RD(m = 1) = 0.623$.

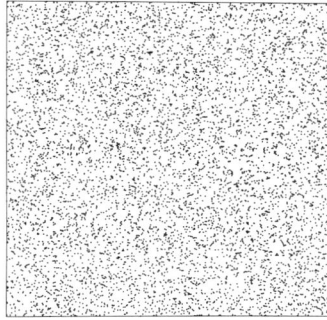

Figure 4.4. An array of points spread over the unit square. 8192 points were generated by a uniform random number generator. What is the variation in the density of the points? What is the variation in distance between nearest neighbors? (From King et al., 1990, their Fig. 1, with permission.)

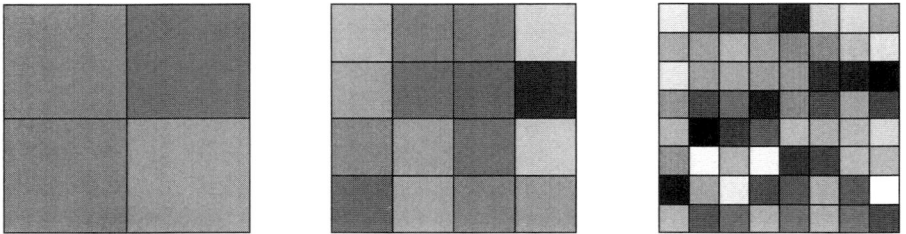

Figure 4.5. Overlays of grids with differing mesh sizes on the array of uniform random points from Fig. 4.4. The number of points per pixel is shaded according to the pixel mean, where white indicates values 20% below the mean for the whole and black 20% above the mean. From King et al., 1990, their Fig. 2, with permission.)

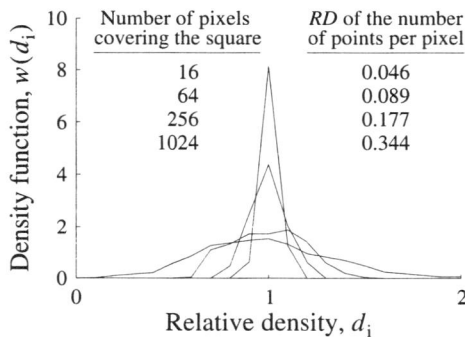

Number of pixels covering the square	RD of the number of points per pixel
16	0.046
64	0.089
256	0.177
1024	0.344

Figure 4.6. Probability density functions of numbers of points per pixel for various-size pixel grids covering the square in Fig. 4.4. For example, the average density per pixel for the sixteen-pixel covering is shown as the middle panel of Fig. 4.5.

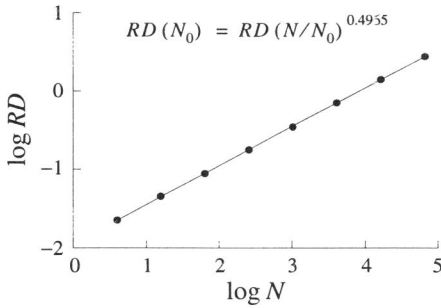

Figure 4.7. Relative dispersions obtained from the uniform random distribution of Fig. 4.4 as a function of number of pixels (N) in the grid. The plot is $\log_{10}N$, for N up to 65,536 pixels. The area of each pixel is inversely proportional to the square of the number of pixels in the grid. (From King et al., 1990, their Fig. 3, with permission.)

Graphs of the probability density functions of points per unit area are shown in Fig. 4.6. With smaller pixels the distributions are broader. The plot of the logarithm of RD versus the logarithm of the number of pixels is shown in Fig. 4.7. The expected slope is such that with each quadrupling of the number of pixels the RD should increase by a factor of two. The fit of a linear regression to the logarithms gave an estimated slope of 0.4965 per quadrupling. (Admittedly, this may not be a completely appropriate regression in view of the fact that the original data were in a linear, not logarithmic, domain as critiqued by Berkson, 1950.) The fractal, or power law expression, is

$$RD\,(N)\,/RD\,(64)\;=\;(N/64)^{0.4965}\,,\tag{4.12}$$

where $N = 64$ was the arbitrarily chosen reference number of pixels, near the middle of the range. The estimated fractal D was thereby 1.4965 compared to the theoretical value of 1.5 for uncorrelated events. (This result can be obtained also from Eq. 4.39 when the covariance is zero.) The nonuniformity observed in Fig. 4.5 by visual inspection is a reflection of this randomness.

Results with Nonrandom Arrays

Performing the same exercise with nonrandom nonfractal arrays leads to an interesting result. Strong correlation between the local densities of neighbors was obtained by generating a function whose densities, while still probabilistic, were distinctly shaped over the field. Gaussian, two-dimensional profiles resulted from generating the sum of two sets of points, each from a random number generator producing a Gaussian density function with standard deviation 0.15, centered at

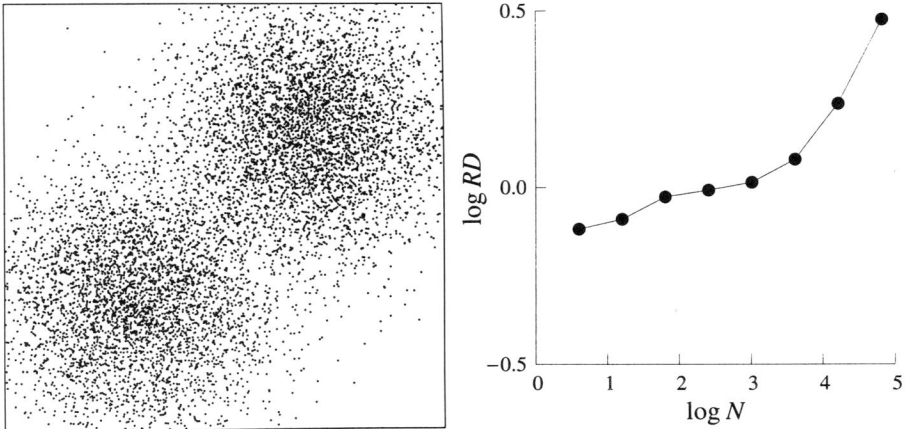

Figure 4.8. Nonrandom array. *Left panel:* Two overlapping Gaussian distributions of points (each with $SD = 0.15$) centered at $X,Y = 0.3, 0.3$ and $0.7, 0.7$. The total number of points is 8192. *Right panel:* Relative dispersion of densities of points at different pixel areas. (From King et al., 1990, their Figs. 4 and 5, with permission.)

$(X,Y) = (0.3, 0.3)$ and $(0.7, 0.7)$. The number of points totaled 8192, as in Fig. 4.4, so that the mean densities are identical in Fig. 4.4 and Fig. 4.8.

The same sequence of grids were placed over this correlated array, and RD calculated for each mesh size as before. The results, shown in Fig. 4.8 (right panel), differ in two respects from those obtained for Fig. 4.4: the line is no longer a simple power law relationship and, at mesh sizes below 64×64 pixels, the slope of the line of RD versus mesh size is less than for uniformly distributed data. A single Gaussian profile would give a similar result. The lower slope with large pixel sizes is about 0.062, giving an apparent fractal D of 1.062. The slope steepens toward 0.5 at large N and small pixel sizes, as it would for any random signal.

An Exploration of Continuous Distributions

Instead of generating the positions of points with a statistical distribution of locations in two-dimensional space, we took the average point density in each of 4096 pixels and by using these average densities within a pixel generated the continuous distribution shown in Fig. 4.9. It has the same mean probability densities, or concentrations, as Fig. 4.8 (left panel). The mesh was 64×64 for a total of 4096 squares within the original unit square. Within each pixel the concentration was considered uniform. The grids were overlaid, the average concentration calculated for each grid unit, and the RDs calculated for the several different levels of resolution as before. The result is strikingly different. The relationship between RD and grid element size, shown in Fig. 4.9, now has a clearly defined plateau beyond which further refinement of the grid element size, or resolution of the

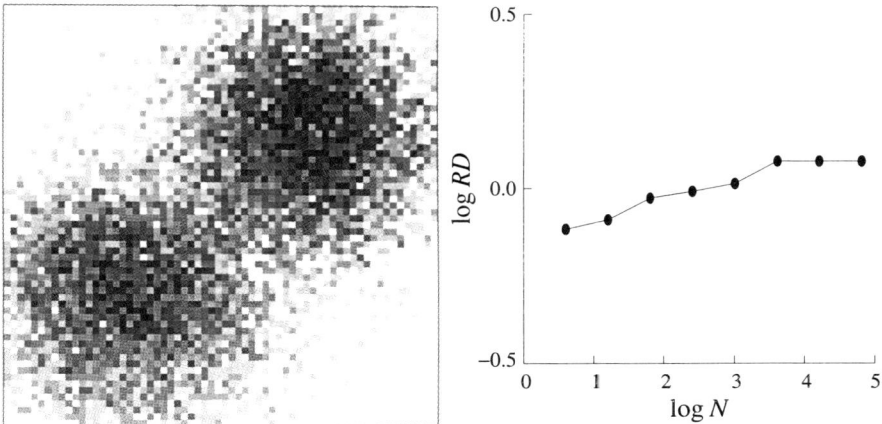

Figure 4.9. Continuous representation of the Gaussian distributions shown in Fig. 4.8. *Left panel:* Each of the 4096 pixels is internally uniform in concentration. *Right panel:* Relative dispersions over the unit square at different grid sizes for the continuous distribution. (From King et al., 1990, their Figs. 6 and 7, with permission.)

observation element size, gives no further increase in the observed heterogeneity. The plateau begins at the grid resolution that matches the pixel size of the original continuous distribution. The result is inevitable, in that subdividing internally uniform units can produce no increase in the apparent dispersion of the concentrations.

Continuous Distributions with Subunits Containing Negatively Correlated Subunits

In dichotomous branching systems where a parent source divides into two daughter branches, there is an opportunity for negative correlation between the daughter branches. An example is the twinkling of local flows in the microvasculature where, if the flow in the parent source is held constant but that in one daughter branch increases, then the flow in the other daughter branch must decrease. Here is a simple example extending that of Fig. 4.9.

Each of the pixels of Fig. 4.9 was further subdivided using a pattern of 16 subunits on a square grid (Fig. 4.10, left panel) where the concentrations in the dark squares were set to be twice the mean concentration of the pixel, and those in the light squares were set to zero. Again the RDs were calculated for the various grids. The results, shown in Fig. 4.10 (right panel), are identical to those in Fig. 4.9 up through the division into 4096 units, $\log(N) = 3.6$. The next division into 16,384 units produces no increase in observed dispersion, since each of the quarters of Fig. 4.10 (left panel) has the same mean as the whole square. The next division,

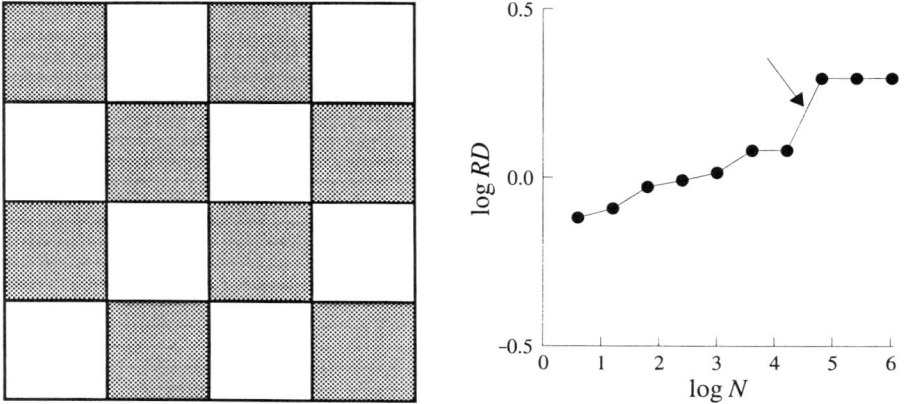

Figure 4.10. *Left panel:* Checkerboard pattern used to further subdivide each of the internally uniform 4096 pixels of Fig. 4.9 into 16 subunits with densities alternately at twice the mean for the unit or zero. *Right panel: RD*s as a function of number of elements into which the unit square is divided. The slope is steepest at the division where the negative correlation appears between the subunits, that is, in dividing 16,384 units into 65,536 units (arrow). With further subdivision of the 65,536 internally uniform pixels the estimates of dispersion are at a plateau. (From King et al., 1990, their Figs. 8 and 9, with permission.)

however, shows a sudden increase in observed dispersion, since the heterogeneity of the units is now revealed. As in Fig. 4.10 (left panel), further division of these now uniform units does not increase the observed dispersion beyond the second plateau.

The negative correlation produces a segment of the slope of $\log RD$ versus $\log N$ with a slope having a fractal D greater than 1.5, that is, a fractal D greater than that of random noise. Because these possibilities for negative correlation exist in many biological situations, combinations of positively and negatively correlated features in tissue structures or functions should not be thought strange. See Table 4.2.

For a function to be considered fractal, its characteristics should show self-similarity or self-affinity. This means that the estimate of the fractal dimension

Table 4.2. Fractal D and correlation in one dimension

Fractal D	Hurst, H	Type of correlation
1.0	1	Exactly correlated or uniform signal
$1<D<1.5$	$0.5<H<1.0$	Positively correlated elements
$D=1.5$	0.5	Uncorrelated, random signal
$1.5<D<2.0$	$H<0.5$	Negatively correlated elements

should be independent of the level of resolution or scale. Consequently, it may be useful, even if less accurate, to calculate D from each pair of points on graphs of $\log RD$ versus $\log$ (piece size) such as might be done for the data in Fig. 4.3:

$$D = 1 + \frac{\log [RD(N_2)/RD(N_1)]}{\log [N_2/N_1]}, \tag{4.13}$$

which is the same as Eq. 4.6 since $\log[N_2/N_1] = -\log(m/m_0)$.

This would exaggerate the influences of noise, but a gradual diminution or a sudden plateauing in D at small pixel size would be revealed.

Application to Densities in Three Dimensions

A one-dimensional analysis can be applied to the variation in the intensities of a property or the concentration of a substance in a tissue, if one can justifiably assume that there is no directionality in the variability of the property. On the basis that there were no systematically located high flow or low flow regions in the heart or endocardial/epicardial gradients (King et al., 1985), Bassingthwaighte et al. (1989) applied the one-dimensional analysis described just above. The method was to determine the blood flow per gram of tissue in about 200 spatially mapped small pieces of the heart to obtain the probability density function at the highest level of resolution. In Fig. 4.1 this gives the *pdf* with the lowest peak and most spread tails of the distribution, with an average piece size $\overline{m}$ of 0.22g and a relative dispersion of 29.4%. Adjacent pieces were then lumped together (mathematically) in various ways (left, right, forward, back, above, below) and the average flow per gram obtained for each piece pair. For each way of pairing all the pieces were used and the RD was estimated; the average RD for the several ways of pairing was 26.5% in pairs weighing 0.43 g. Successively larger aggregations of nearest neighbors gave the *pdf* values and RDs shown in Fig. 4.1.

In order to test whether or not the relationship was fractal, $RD(m)$ was plotted versus m using log-log scaling, Fig. 4.11. A straight line fitted to the data of Fig. 4.1 has the logarithmic slope $1 - D$, in accord with Eq. 4.5, rephrased,

$$\log RD(m) = \log RD(m_0) + (1 - D)\log(m/m_0). \tag{4.14}$$

The values of $RD(m)$ for the largest aggregates are smaller than predicted by the equation. The equation is designed for an infinite size of domain where the self-similarity holds for all scales; in an organ of finite size the fractal relationship cannot hold for large portions of the organ. The idea that a fractal D between 1.0 and 1.5 indicates correlation between near neighbors has a corollary, namely that in a finite domain where there is near-neighbor positive correlation (flows correlated, in this case), there must be negative correlation between the flows in regions separated by large distances.

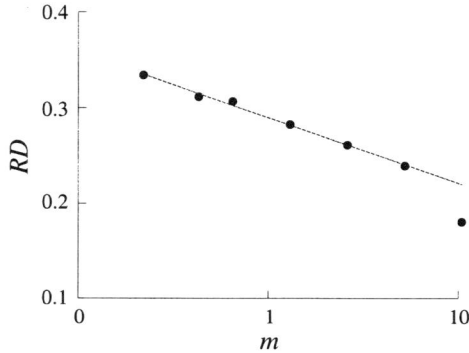

Figure 4.11. Log-log plot of RD versus m for the myocardial blood flow distributions of Fig. 4.1. The logarithmic slope of the fitted line (ignoring the largest aggregates) is -0.137, and the fractal D is 1.137.

4.3   Rescaled Range Analysis: The Hurst Exponent, H

Methods for assessing the fractal characteristics of time-varying biological signals are useful for heart rates, respiratory rates, neural pulse trains, and so on. Such biological signals which vary, apparently irregularly, have been considered to be driven by external influences which are random, that is to say, just "noise." However, in the spirit of distinguishing truly random influences from other types of coordinated irregularity, we will examine methods commonly applied to one-dimensional signals, or signals which are recorded over time. The dispersional analysis defined above can be applied, but other approaches are valuable from the point of view of understanding the phenomena and of making estimates of characterizing parameters. For fractal signals three methods are the rescaled range analysis (Hurst), autocorrelation analysis, and spectral analysis. These give statistical insight. Irregular signals may also be characterized with the methods of chaos analysis explicated in later chapters, which are better for gaining insight into underlying determinism due to nonlinearities, not statistical variation.

Hurst (1951) defined an empirical descriptor of temporal signals describing natural phenomena. This was based on the statistical assessment of many observations of many phenomena. His original paper and a lovely little monograph (Hurst et al., 1965) extending the work contain a wealth of information gathered during his career as a civil, geographical, and ecological engineer working in the Nile River basin. Hurst needed to determine what minimum height of the then proposed Aswan dam would provide sufficient storage capacity to govern the flow of the Nile downstream. The reservoir volume is the integral of the difference between inflow and outflow as diagrammed in Fig. 4.12, where R represents the range between the maximum and the minimum volume over a defined time period.

This project used data covering over a millennium of annual flows, and required prediction of what reservoir capacity would be needed to smooth out the flows.

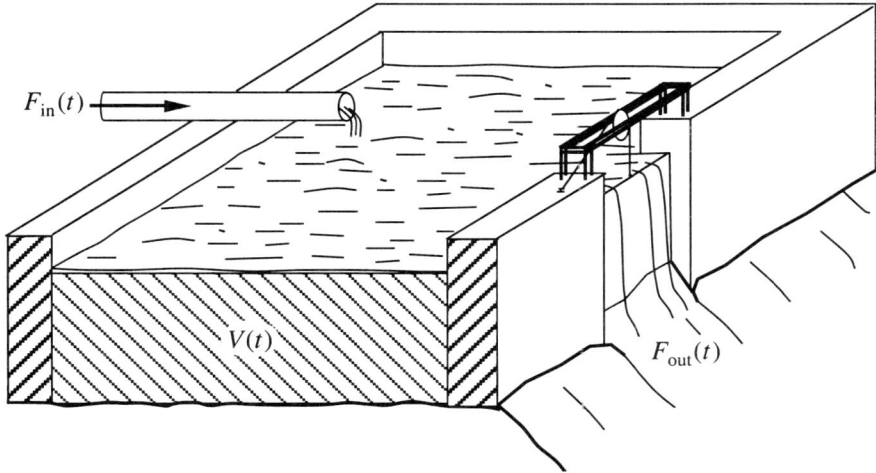

Figure 4.12. Reservoir volume changes over time are a measure of the cumulative differences between inflow and outflow. If the inflow each year is $F_{in}(t)$, then the cumulative inflow over t years will be $\int_0^t F_{in}(\lambda)d\lambda$. The outflow is set at a constant value $\overline{F}_\tau$, the average value of F_{in} over the period τ. Then the volume accumulated in the reservoir at any time t is $V(t) = \int_0^t (F_{in}(\lambda) - \overline{F}_\tau)d\lambda$. The range $R = V_{max} - V_{min}$.

Hurst observed that records of flows or levels at the Roda gauge, near Cairo, did not vary randomly, but showed series of low-flow years and high-flow years. Even though river flows might be expected to fluctuate less widely than the rainfall, the "memory" or correlation between successive years created a serious problem: the dam would need to be much larger than it would if annual rainfalls and river flows were random. The correlation between successive years meant that the reservoir would fill or empty more completely than would occur with random variation. He made comparisons of random signals and of biased signals with his data, and in so doing developed an approach to examining accumulations or integrals of naturally fluctuating events.

This led Mandelbrot and Wallis (1968) to see unusual events such as extended rainfall (stimulating the building of Noah's ark, or producing the Mississippi flood of 1993) or prolonged drought (predicted by Joseph) as natural phenomena that are likely to occur when there is "memory" in the system.

Definition of the Hurst Exponent H

The Hurst exponent H gives a measure of the *smoothness* of a fractal object, with $0 \leq H \leq 1$. A low H indicates a high degree of roughness, so much that the object almost fills the next-higher dimension; a high H indicate maximal smoothness so that the object intrudes very little into the next-higher dimension. A flat surface has Euclidean dimension $E = 2$ and $H = 1$; a slightly roughened surface with $H = 0.98$, for example, and fractal dimension $D = 2.02$ intrudes only a little into three-dimensional space. The general relationship is

$$H = E + 1 - D, \tag{4.15}$$

where $E = 0$ for a point, 1 for a line, and 2 for a surface. Whenever H is less than 1, then the object intrudes into the next dimension, just as a curved or rough surface requires embedding in three-dimensional space to be visualized.

Hurst envisioned an "ideal" reservoir performance, where (1) the outflow is uniform, (2) the reservoir ends up as full as it begins, and (3) the reservoir never overflows. See Fig. 4.12. His goal in designing the Aswan Dam was to minimize the capacity and yet fulfill these conditions. The concept is retrospective only since the setting of the outflow to the average can only occur when all of the data have been collected. Nevertheless, retrospective analysis is valuable in predicting future behavior, and as for all correlated events, the best indicator of next year's situation is this year's.

In the "ideal" reservoir situation we follow the time course of the inflow $F(t)$ and the integral of its differences from the mean, which is the hypothetical accumulated volume $V(t)$, during a period of duration τ. We use a variable u to represent time, beginning with $u = 0$ at the start of the interval as diagrammed in Fig. 4.13 so that

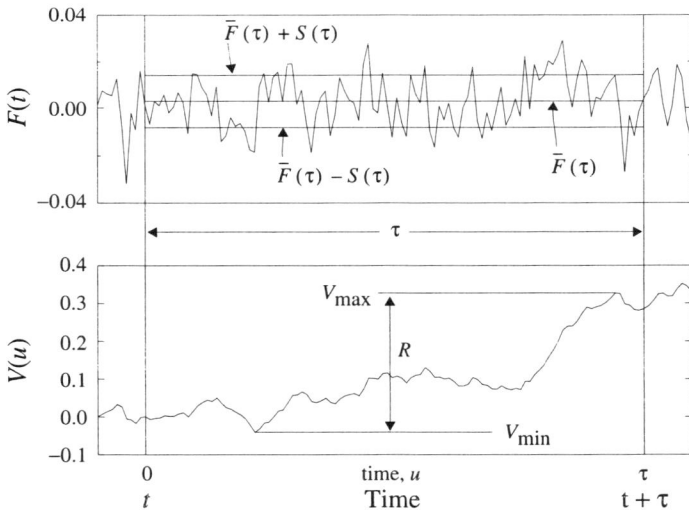

Figure 4.13. An example of a one-dimensional signal $F(t)$ and $V(u)$, the integral of the difference between $F(t)$ and its mean $F(\tau)$ over a period τ. One standard deviation of $F(t)$ over the period τ is $S(\tau)$, and the range $R(\tau)$ is $V_{max} - V_{min}$. Time u within the interval runs from 0 to τ or from t to $t + \tau$. (This fractional Brownian noise signal was generated with a spectral synthesis method adapted from Saupe (1988), using $H = 0.8$.) (From Bassingthwaighte, 1994, with permission.)

the volume accumulated is the same at the beginning and end of the interval, $V(u = 0) = 0 = V(u = \tau)$:

$$V(u) = \int_0^u (F(\lambda) - \bar{F}(\tau)) \, d\lambda,$$
(4.16)

where $0 < u < \tau$, $\bar{F}(\tau)$ is the retrospectively calculated mean inflow over the period τ and equals the constant but hypothetical outflow, and λ is a dummy variable for the integration up to time u. Next, we define the range R over the interval τ beginning at time t, as $R(t, \tau)$, and it is calculated as the difference between the maximum and minimum values of $V(u)$ over the period τ:

$$R(t, \tau) = V_{max} - V_{min}$$
$$= \max(V(u)) - \min(V(u)) \quad \text{over } 0 < u \leq \tau.$$
(4.17)

There will be different values of $R(t, \tau)$ for each interval starting at different times, t. To normalize the range relative to the fluctuations in $F_{in}(t)$, Hurst used the standard deviation $S(t, \tau)$ of $F(t)$ over the same period, τ:

$$S(t, \tau) = \left[\frac{1}{\tau} \int_0^\tau (F(t) - \bar{F}(\tau))^2 \, dt \right]^{1/2};$$
(4.18)

thus he obtained for each τ several estimates of the "rescaled range," R/S:

$$R/S = R(t, \tau)/S(t, \tau).$$
(4.19)

For any long data set where τ is less than some τ_{max} there are many subsets of length τ and therefore many estimates of R/S for any chosen τ. Hurst wisely chose to use nonoverlapping subsets, gaining independence between subsets, but he thereby got fewer estimates of R/S when τ was near τ_{max}, for example only two when $\tau = \tau_{max}/2$.

He found that many natural phenomena gave simple log-log relationships. He used a graphical analysis, plotting the average of the estimates of R/S at a given τ versus the interval length τ on double logarithmic paper, doing this for each of many values of τ. These gave approximately straight lines. (Hurst used the average R/S at each τ; Mandelbrot and Wallis (1969d) advocate the plotting of each of the estimates of R/S, as we will discuss below, producing the so-called "pox plot.") The slope of the straight line best fitting log R/S versus log τ is a power law exponent, H. *The exponent H is calculated from the slope of the log−log relationship between the ratio of the range R to the standard deviation S and the length of the interval observed.* The descriptor "rescaled range analysis" is appropriate for two reasons: dividing by S is a method of normalization allowing one range to be compared to another. The equation for R/S versus τ is

$$R/S = R(\tau)/S(\tau) = (a\tau)^H,$$
(4.20)

where a is a constant, τ is the period over which the calculation is made, and H is the Hurst coefficient (which Hurst called K). For a random Brownian motion, $H = 0.5$, the value of a is $\pi/2$, so the equation is $R/S = 1.253\tau^{H}$ (Feller, 1951).

[Aside]

Steps in estimating H from F(t) (without trend correction). This was Hurst's original method. The basic idea is to determine how the range of the cumulative fluctuations depends on the length of the subset of the data analyzed. In this section we consider the data to consist of values $F(t_i)$ sampled at uniform intervals of Δt.

1. Start with the whole observed data set that covers a total duration $\tau_{max} = N_{max}\Delta t$, and calculate its mean over the whole of the available data collected, $\bar{F}(\tau_{max})$, namely for $\tau = \tau_{max}$ and $N = N_{max}$,

$$\bar{F}(\tau_{max}) = \frac{1}{N}\sum_{i=1}^{N} F(t_i) . \tag{4.21}$$

2. Sum the differences from the mean to get the cumulative total at each time point, $V(N, k)$ from the beginning of the period up to any time $k\Delta t$, where we equate $V(N, k)$ to $V(\tau, u)$, and $u = k\Delta t$, and $\tau = N\Delta t$:

$$V(N, k) = \sum_{i=1}^{k} [F(t_i) - \bar{F}(\tau_{max})] \quad \text{for} \quad 0 < k \leq N. \tag{4.22}$$

3. Find V_{max}, the maximum of $V(N, k)$, and V_{min}, the minimum of $V(N, k)$ for $0 < k \leq N$ and calculate the range:

$$R(\tau) = V_{max} - V_{min} . \tag{4.23}$$

4. Calculate the standard deviation, S, of the values, $F(t_i)$, of the observations over the period τ, during which the local mean is $F(\tau)$:

$$S(\tau) = \left\{ \frac{1}{N}\sum_{i=1}^{N} [F(t_i) - \bar{F}(0)]^2 \right\}^{1/2} . \tag{4.24}$$

5. Calculate $R/S = R(\tau)/S(\tau)$ and plot the value on log-log paper. (At this first stage, N equals the total number of all the values in the time series and is the point at τ_{max} in Fig. 4.14.)
6. For the next stage, N will only cover a fraction of all the values in the data set, typically $1/2$ for $N = N_{max}/2$, and a new $\tau = N\Delta t$. Then repeat the entire procedure, steps 1–5, and determine R/S for each segment of the data set.

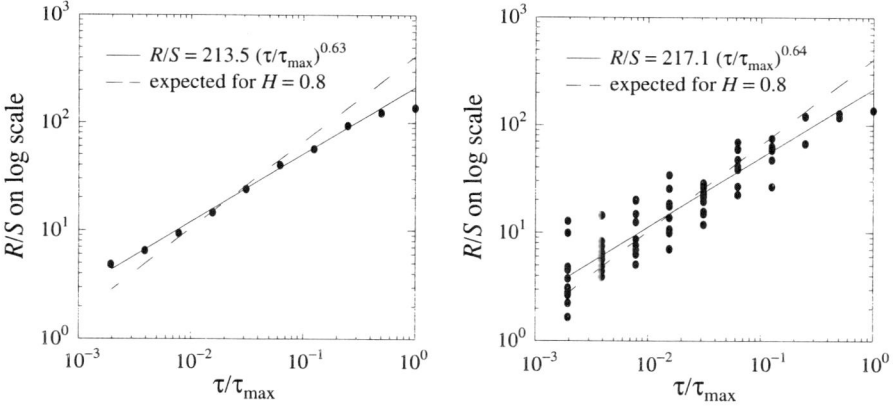

Figure 4.14. Hurst plots for rescaled range analysis without local trend correction. The abscissa is the duration of the interval for the calculation divided by the total time period. The data are a 2048-point array generated using a spectral synthesis method for $H = 0.8$. τ/τ_{max} is the fraction of the total points in an interval of duration τ. *Left panel:* The means of the estimates of R/S for each interval length τ are plotted. Each point is the average R/S of $2048/(\tau/\tau_{max} \cdot 2048) = \tau_{max}/\tau$ estimates. The regression calculation omits the values of R/S at the three largest values of τ/τ_{max}. *Right panel:* Individual estimates of R/S for nonoverlapping intervals are plotted. This is the so-called "pox plot." Each of the ten points is the estimate of R/S for $(\tau/\tau_{max} \times 2048)$ points. The evenly weighted log-log regression uses up to ten points at each τ/τ_{max}, but omits the points at the three highest values of τ/τ_{max}, where less than ten points are available. (From Bassingthwaighte, 1994, with the permission of BMES.)

This gives N_{max}/N estimates, which are averaged and plotted as the value of $\tau_{max}/2$ in Fig. 4.14. Repeat, using successively shorter τ's, at each stage dividing the data set into nonoverlapping segments and finding the mean R/S of these segments. The result, with the estimates of R/S being the average value for each τ, is shown in the left panel of Fig. 4.14. For the pox plot in the right panel that is preferred by Mandelbrot and Wallis (1969a), the individual values of $R(\tau)/S(\tau)$ are plotted for each τ. For the shorter periods, τ, there are τ_{max}/τ estimates, so in order to reduce computation and obtain a reasonable sampling Mandelbrot and Wallis (1969b) chose starting points for each τ so that the intervals sampled did not overlap and were spread over the data set. They did feel that it was worth overlapping the intervals to get more estimates when τ_{max}/τ was large (for example, > 0.15), in order to get 10 or so estimates over the whole data set.

7. The functional relationship over intervals of lengths τ compared to a chosen reference length τ_0 is

$$R(\tau)/S(\tau) = R(\tau_0)/S(\tau_0) \cdot (\tau/\tau_0)^H, \qquad (4.25)$$

where the Hurst coefficient, $H = 2 - D$. This form of the equation is similar to that of the relative dispersion in the preceding section. (In Fig. 4.14 we use $\tau_0 = \tau_{max}$.) The equation for the best fitting straight line is determined most simply by using the linear Y on X regression with $Y = \log (R/S)$ and $X = \log (\tau/\tau_0)$:

$$\log [R(\tau)/S(\tau)] = \log [R(\tau_0)/S(\tau_0)] + H \cdot \log [\tau/\tau_0], \qquad (4.26)$$

where the *exponent H is the slope of the regression line*. A value of $H \neq 0.5$ indicates autocorrelation in the signal. $H > 0.5$ means positive correlation, $H < 0.5$ means negative correlation or antipersistence. Mandelbrot and Wallis (1969b) preferred the "pox plot" because this gives a more representative weighting of the data. They limited the number of estimates of R/S at each τ to a small number, 10, not merely to reduce computation but to avoid overweighting the data at short τ's where N is very large.

A straight line on this plot indicates that there are self-similar correlations at all scales. Mandelbrot (1977, 1980) calls $H > 0.5$ *persistence* because it means that increases are more likely to be followed by increases at *all* distances, and $H < 0.5$ *antipersistence* because increases are more likely to be followed by decreases at *all* distances. Thus, $H < 0.5$ signals are considerably more irregular than those with $H > 0.5$. What one usually finds is that the line does not fit the data precisely, for the data at short τ tend to be scattered and to lie below the straight line, while the data for long τ also tend to lie below the line. This is to say that the values of R/S are usually concave downward. The slope is often artifactually too high for short intervals, presumably because observations are taken at unnecessarily short intervals and a high estimate of the degree of local correlation is thereby obtained; this can occur when the data are prefiltered with a low-pass filter before digitization. One is caught on the horns of a dilemma, for the requirements of Shannon sampling force one to do prefiltering so that frequencies higher than twice the sampling frequency are not represented by lower frequency power, but such prefiltering produces false correlation in truly fractal time signals. The slope is also too low at long intervals, presumably because of a real reduction in correlation at very long intervals such that the slope tends toward an H of 0.5, which is what it would be for a random signal.

The measure of dispersion $S(\tau)$ is in the denominator; if the signal characteristics did not change over the extent of the observations, and the shortest intervals were long enough that good estimates of $S(\tau)$ were obtained, then $S(\tau)$ would not be expected to change with interval length. Why then is this useful? If the signals were random then it would be useless, for the range would be predictable and would grow with interval length in a precise way: for random signals,

$$R/S = (0.5\pi N)^{1/2} = 1.253\sqrt{N}, \qquad (4.27)$$

where N is the number of intervals combined to form longer intervals, as shown by Feller (1951). Or, for intervals of length τ,

$$R(\tau)/S(\tau) = (0.5\pi\tau)^{1/2} = 1.253\sqrt{\tau}. \tag{4.28}$$

When N is small, Hurst noted that the distinction between proportionality to $\sqrt{N}$ and to N could not easily be made, which is the usual problem with short intervals and the statistics of small numbers.

Hurst's actual technique was improved by Mandelbrot and Wallis (1969d) accounting for local trends during each time period τ. The modified approach was intended to correct for long-term trends or periodic fluctuations when using short τ, and is shown in Fig. 4.15. Over short intervals of time or space there may be local trends strong enough that the value of $V(t)$ at the end of the interval is quite different from that at the beginning, thus causing a deviation from "ideal" reservoir behavior. Mandelbrot and Wallis (1969d) give a thorough evaluation of trend correction. To handle this appropriately, that is, to get a more sensible estimate of the range, R, they accounted for the trend by assuming that it was locally linear, and calculated the maximum and minimum of the differences, V_{max} and V_{min}, from the trend line, the straight line between $V(t)$ and $V(t+\tau)$. The trend correction, applied to the reservoir problem, is equivalent to a pretense that the average outflow was regulated so that the reservoir height would have been the same at the end as at the beginning of any chosen period. The standard deviation, S, is calculated for $F(t_i)$ as for the simple method. This prevents exaggerating the range, R, which is a problem at short intervals, by redefining it as in Fig. 4.15. The difference between the two measures of R/S is greatest at the short interval lengths.

The results of the simple versus the trend-accounting method are shown in Fig. 4.16. Trend correction steepens the slope, giving higher estimates of H. The

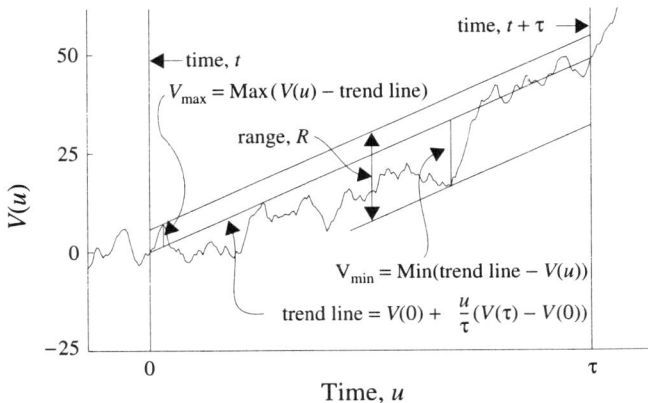

Figure 4.15. Calculation of the trend-corrected estimate of the range $R(t, \tau)$, which is $R(\tau)$ for the interval of duration τ beginning at time t. $R(t, \tau)$ is the sum of the maximum difference of $V(u)$ above the local linear trend line and the greatest excursion below the trend line. In this case the trend-corrected $R(\tau)$ is less than would be obtained from the difference between the maximum and minimum of $V(u)$. (From Bassingthwaighte, 1994, with the permission of BMES.)

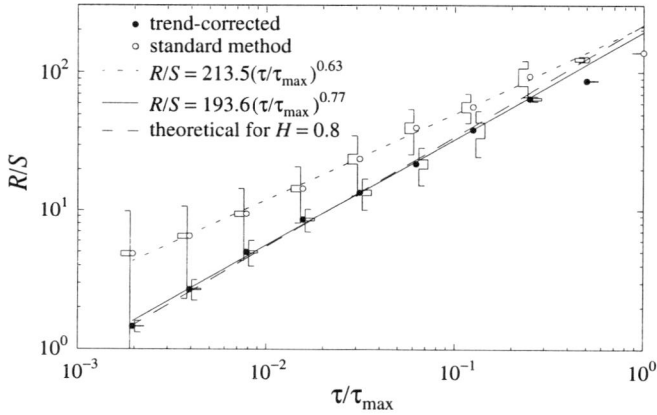

Figure 4.16. Accounting for local trends reduces the slope compared to the original Hurst method for a signal generated with $H = 0.8$. (From Bassingthwaighte, 1994, with the permission of BMES.)

R/S versus τ relationship for the trend-corrected data gave higher estimates of H than did the analysis without trend correction. Mandelbrot and Wallis (1969d) argue that Hurst's estimates from the averages at each τ covered an artifactually narrow range, being too large when $H < 0.72$ and too small when $H > 0.72$. Mandelbrot and Wallis (1969e) tested the adequacy of the rescaled range technique on non-Gaussian correlated processes with large skewness and kurtosis and found it to work well.

A comment on the "pox plot" preferred by Mandelbrot and Wallis (1969 papers) is merited. They choose to use ten sample intervals, nonoverlapping and taken from different parts of the data set. When there is a lot of data, as in the case shown in Fig. 4.17, the enormous number of points at $\tau/\tau_{max} = 2^{-9}$ and 2^{-8} actually has very little tendency to bias the regression line toward being overly steep.

Hurst observed river flows, the thicknesses of annual layers of mud on lake bottoms (varves), tree ring thicknesses, wheat prices, and many other processes, and calculated the slopes of R/S versus τ for each. The relationships had varied slopes, mostly with $H > 0.5$. Almost none showed $H < 0.5$, which would be negatively correlated. Some of his data are shown by Fig. 4.18. Hurst made a most remarkable collection of 837 sets of data, some extending over hundreds of years, with one set (Temiskaming varves) of 1809 years' duration. He found that relatively few phenomena were random; the Hurst coefficients from these 837 data sets are gathered into the histogram shown in Fig. 4.19. From this graph, one would estimate that perhaps 90% of the observed time series were distinctly different from random signals, since about 90% give estimates of H greater than 0.55. The fluctuations in these time series have a fractal form, because they have approximately linear plots of $\log (R/S)$ versus $\log \tau$. Moreover, they mostly have $H > 0.5$, indicating that there are positive correlations at all time scales.

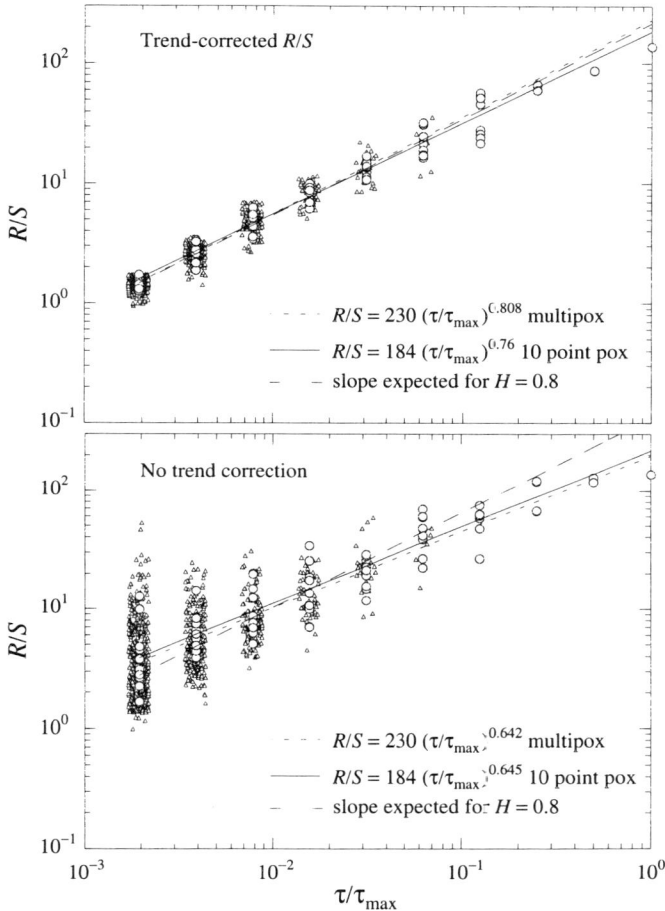

Figure 4.17. Pox plot for the trend-corrected rescaled range analysis on a 16,384-point signal for $H = 0.8$ generated by the spectral synthesis method. Using ten sample intervals gives representative points. Using all the points weights the estimate of the slope toward that given by only short τ's but does not change the slopes significantly. The regressions do not include the largest three τ's at $\tau/\tau_{max} = 1$, 0.5, and 0.25. (From Bassingthwaighte, 1994, with the permission of BMES.)

The following are problems to be worked out for the Hurst analysis:

1. What is the fundamental theory for estimating R? One would like a theory defining the expectation of the range in the presence of a given degree of correlation. (How can the a in Eq. 4.20 be obtained *a priori*? Mandelbrot and

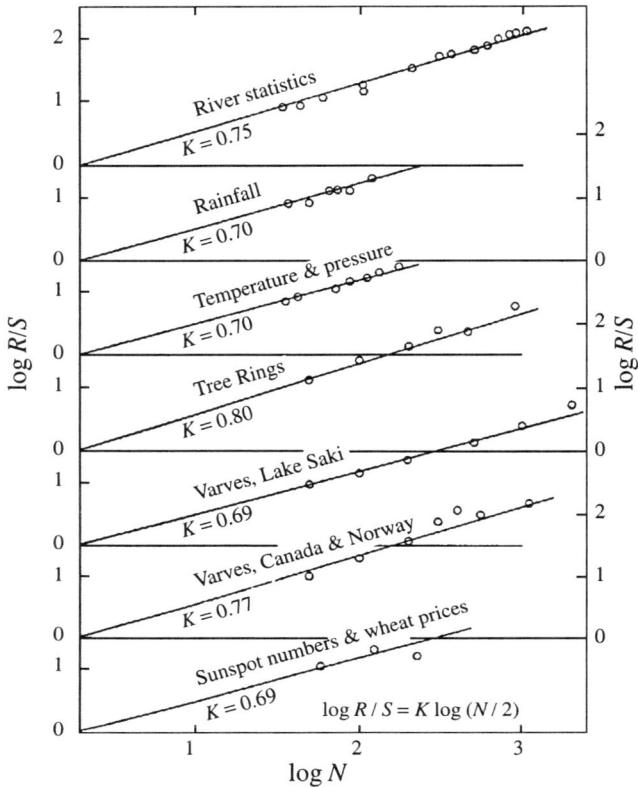

Figure 4.18. This figure from Hurst (1951) illustrates the precision with which R/S behaved as a simple fractal. Hurst's K = our H, the power law slope. The equation at the bottom is the expectation for uncorrelated (white) noise; N is number of data points.

Wallis (1968) derived an asymptotic expression for the correlation between observations separated by a time s, $r(s) = [2H(2H - 1)k]s^{2H - 2}$, but could not define a value for k, so the meaning of the overall coefficient, $2H(2H - 1)k$ is not well defined. The exponent, $2H - 2$, is derived in the next section.)

2. What are the finest intervals that are useful?
3. Where are the effects of trend correction most important? How big are the effects of trend correction?
4. Are the estimates of R/S biased, or is S naturally larger when R is excessive, so making the ratio self-correcting? (Mandelbrot and Wallis, 1969e, state that the method works for both long-tailed and skewed distributions.)
5. What statistical tests can be used to distinguish estimates of H from two data sets, or to signify H different from 0.5? (What are the confidence limits on the estimate of H?)

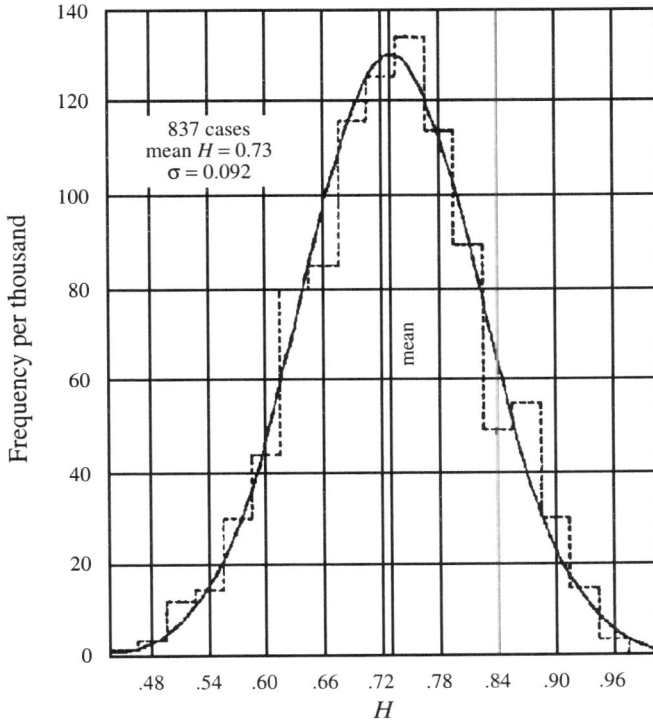

Figure 4.19. Rescaled range analysis for 837 different natural phenomena for which annual values were obtained. The mean value of H was 0.73, with a more or less Gaussian spread with the $SD = 0.09$. (From Hurst et al., 1965.) The spread of these 837 estimates of H is somewhat narrower than if the trend-corrected estimates of R were used.

Consequently, we summarize by saying that the rescaled range analysis is useful but according to Schepers et al. (1992) may not be as good as the dispersional analysis. Clearly, thorough testing of the method is needed. Feder (1988) tested it on Gaussian noise signals ($H = 0.5$) and concluded that the technique worked well when the number of intervals was 50,000 and $\tau > 20$. Data are needed on analyses of known signals with $H \neq 0.5$ and for shorter data sets. Schepers et al. (1992) present analyses illustrating that rescaled range analysis worked best for signals with H near 0.7, while it tended to give overestimates with low H, particularly for signals with $H < 0.5$. There was a tendency to underestimate H for signals with $H > 0.85$. These biases in the technique have been quantitated by Bassingthwaighte et al. (1994), and affirm that range of Hs portrayed in Fig. 4.19 are significantly too narrow. Nevertheless, the reinterpretation of this figure would still lead us to conclude that over two-thirds of these natural phenomena are nonrandom, positively correlated fractal time series.

4.4   Correlation versus Distance

Diminution in Correlation with Distance (in space or time)

That there might be a diminution in correlation with increasing distance between sites of observation evokes no surprise. In considering the structure and function of an organ, it is reasonable to expect that neighboring regions must be more alike with respect to a property than are regions more widely separated. The branching characteristics of excretory ducts and vascular structures, and the directional spreading of neural connections all fit in with the general concept of monotonic decrease in likeness with increased distance.

Under special circumstances, such as in branching structures, there is a basis for *a fractal relationship showing correlation extended over longer distances than would ordinarily be expected*. Van Beek et al. (1989b) showed that the correlation between nearest neighbors in a spatially heterogeneous system (the three-dimensional profiles of regional blood flow in the heart) was defined by the fractal dimension describing the degree of heterogeneity in a size-independent fashion. This is expressible directly in terms of the correlation coefficient, r_1, between nearest-neighbor, equal-size regions (independent of the actual size):

$$r_1 = 2^{3-2D} - 1 .$$

(4.29)

This translates into an analogous expression using the Hurst coefficient where $H = 2 - D$,

$$r_1 = 2^{2H-1} - 1 .$$

(4.30)

For events that are separated further in time or space, we expect that correlation will diminish. For random events there is no correlation even with the nearest neighbor. What we show here is that when the correlation extends to more distant neighbors, then the falloff in correlation with separation is slower than exponential, and the relationship itself is fractal, showing self-similarity on scaling.

A major feature of this correlation is that *the correlation is exactly the same between neighbors of one unit size as it is between neighbors of smaller unit size even when the smaller element is an internal component of the larger element*. This is in accord with self-similarity independent of scale. It cannot occur unless the units of any chosen size are internally heterogeneous. The idea is portrayed in Fig. 4.20: a domain, such as a piece of tissue, is cut into large pieces, as in the upper panel, and the autocorrelation, r_n, determined for the intensity of some property, for example, lactate dehydrogenase levels. The slope of r_n for $n = 1$ to 4 is suggested in the upper panel. By cutting each of the initial large pieces into smaller pieces of 1/2, 1/4, 1/8, etc., the autocorrelation function, r_1 to r_n, can be found for each of the smaller-size

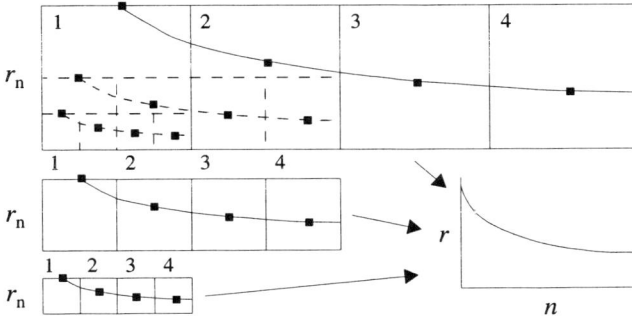

Figure 4.20. Self-similarity in autocorrelation shows scale independence of r_n versus n. The nth neighbor correlation is independent of the size of the regions, and r_n versus n is the same for the largest samples (*upper*), for samples of 1/4 size (*middle*), and for samples of 1/16 size (*lower*), so that the three curves of r_n versus n can be superimposed (*right lower*). (From Bassingthwaighte and Beyer, 1991.)

sets of pieces. If the intensities of the signal are fractal, then r_n versus n has the same H at every level of resolution.

The finding of a fractal fall-off in correlation with distance (measured in terms of numbers of neighboring units) and its affirmation by finding self-similarity in correlation (showing that the fall-off in correlation is independent of piece size) is powerful evidence of the fractal nature and provides a strong incentive to seek the underlying basis in causal processes.

Correlation with the Nearest Neighbor, r_1

Van Beek et al. (1989) gave the key to the derivation of the result that nearest-neighbor correlation was expressible directly by the correlation coefficient in terms of the fractal D. They were inspired by the observation that *high-flow regions of the myocardium tended to have high-flow neighbors*, and *low-flow regions tended to have low-flow neighbors*. Their approach brought together two ideas: the idea that correlation is expressed via the mutual variation (covariance) between pairs of observations; and the idea that the relative dispersions of aggregates of nearest neighbors obey a fractal relationship.

[Aside]

Derivation of r_1. The method is based on calculating covariance from the fractal relationship. For convenience we use these definitions for application to analysis of a one-dimensional signal at even intervals in space or time, F_i with $i = 1$ to N.

$$\text{The mean is } \bar{F} = \frac{1}{N} \sum_{i=1}^{N} F_i. \tag{4.31}$$

$$\text{The variance is } \mathrm{Var}(F) = \frac{1}{N} \sum (F_i - \bar{F})^2. \tag{4.32}$$

The covariance between nearest neighbors is

$$
\begin{aligned}
\mathrm{Cov}_1 &= \mathrm{Cov}(F_i, F_{i+1}) \\
&= \frac{1}{N-1} \sum_{i=1}^{i=N-1} (F_i - \bar{F})(F_{i+1} - \bar{F}) \quad.
\end{aligned} \tag{4.33}
$$

The covariance between n^{th} neighbors is

$$
\begin{aligned}
\mathrm{Cov}_n &= \mathrm{Cov}(F_i, F_{i+n}) \\
&= \frac{1}{N-n} \sum_{i=1}^{i=N-n} (F_i - \bar{F})(F_{i+n} - \bar{F}) \quad.
\end{aligned} \tag{4.34}
$$

The correlation coefficient for n^{th} neighbor correlation is

$$r_n = \mathrm{Cov}_n / \mathrm{Var}(F), \tag{4.35}$$

and is written out as Eq. 4.3.

One can gather pairs of nearest neighbors together to form $N/2$ aggregates of double the original element size. For signals in time this means summing each pair of elements and plotting the sum (or the average) at intervals of $2\Delta t$.

The general expression for the variance of the mean of the paired elements of the signal is

$$
\begin{aligned}
\mathrm{Var}\,(2F) &= \mathrm{Var}\,(F_i) + \mathrm{Var}\,(F_{i+1}) + 2\mathrm{Cov}\,(F_i, F_{i+1}) \\
&= 2\mathrm{Var}\,(F) + 2\mathrm{Cov}_1 \quad.
\end{aligned} \tag{4.36}
$$

Likewise, for aggregates of n adjacent samples,

$$\mathrm{Var}\,(nF) = n\mathrm{Var}\,(F) + 2 \sum_{i=1}^{i=n-1} (n-i)\,\mathrm{Cov}_i. \tag{4.37}$$

As in Section 4.2, the relative dispersion RD is defined as

$$RD(F) = \sqrt{\text{Var}(F)}/\bar{F}, \qquad (4.38)$$

and for paired adjacent elements,

$$RD(2F) = \sqrt{\text{Var}(2F)}/2\bar{F}$$
$$= \sqrt{\text{Var}(F) + \text{Cov}_1}/\sqrt{2}\bar{F}. \qquad (4.39)$$

When there is a fractal heterogeneity over a range of element sizes, as given by Eq. 4.5 in Section 4.2, this is equivalent to

$$RD(2F) = RD(F)\,2^{H-1}, \qquad (4.40)$$

where the right side of the equation (using D or H) comes from the observed power law relationship.

From this, inserting the definitions for $RD(F)$ and $RD(2F)$, reducing and squaring both sides, and substituting r_1 for $\text{Cov}_1/\text{Var}(F)$, one gets

$$1 + \frac{\text{Cov}_1}{\text{Var}(F)} = 2^{3-2D}, \qquad (4.41)$$

and since $r_1 = \text{Cov}_1/\text{Var}(F)$,

$$r_1 = 2^{3-2D} - 1 \quad \text{or} \quad r_1 = 2^{2H-1} - 1. \qquad (4.42)$$

* * *

As van Beek et al. (1989) put it, "this little expression [for r_1] is an important statement because it summarizes the whole situation." If there is no spatial correlation, $r_1 = 0$, then the local flows are completely random and the fractal $D = 1.5$, or the Hurst coefficient $H = 0.5$. With perfect correlation, $r_1 = 1.0$ and the flows are uniform everywhere. For regional myocardial blood flows van Beek et al. (1989b) found fractal dimensions of about 1.2, or correlation coefficients of about 0.52 between adjacent pieces. We caution that while discussing flow variation in a three-dimensional organ, we applied one-dimensional analysis (dispersional analysis and correlation analysis), assuming variation was identical in all directions. Formally, for the densities of a property in three-space, instead of reporting a fractal D of 1.2, it might be reported as 3.2. However, the Hurst coefficient is 0.8 in either case. For $H = 0.5$, the correlation is zero, so whether one calls this a random fractal or noise due to independent events is a matter of choice; in any case the self-affinity remains.

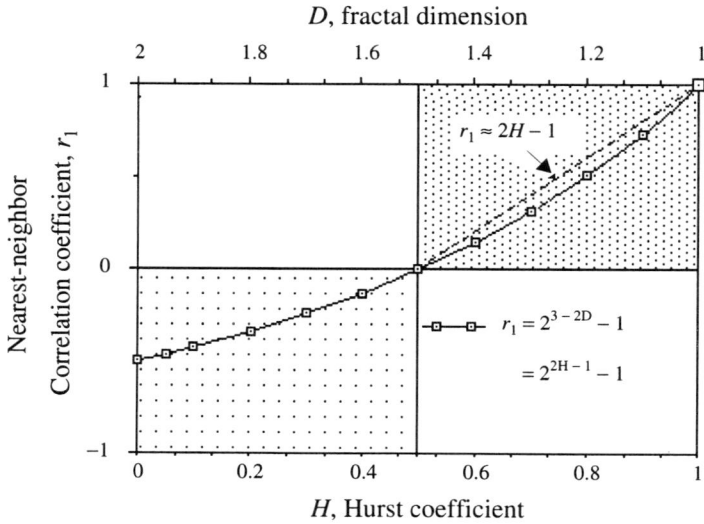

Figure 4.21. Plot of the nearest-neighbor correlation coefficient r_1 versus H and D. Over the range $0.5 < H < 1.0$, the dark-shaded region, the curve is approximated by a straight line, $r_1 \approx 2H - 1$ or $r_1 \approx 3 - 2D$. (In the light-shaded region of negative correlation, $r_1 \approx H - 0.5$ or $r_1 \approx 1.5 - D$, which doesn't make good general sense and wouldn't work for a fractal object that is nearly space-filling.)

The relationship between r_1 and the fractal exponents, D and H, is shown in Fig. 4.21. An approximating linear relationship is not too bad in the range of positive correlations:

$$r_1 = 2H - 1 \quad \text{for} \quad 0.5 < H < 1 \text{, or} \tag{4.43}$$

$$r_1 = 3 - 2D \quad \text{for} \quad 1.0 < D < 1.5 . \tag{4.44}$$

The relationship is not continued into the range of negative correlation, and the concept breaks down at $r_1 = -1$, perfect inverse correlation.

Extended Correlation with Nonadjacent Neighbors

The correlation with the piece one beyond the immediate neighbor is found by grouping pieces in threes and following the same arguments. Here we are deriving an interpretation for the two-point autocorrelation function between n^{th} neighbors. As before, this stems from the observed fractal relationship given by Eq. 4.5:

$$RD\,(nF) = RD\,(F)\,n^{\text{H}-1} . \tag{4.45}$$

[Aside]

Derivation of r_n. Eq. 4.44 gives us directly

$$RD\,(F)\,n^{H-1} = \sqrt{\text{Var}\,(nF)}\,/n\bar{F}$$

$$= \sqrt{n\,\text{Var}\,(F) + 2\sum_{i=1}^{i=n-1} (n-i)\,\text{Cov}_i/n\bar{F}}$$

$$= \frac{1}{n}\sqrt{\frac{\text{Var}\,(F)}{\bar{F}}}\sqrt{n + 2\sum_{i=1}^{i=n-1} (n-i)\,\text{Cov}_i/\text{Var}F}\,. \qquad (4.46)$$

Since $RD(F) = \sqrt{\text{Var}\,(F)}\,/\bar{F}$ and $r_i = \text{Cov}_i\,/\,\text{Var}(F)$, when one squares both sides

$$n^{2H} = n + 2\sum_{i=1}^{i=n-1} (n-i)\,r_i\,, \qquad (4.47)$$

and, equivalently,

$$\sum_{i=1}^{i=n-1} (n-i)\,r_i = \frac{1}{2}\,(n^{2H} - n)\,. \qquad (4.48)$$

Eq. 4.48 provides a recursive expression defining r_{n-1} in terms of $r_{n-2}, r_{n-3}, \cdots$ r_2, and r_1. To find an expression that avoids the recursion, we apply Eq. 4.48 for $n = 1, 2, \ldots, m + 1$, following the strategy of Bassingthwaighte and Beyer (1991), and get the following set of equations:

$$r_1 = \frac{1}{2}\,(2^{2H} - 2)$$

$$2r_1 + r_2 = \frac{1}{2}\,(3^{2H} - 3)$$

$$3r_1 + 2r_2 + r_3 = \frac{1}{2}\,(4^{2H} - 4)$$

$$\cdot$$
$$\cdot$$
$$\cdot$$

$$mr_1 + (m-1)\,r_2 + 2r_{m-1} + r_m = \frac{1}{2}\,\{\,(m+1)^{2H} - (m+1)\,\}\,. \qquad (4.49)$$

This can be rewritten in matrix form:

$$\begin{bmatrix} 1 & 0 & 0 & \dots & 0 & 0 \\ 2 & 1 & 0 & \dots & 0 & 0 \\ 3 & 2 & 1 & \dots & 0 & 0 \\ \cdot & \cdot & \cdot & \dots & \cdot & \cdot \\ \cdot & \cdot & \cdot & \dots & 1 & 0 \\ m & m-1 & m-2 & \dots & 2 & 1 \end{bmatrix} \begin{bmatrix} r_1 \\ r_2 \\ r_3 \\ \cdot \\ r_{m-1} \\ r_m \end{bmatrix} = \frac{1}{2} \begin{bmatrix} (2^{2H} - 2) \\ (3^{2H} - 3) \\ (4^{2H} - 4) \\ \cdot \\ \cdot \\ (m+1)^{2H} - (m+1) \end{bmatrix} . \tag{4.50}$$

The coefficient matrix is lower triangular, so by inversion we get for $n = m$:

$$r_n = \frac{1}{2} \{ (n+1)^{2H} - 2n^{2H} + (n-1)^{2H} \} , \tag{4.51}$$

which is valid for $n \geq 1$. (This simplification of the derivation given by Bassingthwaighte and Beyer (1991) is due to Dr. Spiros Kuruklis.)

＊   ＊   ＊

The values of r_n for various H and n values are shown in Fig. 4.22, left panel. With increasing n, the correlation initially falls off rapidly, but at large n the

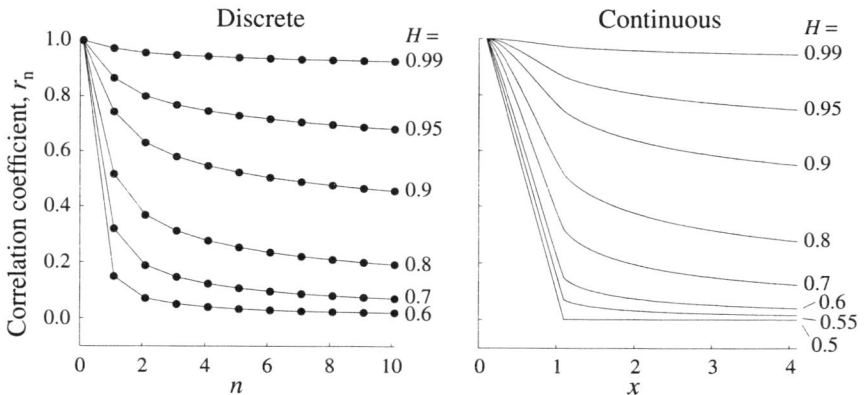

Figure 4.22. Fractally extended correlation. *Left panel:* Eq. 4.51 gives discrete steps in correlation coefficient between units of a discrete size separated by $n - 1$ intervening units. Symbols for a given Hurst coefficient (H) where $H = 2 - D$ for one-dimensional signals. *Right panel:* Eq. 4.53 gives fractional or continuous correlation over distance x. The shapes of $r(x)$ are continuous curves for any values of x, be they integer or fractional. (Figures adapted from Bassingthwaighte and Beyer, 1991, their Figures 2 and 3, with permission from IEEE.)

diminution is slow. In fractal signals with $H < 0.7$, the values of r_n for $n > 2$ are so low that it is easy to understand that fractal signals would be regarded as noise.

From this follows an asymptotic relation:

$$\frac{r_n}{r_{n-1}} = \left(\frac{n}{n-1}\right)^{2H-2}, \tag{4.52}$$

which is a simple power law relationship.

Some further algebra (Bassingthwaighte and Beyer, 1991) demonstrates that Eq. 4.52 can be put in a continuous rather than a discrete form. Though the unit size remains at 1, the correlation between units is

$$r(x) = \frac{1}{2}\left\{|x+1|^{2H} - 2|x|^{2H} + |x-1|^{2H}\right\}, \tag{4.53}$$

where $r(x)$ is continuous in x and valid for $x \geq 0$. With $x < 1$, the units overlap; with $x > 1$, the units are separated by a part of a unit or more. See Fig. 4.22, right panel. With $H = 0.5$, $r(x) = 0$.

The Shape of Extended Correlation Functions

Fig. 4.22 shows the fractional correlation for various values of H over the range $0 \leq x \leq 4$. The curves are continuous, without discontinuity in the derivative, except in the random case ($H = 0.5$). With $H = 0.5$, the correlation for $0 < x < 1$ is the straight line $r = 1 - x$, and $r = 0$ for $x > 1$; the correlation coefficient equals the fractional overlap between the units. With positive correlation, $0.5 < H < 1$, the curves appear as something summed with the line $r = 1 - x$. So they should, because the fractional overlap makes the same contribution as is made for the random function. The continuity of the derivative has been examined for cases with $0.5 < H < 0.55$, and though it is steepest at $x = 1.0$, there is no break in the derivative and it is smooth for values of x from 0.999 to 1.001.

The shapes of r_n versus n exhibit a falloff in correlation that is surprisingly slow, slower than exponential or Gaussian, as is shown in Fig. 4.23. The "stretched exponential" used by Liebovitch et al. (1987a) for fractal ionic channel kinetics has a position close to the line for r_n, but it falls off more rapidly. The stretched exponential equation is

$$r(x) = Ax^{H-1} \cdot e^{-Ax^H/H}. \tag{4.54}$$

Approximating and Asymptotic Relationships

The relationship of r_n versus n is surprisingly simple for the range $0.5 < H < 1$ and is shown in Fig. 4.24. For $n \geq 1$ the log-log relationships are nearly straight, and for $n \geq 2$ are within 0.1% of the asymptotic expression, Eq. 4.52, with a log-log slope of

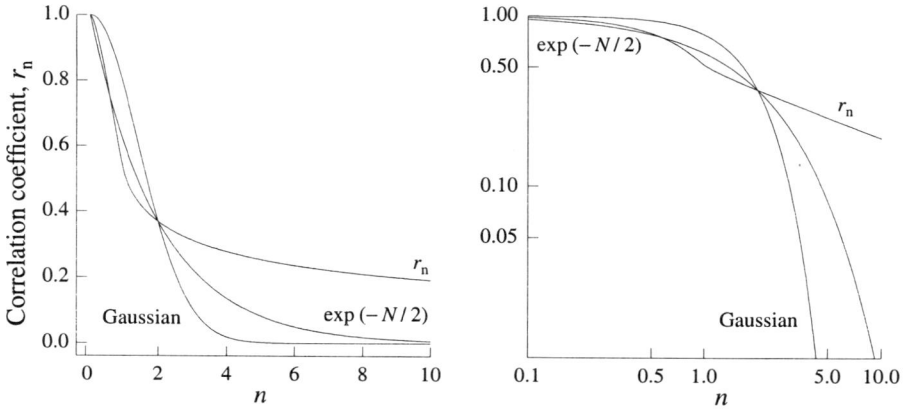

Figure 4.23. Fractal correlation, $r(x)$, as a continuous function of a number of intervening elements of unit size for the case $H = 0.8$. The shape of the relationship is independent of piece size. For comparison are shown exponential and Gaussian functions chosen to give the same value for $r(n = 2)$. *Left panel:* Linear plots. *Right panel:* Log-log plots. The correlation function approaches a power law function for $n > 3$, while the Gaussian and exponential functions fall off more steeply. (Reproduced from Figure 4 of Bassingthwaighte and Beyer, 1991, with permission from IEEE.)

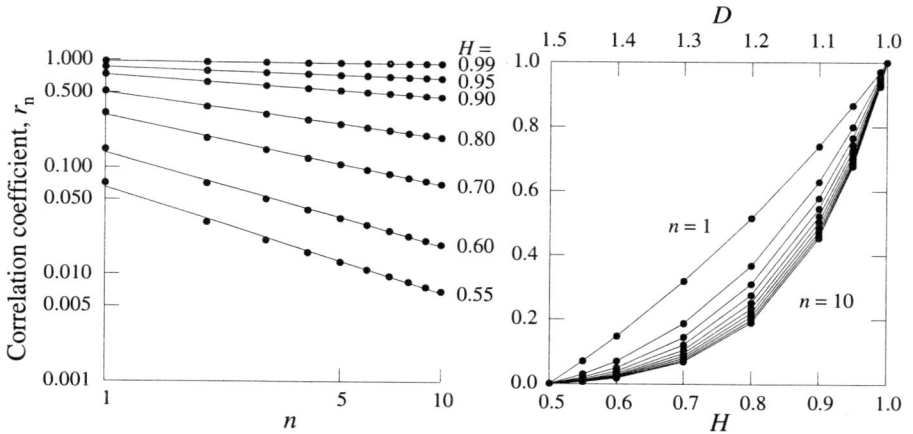

Figure 4.24. Correlation diminution with distance. *Left panel:* Log r versus log n for Hs with positive correlation. The symbols are those of Eq. 4.51 for the specified values of n and H. The curves have been fitted by linear regression to provide the best power law fit. The symbols are not quite in a straight line, for there is a distinct bend in the line occurring between r_1 and r_2. The straight line best fits over the range from 1 to 10 are shown for several Hs. *Right panel:* The r_n versus H curves are increasingly curved at higher values of n. The fractal dimension $D = 2 - H$.

98

$2H - 2$. The lines are close enough to being straight to suggest an empirical next step, shown in Fig. 4.25, where the exponents of the best fitting power law functions in Fig. 4.25 are plotted against H. A simple, graphical relationship between r and H for small n approximates the complicated recursion in Eq. 4.47:

$$r_n \approx r_1 \cdot n^{2.2\,(H-1)} .\tag{4.55}$$

Using $r_1 = 2^{2H-1} - 1$ as the starting point, this approximate expression is a poor fit to the general expression Eq. 4.51 because most of the curvature is near r_1. A generalization of the asymptotic expression from Eq. 4.52 is

$$r_n \approx r_m\,(n/m)^{2\,(H-1)} .\tag{4.56}$$

The right panel of Fig. 4.25 shows that for higher n, the exponent $2H - 2$ fits the correlation well, as might be expected because this is the slope for the asymptotic expression. The analytic forms for r_2 and r_3 are reasonably simple, but to avoid the curvature near r_1 they can be used to provide two expressions:

$$r_n \approx r_3\,(n/3)^{2\,(H-1)} ,\text{ or}\tag{4.57}$$

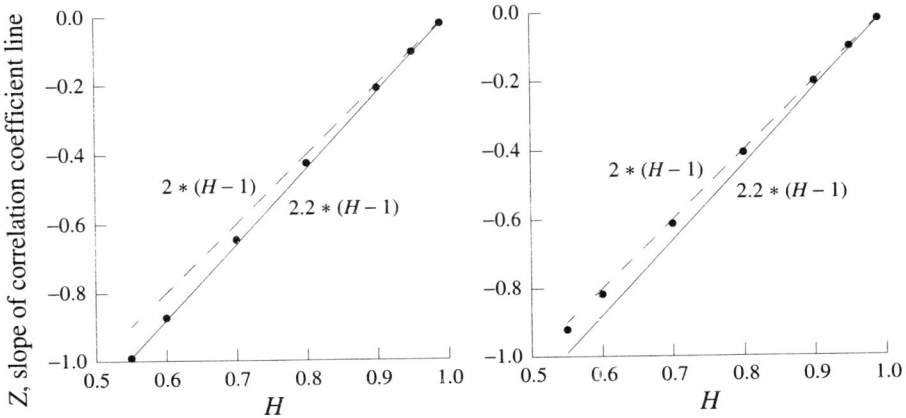

Figure 4.25. Relation of power law exponent, Z, to H. *Left panel:* The relationship is best fitted with the line $Z = 2.21 - 2.228H$, where the power law expressions represented the fits over the range r_1 to r_{10}. It is almost as well fitted by $Z = 2.2(1 - H)$, but is distinctly different from the asymptotic expression, $Z = 2(1 - H)$. *Right panel:* When r_1 is ignored and r_2 to r_{10} are used to derive the power law fits, then the asymptotic expression $Z = 2(1 - H)$ is an excellent descriptor.

$$r_n \approx r_2 \, (n/2)^{2(H-1)}. \tag{4.58}$$

Using the less accurate approximation, Eq. 4.58, with $H = 0.8$, yields an estimated $r_3 = 0.312$ instead of the correct 0.311. For r_{10}, the approximations may still be acceptable; calculating from r_3 (using Eq. 4.57) gives 0.192; and calculating from r_2 (using Eq. 4.58) gives 0.193, instead of the correct 0.191.

Estimating Fractal Dimensions from the Two-Point Correlation Function

Estimating H *or* D *by nonlinear optimization.* The theoretical relationships and approximations above can be used to estimate D and H from data. Because noise superimposed on a fractal phenomenon has a greater effect on more distant relationships, r_n to r_3 usually merit higher weighting than r_n for high n. Nonlinear optimizing techniques (least squares, etc.) are useful for fitting the general expression for r_n, Eq. 4.53, to the observed r_n because they allow weighting the observation of r_n appropriately to the data set. A one-parameter search is performed efficiently by the optimization routine GGOPT (Bassingthwaighte et al., 1988a).

[Aside]

Steps in estimating D *graphically from the two-point autocorrelation,* r_n *versus* n *.*

1. Obtain the two-point autocorrelation function, r_n, for the data set Y_i, $i = 1, N$. For nearest neighbors calculate r_1, as in Eq. 4.33:

$$r_1 = \frac{\Sigma \, (Y_i - \bar{Y}) \cdot (Y_{i+1} - \bar{Y})}{\Sigma \, (Y_i - \bar{Y})^2} = \frac{\mathrm{Cov} \, (Y_i, Y_{i+1})}{\mathrm{Var} \, (Y)} \quad \text{and} \tag{4.59}$$

$$r_n = \frac{\Sigma \, (Y_i - \bar{Y}) \cdot (Y_{i+1} - \bar{Y})}{\Sigma \, (Y_i - \bar{Y})^2} = \frac{\mathrm{Cov} \, (Y_i, Y_{i+1})}{\mathrm{Var} \, (Y)}, \tag{4.60}$$

  for all $n > 1$, Eq. 4.34. Standard computer routines are commonly available.
2. Plot $\log r_n$ versus $\log n$ as in Fig. 4.24 (left panel). Fit a best straight line through the curve, by eye and ruler or by a linear regression calculation. Estimate the slope from any two points on the line:

$$\text{Slope} = \frac{\Delta \log r_n}{\Delta \log n} = \frac{\log r_n - \log r_{n-d}}{\log d}. \tag{4.61}$$

3. Calculate H: Both the slope and intercept may be used, and both should be considered because the autocorrelation function r_n obtained from

experimental data is doomed to be somewhat noisy. For these purposes we chose the "best simple" expression that uses the theoretically correct intercept, the value of r_1, and the empirically simplest slope from Fig. 4.25 (left panel), by taking the logarithms for Eq. 4.55:

$$\log{(r_n)} \; = \; \log{(2^{2H-1} - 1)} + 2.2 \, (H-1) \log{(n)} \; . \tag{4.62}$$

Estimates of H can be made directly from the slope alone. For data sets where n is small or the estimates of r_n are very scattered at large n, use an approximation for small n:

$$H \; = \; \text{slope}/2.2 + 1 \, . \tag{4.63}$$

Eq. 4.63 gives estimates within 1% of the correct values for the curves of Fig. 4.24, as illustrated in the left panel of Fig. 4.25. For data sets where n is large and the $\log r_n$ versus n relationship good, then use

$$H \; = \; \text{slope}/2 + 1 \, , \tag{4.64}$$

omitting r_1 from the data used to obtain the slope.

This asymptotic expression for larger n gives estimates that are nearly correct at Hs approaching 1.0, but underestimate H by linearly increasing percentages to about 10% at $H = 0.55$. Regressions omitting r_1 are very close to the asymptotic expression, and 2.0 should be used in Eq. 4.63 rather than 2.2 (right panel of Fig. 4.25).

4. Calculate the fractal dimension D, which is $2 - H$.

Comment on estimation methods: There is something to be said for using r_1 alone to estimate H. Reformulating Eq. 4.30 we get

$$H \; = \; [\log{(r_1 + 1)} / \log{(2 + 1)}] / 2 \, . \tag{4.65}$$

The values of r_1 are higher than for r_2, etc., and, particularly with small data sets, more data go into the estimate. Both factors ensure that r_1 provides greater discriminating power in estimating H than does any other r_n. While this expression is useful for calculating H or D, observing whether or not there is a straight log-log relationship in r_n versus n, or in RD versus n, is needed in order to know if the overall situation is fractal.

Applications of Fractal Autocorrelation Analysis

The estimation of H and D in practical situations is sensible when the situation is a one-dimensional fractal. The relationship among H, D, and β is summarized in Table 4.3; the β is the exponent describing the slope of the log-log relationship between the amplitude and frequency of the Fourier transform of a fractal signal;

Table 4.3. Fractal D, H, and β for one-dimensional signals that have the form of fractional Brownian motion

	D	H	β
$D =$	D	$2 - H$	$(5 - \beta)/2$
$H =$	$2 - D$	H	$(\beta - 1)/2$
$\beta =$	$5 - 2D$	$1 + 2H$	β

amplitude is proportional to $f^{-\beta}$, where f is frequency. In principle, the method should apply to higher dimensional data such as surface characteristics or fractal variations in properties in three-dimensional tissues. Schepers et al. (1992) tested the autocorrelation, Hurst, and dispersional methods on some artificial signals. Künsch (1986) and Geweke and Porter-Hudak (1983) used the approach in looking at the statistics of correlated phenomena. From the discussion of Haslett and Raftery (1989) an application to the spatial distributions of wind velocities may be envisaged. The growing understanding and utility in statistics is discussed by Hampel et al. (1986). But there is no evidence that the method is at all useful in characterizing chaotic functions of time, even though autocorrelation is helpful in determining how to formulate phase space plots of chaotic signals (Chapter 7).

To test the accuracy of estimation of H on known signals, Schepers et al. (1992) generated a fractal signal of 32,768 evenly spaced points by performing the inverse Fourier transform on a frequency domain spectrum with amplitude $1/f^{\beta}$ with $\beta = 2.6$ or $H = 0.8$ and randomized phases. (Fractal signal generation is described in Chapter 5.) The two-point autocorrelation function was calculated for this artificial data set, and, considering it as an experimental data set, an optimized best fit to the set was obtained using GGOPT (Bassingthwaighte et al., 1988a) by adjusting only H in Eq. 4.53, modifying it to use the asymptotic extension (Eq. 4.52) for $n > 4$. The result is shown in Fig. 4.26 (left panel). The estimated H was 0.8002 ± 0.0034, very close to 0.8. With a similar signal of only 512 points the estimated H was 0.743 ± 0.023 (Fig. 4.26, right panel). With large data sets the H was determined with good accuracy; smaller sets allowed more error.

More work is needed on this method. One incentive is a logical truism: whenever there is short-range positive correlation in a signal of finite length there must be long-range inverse correlation. This is obvious in a situation where an organ's blood flow comes from a single artery: if neighboring regions are correlated, having flows for example greater than the mean flow per gram of tissue, the corollary is that there must be distant regions with flows less than the mean. Vascular "steal," where a vasodilatory increase in local flow occurs at the expense of flow to another region, is an example of reciprocal fluctuations.

For application to some actual data we choose a pair of examples from Hurst et al. (1965) on the fluctuations in flows of the Nile and Niger Rivers. On his pages 99 and 100, Hurst provides data on two-point correlation coefficients on the Niger from Yevdjevich, and on the Nile from his own collected data. The periods are one year. The fit of Eq. 4.51 to these data is shown in Fig. 4.27. The correlations for r_1 to r_4

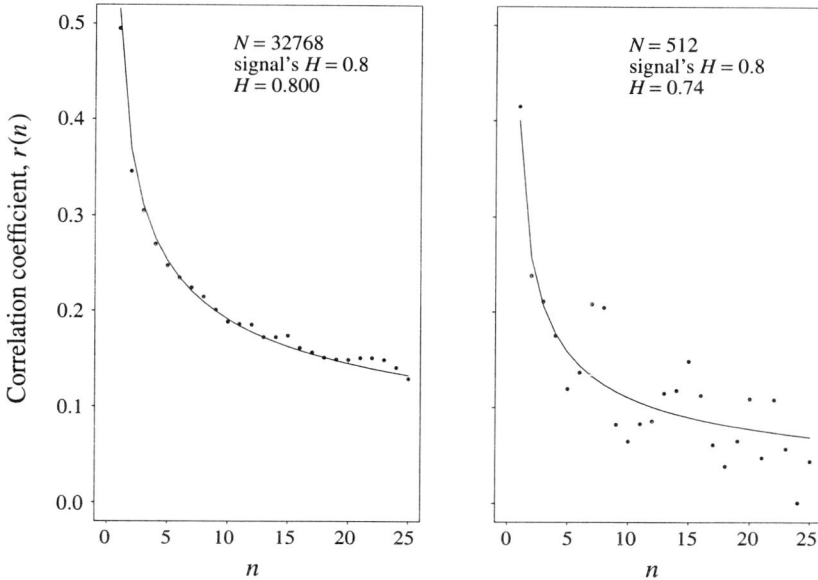

Figure 4.26. Fitting of the model for fractal extended correlation (Eq. 4.51) to the two-point autocorrelation function on artificial fractal signals with $H = 0.8$. *Left panel:* Signal composed of 32,768 points. Estimated $\hat{H} = 0.800$. *Right panel:* Signal composed of 512 data points. $\hat{H} = 0.74$. (Adapted from Figure 5 of Bassingthwaighte and Beyer, 1991, with permission from IEEE.)

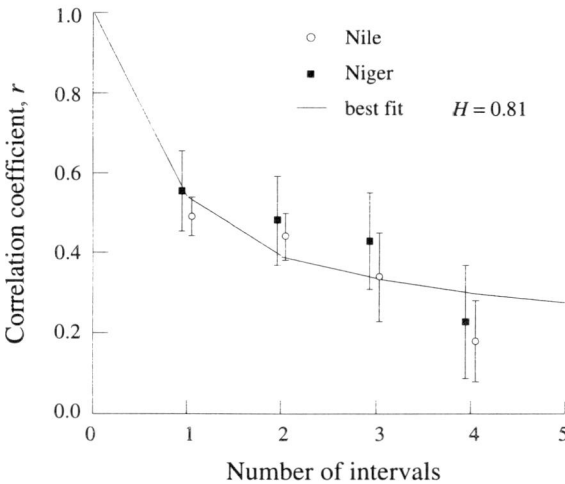

Figure 4.27. Fluctuations in annual flows in the Niger and Nile rivers from Yevdjevich and from Hurst. The nth annual correlations for both rivers have fitted approximately with the general expression for extended correlation with $H = 0.81$. (From Bassingthwaighte and Beyer, 1991, their Fig. 6, with permission.)

are fitted reasonably well with Eq. 4.51 using $H = 0.81$. However, the good fit of Hurst's Nile river flows data is of questionable consequence: he chose two subsets of the same data and came up with correlations giving r_1s of just over 10% and effectively zero for longer lags. Therefore, Hurst felt the correlation technique had little virtue. Because we do not have the original data on the 87 years involved, the source of the apparent discrepancy remains unresolved. Because the rescaled range method and the extended correlation method should give the same result, apart from biases or discrepancies in the analytic approaches, Hurst's observations raise the question of how good any of these methods are when the data sets are short.

The original analyses of relative dispersions at various volume element sizes (Bassingthwaighte, 1988; Bassingthwaighte et al., 1989) implicitly took into account that the subelements were nearest neighbors, and that continuing to divide a region ever more finely would reveal more and more of the underlying heterogeneity. The algorithm for extended correlation allows the testing of the same data sets from a very different point of view, that of local correlation. Furthermore, the test to determine whether or not the signal is fractal can be expressed in two interdependent ways: first, in a way that will express the compatibility of the two-point correlation with the fractal curve of Eq. 4.53; and second, in a way that will express the compatibility of the relationship given by halving or doubling the unit size with the same estimate of H or D. The fractal curves are noise-free model solutions which are fitted through noisy data, and the standard statistical tests of the goodness of fit of the model to the data should be appropriate. The basic statistical assumption that the error in the measurements has a Gaussian distribution should have the same degree of questionability in this setting as in other statistical assessments of the data. Given that the two-point autocorrelation function on the data set has Gaussian error, a nonlinear optimization routine to fit Eq. 4.53 should be used rather than fitting the logarithms graphically. The difference is slight when there are large numbers of points.

The "Double Bind" Test for Fractal Correlation

A test that is based on determining the self-similarity in the variances in the signal is more than a play on the phrase "double blind." The test is to see whether or not the autocorrelation functions superimpose, independent of piece size, as in the right lower panel of Fig. 4.20. We call this the "double bind" test since one will normally start with the smallest pieces, determine r_n (singles) versus n, then bind or combine pairs of neighboring pieces together, determine r_n (doubles), and so on.

In Fig. 4.28 are shown the correlation data for a fractal signal of 8192 points generated for $H = 0.8$ by the spectral synthesis method (Voss, 1988; Schepers et al., 1992). Four levels of resolution are shown; the correlation falloff for all four sample spacings was close to the theoretical line, Eq. 4.53.

The fact that the curvature for the artificial data was not distinctively different from that of the model function was reassuring. The correlation extended farther than for a comparable exponential, as in Fig. 4.23. This test would allow one to say that the fractal model could not be refuted by the comparison, within a specified

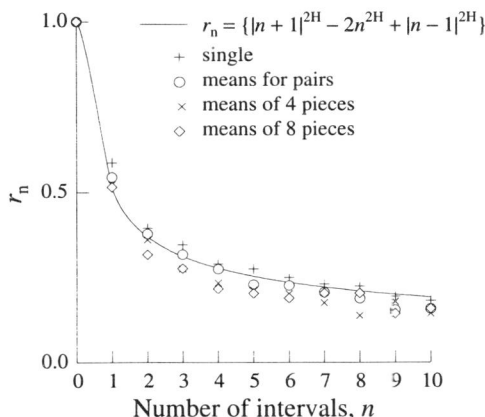

$$r_n = \{|n + 1|^{2H} - 2n^{2H} + |n - 1|^{2H}\}$$

+ single
○ means for pairs
× means of 4 pieces
◇ means of 8 pieces

Figure 4.28. "Double bind" test for self-similarity on 8192 point fractal array with $H = 0.8$ generated using the spectral synthesis method. Four analyses are shown with group sizes of nearest neighboring sections of the signal with grouping of 1,2, 4, and 8 elements. The solid line is Eq. 4.53 with $H = 0.8$. The near superimposition of the four analyses is evidence for self-similarity.

probability. The second and third tests use pieces of double and quadruple the size and confirm that there was self-similarity: this not only reaffirms the applicability of the fractal extended correlation as an appropriate model, but also affirms the estimate of the value for H or D.

4.5   History of Fractal Correlation Analysis

Feder's (1988) very readable book gives an outline of Hurst's "rescaled range analysis," the hugely important idea proposed in 1951, which has been almost completely ignored. In passing, we note that Hurst et al. (1965) denigrated the use of the correlation over extended range, because of some observations regarding analysis of a fairly small set of data; it appears that they were wrong, and we ought to keep in mind that the sensitivity of correlation analysis to the size of the data set has not been established.

Mandelbrot was among the first to recognize the importance of Hurst's work, and that a Hurst coefficient between 0.5 and 1.0 meant that there was correlation over a range. Mandelbrot and Wallis (1968) mention in their discussion the asymptotic relationship given by Eq. 4.52. The series of papers by Mandelbrot and Wallis (1968, 1969) focused on the Hurst approach, and strengthened it by improving the statistical treatment. They provided many tests of different fractal signals (always with the provision that they were not certain how far the fractal characteristics of the signal could be trusted) and illustrated that fitting the R/S estimates with a regression should be done on the individual estimates within each segment of the data (making a "pox" diagram), rather than running the regression through the means of the estimates at each segment length. The great statistician and theorist W. Feller, in his 1951 paper, clearly recognized the huge implications of Hurst's work, but even he failed to follow up on the ideas and doesn't mention it in his 1968 treatise.

Mandelbrot (1965) related measures of covariance to the Hurst measure and gave a formula seemingly for covariance but that we now recognize to be a formula for the ratio of two covariances. His paper, however, seems to have been generally ignored. This is somewhat understandable, because the paper appeared in *Compt. Rendues Acad. Sci.*, Paris, rather than in a journal more widely read by statisticians. Further, because Mandelbrot did not give the derivations for his formulas, the implications and power of the approach were not brought out. What it does illustrate is Mandelbrot's marvelous ingenuity and insight. The logic and rationale for the equations were developed by Mandelbrot and van Ness (1968) using a Fourier transform approach quite different from what we described here. (The Bassingthwaighte and Beyer (1991) effort is yet another example of *re*search, that is, the reinvention of a previous discovery, albeit by a different route than Mandelbrot and van Ness, 1968.)

The limitations of the techniques must yet be worked out. As with other methods for spatial or temporal signals, the number of observations is of the highest importance, just as for Fourier analysis. The quality or precision of the observations is also vital, for it is evident that a small amount of random noise added to a beautiful fractal signal with $H = 0.6$ will totally obscure the fractal nature, and the estimate of H will not be distinguishable from 0.5. The relative reliance on near versus distant neighbor correlations must differ with different situations. Although nothing is fractal forever, and the relationship can hold only over a range, there arises the question of whether the correlation is "oversampling" in the sense that the observation element is so small the signals are doomed to be correlated. At the other extreme, too sparse observations will leave one in the noise that must dominate all at extreme distances.

4.6   Summary

Properties measured over space or time may have correlations with fractal properties. If this is the case, then the dispersion (such as the variance) depends on the size of the element used to make the measurement. Three methods that can test for the existence of fractal correlations were described:

1. The analysis of the relative dispersion (standard deviation divided by the mean) as a function of the element size used to make the analysis, Eq. 4.5.

2. The Hurst analysis of the range of the cumulative deviations normalized by the standard deviation as a function of the length of the signal, Eq. 4.20.
3. The analysis of the correlation coefficient between ever more distant neighbors, Eq. 4.53.

The fractal dimension D of the signal, and the Hurst parameter H, which quantify the degree of correlation, can be determined from each of these methods of analysis.

The question of which method gives the best measure of the fractal dimension under given circumstances is not yet known. These three methods and Fourier spectral analysis were tested by Schepers et al. (1992) on synthetic Brownian fractal signals; the results favored the spectral analysis and the dispersional analysis. Bassingthwaighte et al. (1994), in a more detailed appraisal of Hurst analysis, found systematic biases in the estimates of H for synthetic fractal time series, and observed that dispersional analysis gave much less bias. Since one can correct for known biases, the question shifts to "how accurate are the estimates?", and "what are the effects of additional random noise on the estimation of H or D?" The Hurst analysis requires data sets of 1000 noise-free points to provide 5% accuracy in H, using bias correction. Thus, all that is clear at this point is that further evaluation of these methods is required.

Background References

Hurst's original publication (1951) and Mandelbrot and Wallis (1968) give a good approach to, and in later papers a critique of, the rescaled range analysis. Feder (1988) gives a straightforward description, but he omits the trend correction. The correlation analysis was developed by Mandelbrot and van Ness (1968), mentioned by Mandelbrot (1983, p. 353), and is alluded to by Hampel et al. (1986). The statistical approaches using relative dispersions were introduced by Bassingthwaighte (1988) because of a specific need to describe heterogeneity in a scale-independent fashion. Bassingthwaighte and Beyer (1991) developed the correlation analysis from the fractal statistical viewpoint. This will undoubtedly lend itself to the assessment of variation in sparsely sampled data, as in "kriging" (Krige, 1951), but the fractal model is only one of many. Voss (1988) gives an excellent start on understanding fractional Brownian noise.

5

Generating Fractals

So Nat'ralists observe, a Flea
Hath smaller Fleas that on him prey,
And these have smaller fleas to bite 'em
And so proceed ad infinitum.

Jonathan Swift

5.1   Introduction

Examples provided earlier as introductions to fractal ideas, such as the Koch snowflake, fall into the class of geometric fractals. They are simple, beautiful, and powerful. They startle us: so much diversity is captured in such simple beginnings. The power is not so much in the "beginning," but in the process of recursion. A simple act, repeated sufficiently often, creates extraordinary, often unsuspected results. Playing with recursive operations on the computer is the key to the revelation; reading the book spoils the story when the result is before you. Create your own monsters and beauties, and the insight comes free!

The application of fractal rules to simulate biological systems is a new art form. It has been better developed for botanical than zoological forms. Lindenmayer's pioneering work on plants has been extended so effectively (e.g., Prusinkiewicz and Hanan, 1989; Prusinkiewicz et al., 1990) that one has difficulty distinguishing computer generated plants from photographs. The use of computer generated trees as background in films attests to the degree of realism provided by recursively applied rules.

Perhaps in mammalian physiology we are too hesitant in our attempts at pictorial simulation. There are no books on simulating lungs, glands, brains, or blood vessels to levels of artistry that one can enjoy. There are, however, branching stick diagrams of vascular trees, usually pictured to show the rule or the first generation or two. The remarkable thing is that models formed from recursive application of a single rule have been accurately reflective of the variability, or texture within an organ, of regional flow distributions (see Chapters 10 and 11). Just think how much better these algorithms might be if they used three or four rules recursively, as is done for some of the simulated plants in this chapter.

The engine of creation is the recursively applied "copy machine." The simplest single-level recursion rule is

$$x_{n+1} = f(x_n) \,, \tag{5.1}$$

and the commonest variant is $x_{n+1} = f(x_n) + c$, where x_{n+1} is the output of a copy machine $f(\cdot)$, meaning that x_{n+1} is a function of whatever is provided to the operator in the place of the dot; in this case the dot is replaced by x_n. The c is of great importance, and can be regarded as the influence or initial value provided by the controller for the operation. It may be "just a constant" but it can be the dominant controller; it is the sole controller when the initial value x_0 is zero. The generalization is diagrammed in Fig. 5.1, where the "production rule" or "iterator" is to be defined for the processing unit in the box. The processing unit receives the previous output as one input, and an instruction from a control element C as a second input, c, for each iteration.

Eq. 5.1 is the evolution equation for a discrete system and is also known as a *map*. The evolution of the system is obtained by iterating the map, i.e., by repeated application of the mapping operation to the newly generated data points. Thus, if the function $f(\cdot)$ maps a value of X_j in the interval $[a, b]$ back into itself, then Eq. 5.1 is interpreted as a discrete time version of a continuous dynamical system. The size of the interval is arbitrary since the change of variable $Y = (X - 1)/(b - a)$ replaces a mapping of the interval $[a, b]$ into itself by one that maps $[0, 1]$ into itself. May and Oster (1976) pointed out a number of possible applications of the fundamental equation of Eq. 5.1 for a single variable. In genetics, for example, X_n could describe the change in gene frequency between successive generations; in epidemiology, the variable X_n could denote the fraction of the population infected at time n; in psychology, certain learning theories can be cast in the form where X_n is interpreted as the number of bits of information that can be remembered up to generation n; in sociology, X_n might be interpreted as the number of people having heard a rumor at time n and Eq. 5.1 described the propagation of rumors in societies of various structures. The potential application of this modeling strategy is therefore restricted only by our imagination.

It is useful to point out that the equations of the form of Eq. 5.1 have most often been obtained in the past by numerically solving continuous equations of motion with a computer. It is necessary to discretize such equations to numerically integrate them, and what is of general interest is that the discrete equations have a much

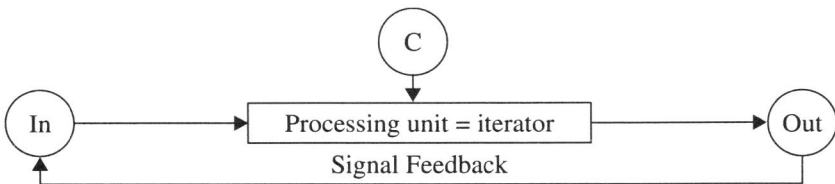

Figure 5.1. The recursion or copy machine, with an input, a control unit, C, and an output. The recursion is the feedback to use the output as the next input.

broader range of solutions than do their continuous counterparts. A continuous equation of evolution with a solution that may monotonically approach a fixed value at long times, for example, a bacterial population approaching the saturation limit of the population that can be supported by the available food supply, can have a corresponding discrete equation with solutions that oscillate in time as well as solutions that are *chaotic*. In the next few chapters we shall reveal how the apparently benign relation described by Eq. 5.1 implies such a richness of structure in its solutions.

5.2   Mandelbrot Set

The Mandelbrot set is the map for creating an infinite variety of Julia sets, using the recursion machine of Fig. 5.1. Here we review some of the geometrical structure that is implicit in certain nonlinear maps. The Mandelbrot set in two dimensions is

$$z_{n+1} = z_n^2 + c, \tag{5.2}$$

where z and c are numbers in the complex plane, $z = x + iy$. Start with $z = 0$, pick a value of c, a complex number that is held constant and used at each step, and iterate, as outlined in Table 5.1, then plot the sequence of values of z in the complex plane. The result is a Julia set, a complex structure with fractal boundaries (Fig. 5.2). The Julia set on the left is "connected," all of its parts are connected internally; that on the right is "disconnected." The famous two-dimensional Mandelbrot set is defined as the set of starting values, c, for which the values of z are bounded after an infinite number of iterations. Starting with a value of c within the central body of the Mandelbrot set (Fig. 5.3) produces Julia sets of the "connected" type. Values of c

Table 5.1. The Mandelbrot set and complex arithmetic

The Mandelbrot set is the set of complex numbers c for which the size of z is finite after infinite iteration, with $z = x + iy$.

The iteration: $z_{n+1} = z_n^2 + c$

Ignore values of $z > 2$, for they enlarge to infinity, rapidly.

Definition: $i = \sqrt{-1}$ or $i^2 = -1$

Addition: $(3 - 2i) + (6 + 4i) = (9 + 2i)$

Multiplication: $(3 - 2i)(7 + 4i) = 21 + 12i - 14i - 8i^2 = 29 - 2i$

Size of z, or Distance from 0 to z: for $7 + 4i$, $size = (7^2 + 4^2)^{1/2} = 8.06$

Figure 5.2. A "connected" and a "disconnected" Julia set. Each is an "attractor" for the set of points iterated by Eq. 5.2 from an initiator of value c.

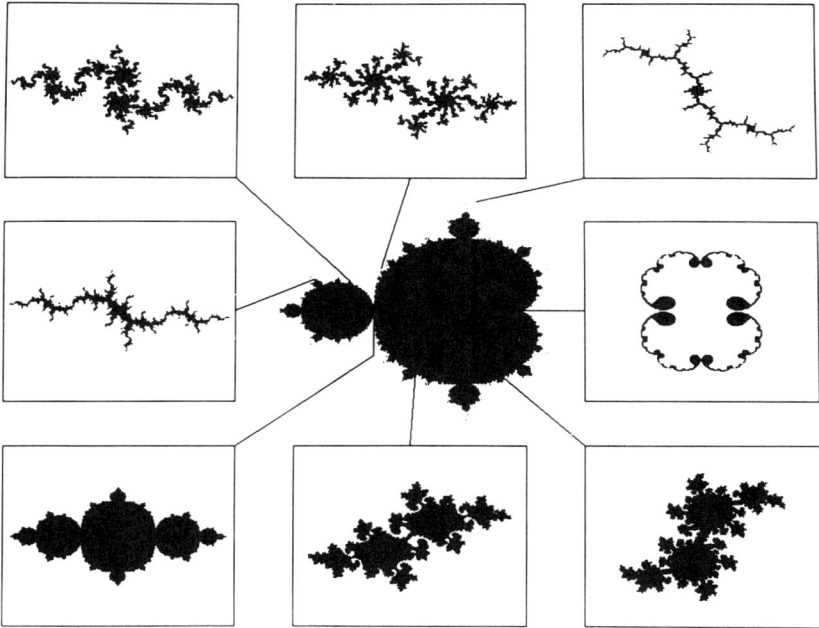

Figure 5.3. The map of Julia sets, the two-dimensional Mandelbrot set (from Figure 8.3.2b of Barnsley, 1988). The varied Julia sets and filled Julia sets tend to look much like the form at the boundary of the Mandelbrot set (in center of figure) from which the initial value was taken.

outside the main cardioid body of the Mandelbrot set produce "disconnected" Julia sets, clusters of points and picturesque objects in the plane. Starting values of c outside the Mandelbrot set do not "converge to settle in a basin of attraction," to form a Julia set, but reach values large enough (e.g., greater than two) which quickly, iteratively move the next generation, z_{n+1}, toward infinity. The Mandelbrot set as a map is shown in Fig. 5.3. Since this odd object was a huge stimulus to the development of the field it deserves our attention and admiration. (Even so, it is not so unique, and you can make up your own, using z^3, z^4, or combinations of all sorts.) The general description of the impact of the choice of c is given by Table 5.2. Peitgen and Richter (1986) give many examples and more detail.

5.3   Line Replacement Rules

Segmental Replacement

The Koch curve, introduced in Chapter 1, is a classic fractal invented by Helge von Koch in 1905 (Fig. 5.4). He did it first by adding an equilateral triangle to the middle third of a line bounding one side of an equilateral triangle (see Fig. 2.1), then repeating the operation, putting the triangles always to one side of the lines. In our line replacement algorithm the middle third of the line is replaced by two lines, each of the same length, so that the line is 4/3 as long. After infinite recursions the line is infinitely long, and covers some of the points in the plane: it is thus an object with topological dimension one, but because of its breadth and denseness of coverage of the adjacent parts of the plane it has a dimension greater than the Euclidean dimension of a "normal" line. It was in those early times another example of a "pathological" function. The program "Designer Fractal" by van Roy et al. (1988) has a very nice replacement rule algorithm to generate such structures.

Table 5.2. The Mandelbrot set as a map of fractal figures

1. **c in interior of main body M:** a fractally deformed circle surrounds one attractive fixed point. Julia sets.
2. **c in a bud:** The Julia set is infinitely many fractal circles each surrounding the points of a *periodic attractor.*
3. **c = germination point of a bud:** Parabolic case. Boundary → tendrils reaching to marginally stable attractor.
4. **c on boundary of cardioid or bud:** Siegel disk. A central attractor is surrounded by circular attractors. A point coming in from a margin rotates around the circle. (See illustrations in Peitgen and Richter, 1986.)
5. **c outside M:** Produces disconnected Julia sets or clusters, then $z \to \infty$.

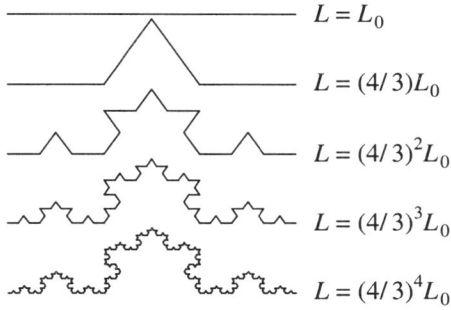

$L = L_0$

$L = (4/3)L_0$

$L = (4/3)^2 L_0$

$L = (4/3)^3 L_0$

$L = (4/3)^4 L_0$

Figure 5.4. Koch curve by replacement of the straight line of length L_0 with a specific crooked line. The line of length L_0 is the "initiator"; the "generator" has the form of the second line, the first iteration. At each stage of the recursion each "initiator" piece is replaced by a "generator" piece. For the infinite recursion the fractal $D = \log 4/\log 3 = 1.2618$; the iterations shown are a "prefractal," and the construction is of the "dragon" type. (From Bassingthwaighte, 1988.)

Another type of construction is called a "trema" type (Edgar, 1990), where trema is defined as that part of the structure which is deleted. In the Cantor set, the middle third of a line is the trema that is deleted, recursively (see Chapter 2, Fig. 2.1). For the Koch curve trema-type construction one starts with a triangle with angles of 120, 30, and 30 degrees, then deletes the equilateral triangle based on the middle third of the base. When this is repeated, the area of the remaining triangles shrinks toward that of the points covered by the dragon-type construction, and the end result is the same, as seen in Fig. 5.5. This is *not* a line replacement algorithm: the message is that there are several ways of creating specific fractals, and all of these are legitimate. Both of these approaches are one-step machines of the type shown in Fig. 5.6.

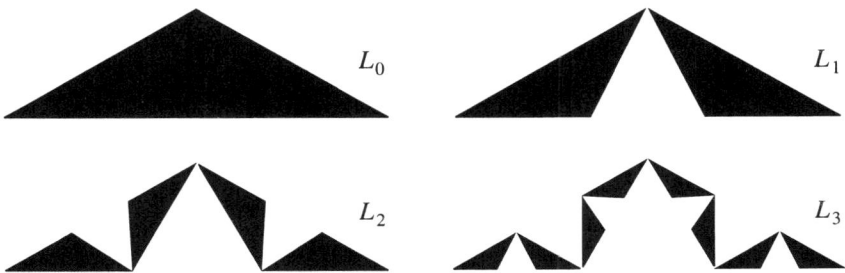

L_0

L_1

L_2

L_3

Figure 5.5. Koch curve constructed by "trema," recursive deletion of a part of a two-dimensional object to create a fractal "line."

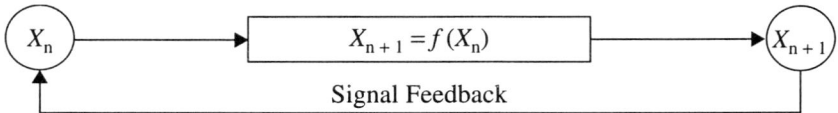

Figure 5.6. One-step feedback machine. A single rule is implemented recursively.

A slight addition to the rules for simple line replacement is to allow a "silent" or "invisible" line. In the algorithm for the bush in Fig. 5.7 there are three "silent" lines: two are clearly marked as shaded lines on the template. The third is really invisible on this picture: it is the initiator or original line; on the template it runs from the bottom to the dot.

Sidedness

Sidedness, a two-dimensional idea, or asymmetry, in any higher dimension, is a feature of the natural world. Vertebrate embryos begin symmetrically, but in early embryonic development begin to show sidedness in the position of body structures. In adult mammals, the left aortic arch is the remnant of two symmetrical arches. The small biological molecules, sugars, and amino acids are sided, and only the biologically selected stereoisomer, e.g., D-glucose and L-histidine, are transported and utilized. The transporter proteins are likewise specific to the metabolized species. The double helix of DNA spirals clockwise, and the structures of its pyrimidine and purine components are so specific that the linkages between the helices are provided by the pairing of the purine adenine to the pyrimidine thymine and the pairing of guanine to cytosine, but not the other pairings. Probably not by

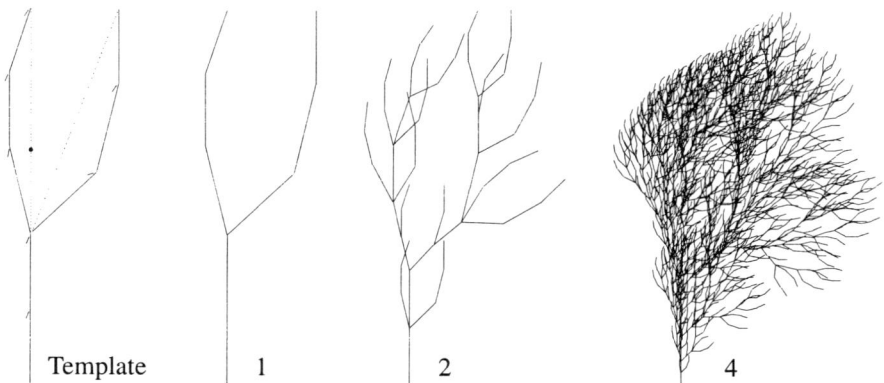

Figure 5.7. Bush, constructed from the template shown, is from van Roy et al. (1988). The form of the object shown is for one, two and four iterations.

coincidence, in the adult cell the membrane transporter for adenine and adenosine also transports thymine and thymidine, but has lower capacity for transporting guanine and cytosine. Receptors, and drugs which interfere with their action, mostly exhibit stereospecificity. While the synthesis of proteins is not seen to be fractal, the processes of transcription to form RNA from the DNA template and translation to form the protein from the RNA are processes which are sided, which are repeated over and over, and which are essential to the growth, development, and maintenance of the cell, tissue, organ, and organism. While a molecule is not generated fractally the processes that follow protein synthesis are usually or at least often fractal. (See Chapter 11 on growth processes.)

Recursive line replacement also involves sidedness. For any generator which is symmetric about the initiator no second rule is needed. For the Koch curve, the implicit definition is that the replacement for each line segment puts the pair of line segments replacing the middle third always on the same side of the line; if the fourth line segment were replaced with a down-going generator, then the picture is quite different (see Fig. 5.8). (Most "fractal" figures are formally "prefractals," so named because the number of generations is less than infinite. By this token nothing in biology or physics is ever fractal, since all except perhaps the universe is finite.)

Now the power of the simple line replacement is more evident. Changes in the sidedness of the replacement, a second part of the rule, have a huge effect on the fractal produced. (The sidedness could be randomized by coin flipping for each segment at each iteration to produce yet greater variety, or biased sequences of coin flips could be used.) Examples of two-segment generators are shown in Fig. 5.9, and the resultant prefractals are shown in Fig. 5.10. In Fig. 5.9 the replacement rule is to replace a line with a pair of lines at right angles (and therefore having total length $\sqrt{2}$ times the original length). Repeat this operation recursively, taking into account sidedness, indicated by the direction of the arrow tip. One may play a game by trying to match the generators with the fractals they produce, i.e., match the numbered fractals with the lettered generators. The answers are given at the bottom

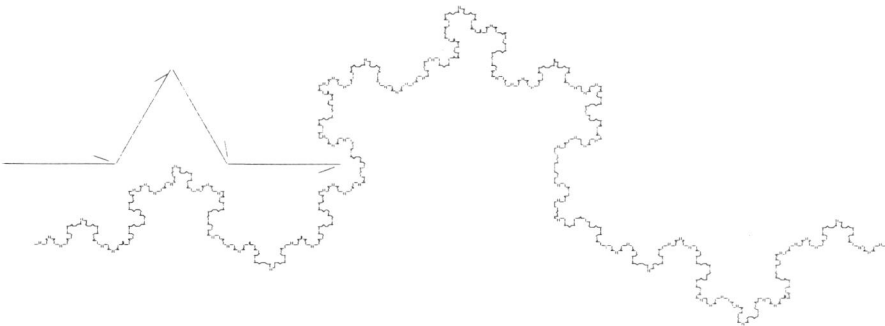

Figure 5.8. A variant on the Koch curve. The fourth line segment is replaced by the same generator, but facing toward the opposite side of the initiator line from the other three line segments. Five generations make up this "prefractal."

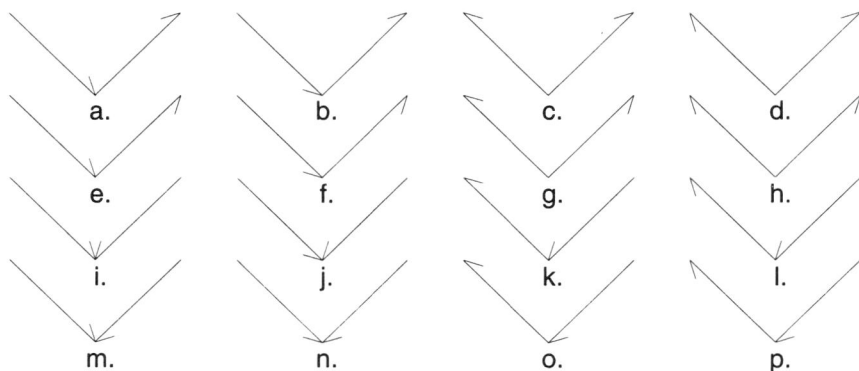

Figure 5.9. Generators or seeds showing a variety of sided replacements. These define the replacement rules for the fractals of Fig. 5.10. (From Garcia, 1991.)

of Fig. 5.10. The miracle is that all of these are produced from a generator composed of two line segments at right angles. Changing the angle slightly greatly changes the form of the fractal, as one can find by exploring the use of "Designer Fractal" (van Roy et al., 1988).

5.4   Area and Volume Replacement Rules

What one does to lines one can do to areas and volumes. One can use a trema-type construction to form the Sierpinski gasket (by removing from a triangle the middle triangle one quarter the size of the original) or the Sierpinski carpet (by removing from a square the middle square one ninth the size of the original), and repeating the operation on each of the residual triangles or squares, as in Fig. 2.1.

Similarly, one creates the Menger sponge out of a cube (Fig. 5.11). The trema method is to remove a square beam of sides one third of the length of the sides of the cube from the center-to-center of opposing faces of the cube, doing this in x, y, and z directions. This removes 7 small cubes from the initial 27 subcubes making up the initial cube. From this the fractal dimension is $D = \log 20/\log 27 = 0.7686$. If the original specific gravity of the sponge were 1.0 g/ml and the cubes removed were replaced by a vacuum, what would be the apparent specific gravity of the cube after 10 recursions? After 1000? (Answers: 0.0498 and ~0.) What is the surface area exposed after 1, 10, and 1000 iterations compared to the original surface area? (Answer: After one iteration the surface is 72/54 times the original surface area.) Now one has to distinguish side from corner cubes for the subsequent iteration. The answer is that the surface area is huge, and the surface-to-volume ratio nearly infinite, *a very efficient object for surface reactions or catalysis.*

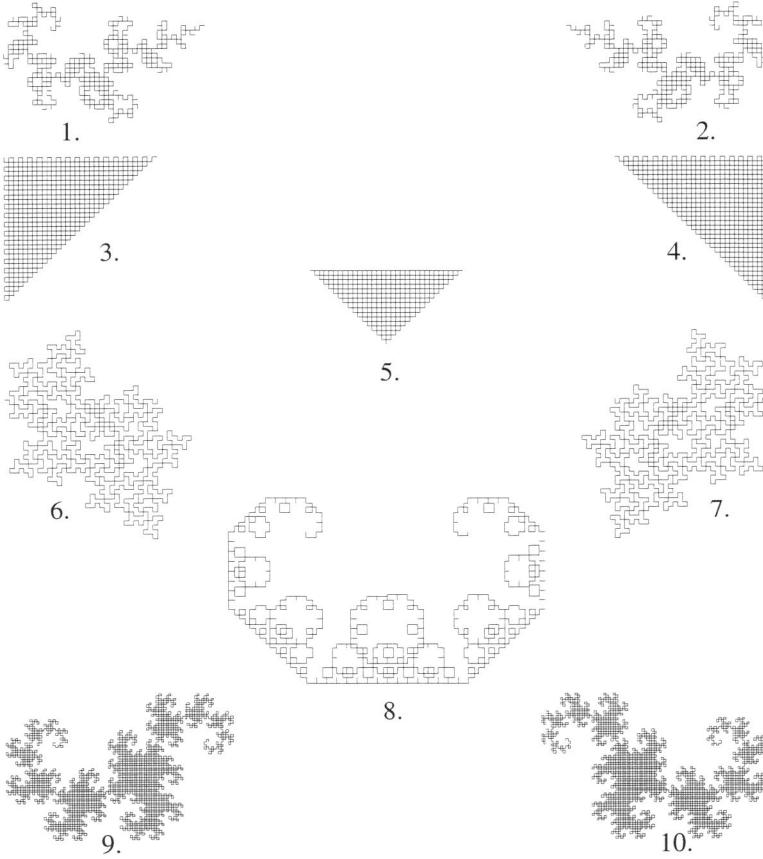

Figure 5.10. Tenth-generation "prefractals" obtained from the generators in Fig. 5.9. There are fewer fractals than generators because some generators are mirror images of another. The match-up is listed at the bottom of the figure. (From Garcia, 1991.)

One-Step Two-Variable Machines

An example of a system where two variables, rather than one, are changed at each step is the Fibonacci series: Beginning with $x_o = 0$ and $y_o = 1$, one constructs sequential values for x_n and y_n according to the two-variable one-step machine form of the type shown in Fig. 5.12:

$$x_{n+1} = x_n + y_n, \tag{5.3}$$

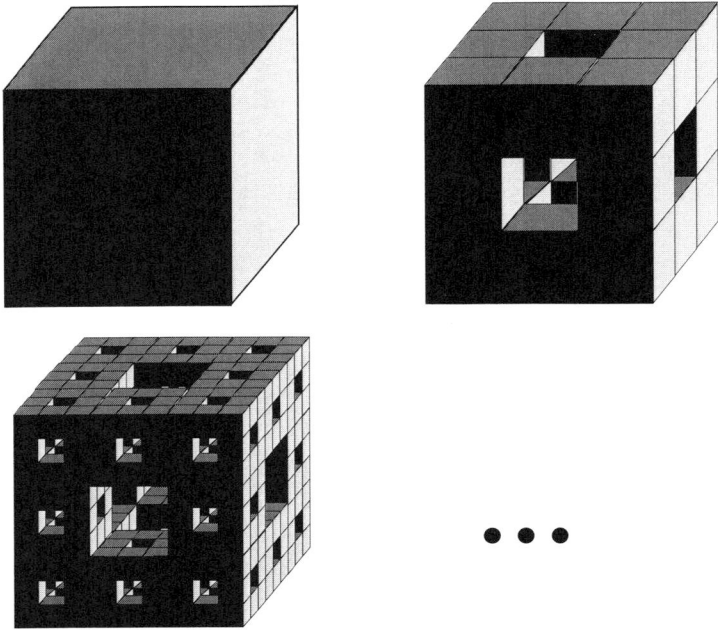

Figure 5.11. The Menger sponge, a fractal with an infinitely high surface-to-volume ratio (from Peitgen et al., 1992).

$$y_{n+1} = x_n. \tag{5.4}$$

The resultant values of x are 0, 1, 1, 2, 3, 5, 8, 13, 21, etc.

Other variants on one-step machines allow composite rules, including decisions based on comparisons or choices based on chance. In the classic case of the Sierpinski gasket the algorithm is not changed, merely the point to which it is applied. The rule is "move half the distance toward a chosen vertex." A role of a die, or spinning the fortune wheel (Fig. 5.13), determines the choice: 1 or 4 on the die

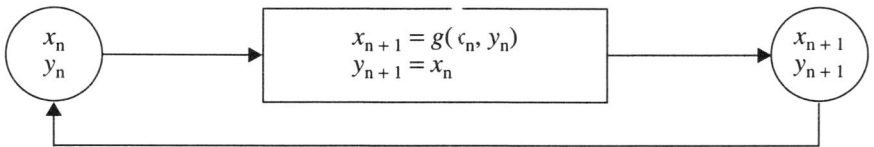

Figure 5.12. A one-step machine may operate on two variables, which may be regarded as composing a vector (x,y).

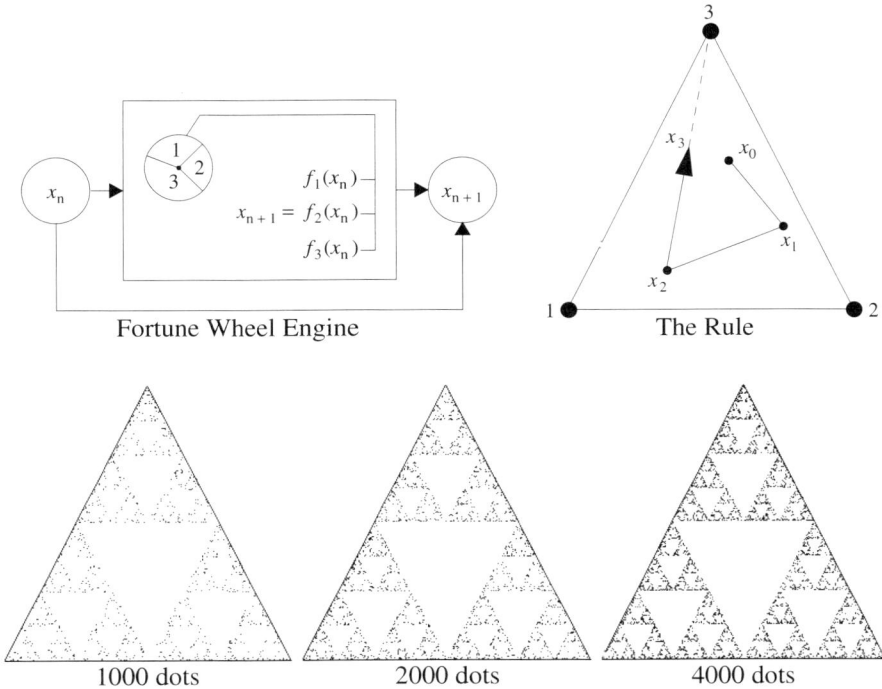

Figure 5.13. A one-step fortune wheel machine. *Left upper panel:* The random wheel makes the choice for each step. *Right upper panel:* For the Sierpinski triangle (gasket) the algorithm is to make a new point halfway between the current point and the vertex of the initiator triangle chosen by the wheel. *Lower panel:* Dot positions after 1000, 2000, and 4000 iterations.

chooses vertex 1, 2 or 5 on the die chooses vertex 2, and 3 or 6 on the die chooses vertex 3. The repetition of the rule, move halfway to a vertex, looks initially like a random movement since the vertex toward which it moves changes almost every iteration. (The first few steps should be ignored, whether starting inside or outside the triangle.) But the pattern gradually emerges, and the dot continues to bounce around the triangle leaving the pattern of its hits. So the random choice leads to very precise patterning; the dot is restricted to the "attractor," which is the Sierpinski gasket itself. Here we now see that this is a "stable attractor" or a stable "basin of attraction" which captures and holds the traveling dot. This attractor is an area of attraction; for others the attractor may be a line, or many lines. Here we use the language of dynamics in that an attractor is the structure toward which the initial choice of values evolves over time. The initial state of the system is attracted to this structure, and once there does not move off it. Dynamically interesting systems are those whose attractors are fractal as we see in the next section.

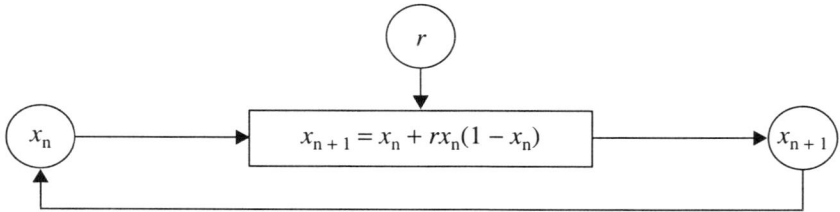

Figure 5.14. The logistic equation as a generator.

5.5   The Logistic Equation

In this section we use the recursion rule approach to generating a one-dimensional function, $x(n)$ or $x(t)$, at uniform intervals. The example chosen is a function which may be nicely stable or "chaotic," depending on one rate parameter. This function suggests some of the types of phenomena to be discussed under the topics of nonlinear dynamical systems. It also illustrates a method for generating a one-dimensional signal that is *not* a simple statistical fractal.

The logistic growth rule, Fig. 5.14, is an example of how a simple equation can produce a wild function. One finds the equation written two ways:

$$x_{n+1} = x_n + r \cdot x_n (1 - x_n) , \quad \text{or} \tag{5.5}$$

$$x_{n+1} = a \cdot x_n + (1 - a) x_n^2 . \tag{5.6}$$

With $a = 1 + r$ they are the same.

Solutions to the equation with small values of r leads to a slow growth in x, plateauing at 1.0. The original reason for looking at this expression was to describe population growth toward stability: the $1 - x$ component represents the impact of the predator or of the limit of the food supply. The growth curves actually fit a good many observations. The predator-prey relationship becomes more striking when r is higher, as for rabbits and foxes. With higher r the population of rabbits fluctuates; more rabbits, more foxes to eat them, fewer rabbits, less food for foxes, fewer foxes, more rabbits, and so on. Of course, with rabbits and foxes it takes some time for a cycle but with instantaneous feedback the excursions are sharp, as in Fig. 5.15. With growth rates greater than a doubling of the population with each period, $r > 2$, there is instability in x, but x has a specified set of values for each r. With $r = 2.3$, there are two values, with $r = 2.5$, four values, with $r = 2.542$, eight values; the break points where the number of values increases are called "period-doubling bifurcations on the route to chaos." "Chaos" occurs with $r > 2.57$. The mapping of these points of x as a function of r is a "map" in the r, x plane, the diagram known as the Verhulst Logistic Map (Fig. 5.16). Like the Sierpinski gasket, there are holes in the map,

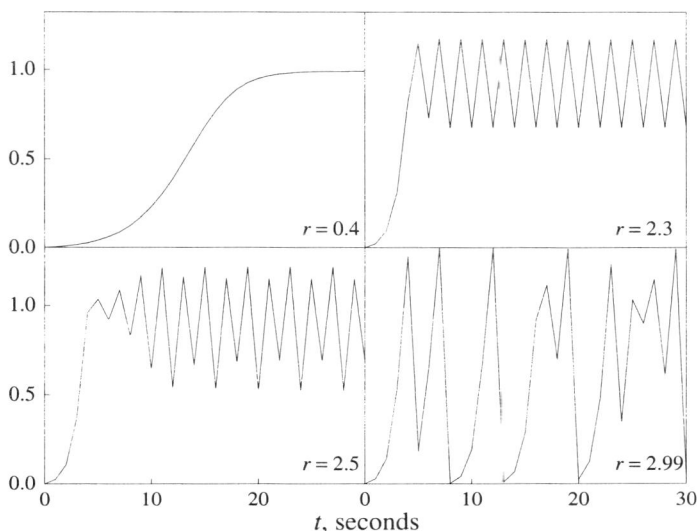

Figure 5.15. Iterations of the logistic equation showing slow growth to a plateau at low growth rates r, but periodicity and apparent chaos at higher rates.

even in the so-called chaotic regime for $r > 2.6$. As an exercise, find the value of r for the window marked with the box.

The Verhulst process map in the r, x plane only tells a little of the story. Fig. 5.16 does show that the density of values of x at a given r is nowhere uniform; in the chaotic regime the values tend to be more dense toward either extreme values or approaching certain boundaries within the chaotic regime. The map also shows self-similar characteristics in repeating patterns. A second part of the story is in the time course of the individual values of x_n as the machine iterates on the function. (As a second exercise, find a nearby region where there is "intermittency," i.e., where the oscillation stops periodically for several iterations and then restarts.) Computing this series of values also show the lack of predictability: a quick demonstration can be done with a hand calculator versus a computer; simply tabulate values for the first 20 iterations with a value of $r > 2.57$ or anywhere in the chaotic regime. A third part is in phase plane mapping, a remarkably revealing method of looking at the function x against previous values, i.e., x_n versus x_{n-1}. In the phase plane the attractor is a parabola, defined by the equation in Fig. 5.14. The mapping in Fig. 5.16 can also be regarded as another version of the attractor; although values will differ on different computers and with different starting values of x, the map always emerges in this form. Another question is whether or not the results of the iterative process are fractals: at $r = 2.3$ it is a periodic function, but at the high end, $r = 3$, the recursion pretty well fills the range from 0 to 1. Is this a fractal with dimension 2?

In this section we have chosen examples mathematically so simple that one might wonder if they deserve much attention. In each case the result of an iterative

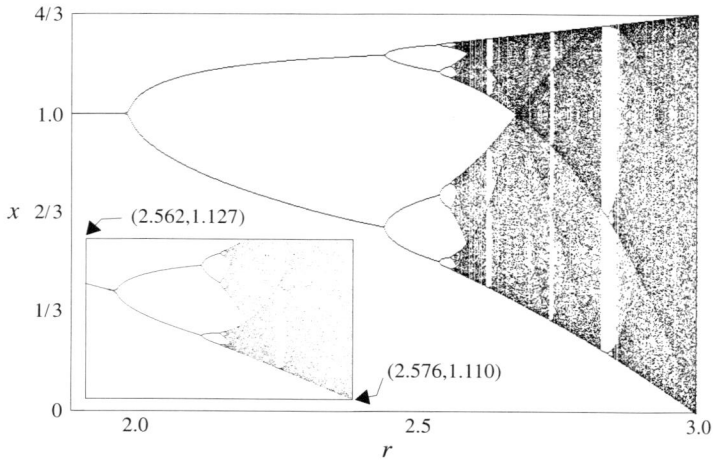

Figure 5.16. The mapping of values that x assures for a given r, for various values of r for the logistic equation or Verhulst growth process. The insert shows the pattern within a window located at $r = 2.562$ to 2.576 and $x = 1.110$ to 1.127; the pattern is faint because only a small fraction of the values for x fall within the window. It is almost an inverted version of the parent map. Within it, of course, are many more of the same at a finer scale.

procedure was to produce something that the unarmed insight could not readily predict. With single line replacement rules the results were structures quite different from the generator. With recursive point plotting from difference equations, we found mapping with structure, and the randomness or inexactness of chaos forming stable attractors, which are very precisely positioned.

5.6   Iterated Function Systems

Barnsley's (1988) *Fractals Everywhere* describes many applications of iterated functions. His fern (Fig. 3.10.1) has become a centerpiece example of what can be done with a few transformation rules, and is the opening graphic for his entertaining Macintosh program "Desktop Fractal Design Systems" (Barnsley, 1990). Its beauty, coming out of mere numbers, is startling. How does this work so well?

To draw in a plane one needs recursion rules for x and y. Purely symmetric recursions in x and y are not too interesting, but affine transformations make for immense variation. These are really two-rule, one-step transformations, with the choice being made by a fortune wheel. The one-step rule

$$x_{n+1} = a \cdot x_n, \text{ given } x_0, \tag{5.7}$$

is a scaling transformation. It moves x along a straight line running through x_0 and the origin. If $0 < a < 1$, then $x_{n+1} < x_n$ and x shrinks toward the origin. If $a > 1$, then the iterates of x expand away from the origin. If a is negative, x jumps to the other side of the origin, but if $-1 < a < 0$, then x jumps back and forth in ever decreasing jumps until it settles at the origin. The origin is the attractor.

Adding translation to the scaling idea is expressed by

$$x_{n+1} = a \cdot x_n + b, \tag{5.8}$$

where b is a real number. With b positive and $0 < a < 1$, then x moves rightward at each jump. It doesn't settle down, and continues with jump size b.

Jumps in x and y can be made functions of both x and y:

$$x_{n+1} = a \cdot x_n + b \cdot y_n + e \quad \text{given } x_0, \tag{5.9}$$

$$y_{n+1} = c \cdot x_n + d \cdot y_n + f \quad \text{given } y_0, \tag{5.10}$$

where the scaling and translation parameters are all real numbers. We can write this in matrix notation to be more general:

$$W \begin{pmatrix} x \\ y \end{pmatrix} = \begin{pmatrix} a & b \\ c & d \end{pmatrix} \begin{pmatrix} x \\ y \end{pmatrix} + \begin{pmatrix} e \\ f \end{pmatrix}, \tag{5.11}$$

where W is an affine transformation. The transformation allows for contraction or expansion, for translation and rotation. The interesting applications are not in moving a point so much as they are in transforming an object. Start with a unit square, S_0, and apply a rule-pair which effects only a reduction of scale by 25% at each iteration:

$$W \begin{pmatrix} x \\ y \end{pmatrix} = \begin{pmatrix} 0.75 & 0.00 \\ 0.00 & 0.75 \end{pmatrix} \begin{pmatrix} x \\ y \end{pmatrix} + \begin{pmatrix} 0.00 \\ 0.00 \end{pmatrix}. \tag{5.12}$$

The rule is applied to an initiator, in this case a square, so that each iteration produces a single square of reduced size, untranslated (Fig. 5.17).

Since normally the initial square would not be written, only the reduced-size copy appears on the screen. To preserve both the original and the copy a second production rule is used, namely the identity transformation. To write a list of transformations compactly a compressed tabular form is used:

	a	b	c	d	e	f
Rule #1	0.75	0	0	0.75	0	0
Rule #2	1	0	0	1	0	0

where the values of the coefficients in Eq. 5.11 are listed for each transformation.

Original

First Iteration

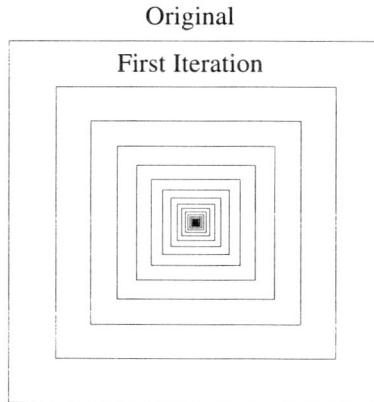

Figure 5.17. Reduction copy machine operating to reduce a unit square by 25% per iteration, and using also an identity transformation to reproduce the original square. The initiator is the square on (−1, −1) to (1, 1), so that the center point of convergence after infinite iterations is (0, 0).

These transformation rules allow great complexity. The information that goes with them to make them possible is equally important. To get the squares in Fig. 5.17 one provides the initiator in the form of a unit square whose center is placed at 0,0. If one started with a circle then the same rules would give a series of concentric circles. From these rules one can guess that the attractor is at 0,0, so that if one were to start at any random point on the initiator then with successive iterations the dot would move at an exponential rate toward the origin. Starting with a single point is therefore uninteresting.

When there are multiple rules the story changes. The compact notation for a variant of Barnsley's fern is shown in Table 5.3. Now with each rule we must also provide a relative probability: the Fortune Wheel directs the production of the next dot from the current position in accord with the choice of rule, where the rule is chosen in accord with the listed probabilities, p. The result is shown in Fig. 5.18 for 50,000 iterations from an arbitrary starting point.

The fern itself is the attractor. Any point will serve as the initiator. This works because the rules force the jumping point to move around the attractor forever! This idea is shown in Fig. 5.19.

An alternative form for the recursion rules is the *scale-rotation form*:

$$W\begin{pmatrix} x \\ y \end{pmatrix} = \begin{pmatrix} r\cos\theta & -s\sin\psi \\ r\sin\theta & s\cos\psi \end{pmatrix}\begin{pmatrix} x \\ y \end{pmatrix} + \begin{pmatrix} e \\ f \end{pmatrix}. \tag{5.13}$$

One can make a tabular version of this for convenience as an alternative to the form shown in Table 5.3, where six parameters per rule, plus the probability, is again required, as shown in Table 5.4.

Table 5.3. Fern variant scaling and translation parameters

	a	b	c	d	e	f	p
#1	0	0.014	0	0.215	−0.024	0.007	0.01
#2	0.85	.04	−.04	.85	.0001	.155	.75
#3	0.2	−.26	.23	.22	0	.183	.12
#4	−0.127	.3	.272	.215	−0.012	.236	.12

Figure 5.18. Fern variant using the production rules of Table 5.3 or 5.4. Display is on a rectangle from (−0.3, 0) to (0.3, 1.0). (Figure produced using the program of Lee and Cohen, 1990.)

Table 5.4. Fern variant: scale, rotation, and translation parameters

	Scale		Rotation		Translation		
	x or r	y or s	x or θ	y or ψ	x or e	y or f	p
#1	0	.215	0	−3.83	−0.024	0.007	0.01
#2	0.851	.851	−2.69	−2.69	.0009	.155	.75
#3	.305	.341	49.0	49.8	0	.183	.12
#4	.30	.369	115	−54.4	−0.012	0.236	.12

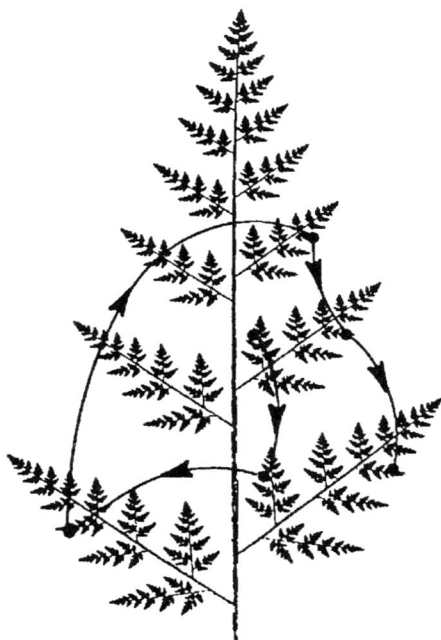

Figure 5.19. The iterated dot position remains on the attractor forever, and by moving over it leaves its trail of blackened pixels in the form of a fern. (Diagram from Barnsley, 1988, p. 142.)

The colloquial parlance is that r is the x-scale and s the y-scale, θ the x-rotation and ψ the y-rotation. In some situations this is easier to use, and in a program like "Fractal Attraction" by Lee and Cohen (1990) one can work with any of the forms, the full matrix or compact matrix forms or the scale-rotation form.

5.7   The Collage Theorem

Barnsley (1988) observed that an iterated function algorithm could be used to produce realistic patterns out of relatively crude initial forms. A guide to positioning the initiators is to make a collage covering the approximate shape of the object desired. Our example is to make a fern out of a fern replica.

Barnsley's nonmathematical statement of the theorem is "The theorem tells us that in order to find an IFS (iterated function system) whose attractor is 'close to' or 'looks like' a given set, one must endeavor to find a set of transformations—contraction mappings on a suitable space within which the given set lies—such that the union, or collage, of the images of the given set under the transformations is near to the given set."

Figure 5.20. The collage theorem is illustrated by using polygonalized reduced copies of a leaf to cover the outline of a leaf. *Left panels:* The target pattern is the large polygon; the replicas are the reduced, transformed copies defining the production rules. In the left lower panel the replicas have been malpositioned so that they do not overlap the target polygon. *Right panels:* The resultant attractors for the two sets of rules portrayed on the left. The malpositioning of the replicas in the lower set leads to a fractal attractor with many separate parts, like a Julia set whose initiator was on the outer fringe of the Mandelbrot set.

In Fig. 5.20 is shown a target set, the polygonalized boundary of a leaf, best seen in the lower left panel. Superimposed on the target are four smaller replicas, four affine transformations of the target set; in the lower left diagram they are poorly positioned and don't cover the target set well. The overall contractivity (the scale contraction with each iteration) is about 0.6, so that the final attractor must be smaller than the target set. The result of random applications of the production rules is the attractor shown in the right lower panel, which is not too close to the original leaf. A better placement of the replicas is shown in the left upper panel; the coverage of the target is quite close and the attractor that results, shown in the right upper panel, is a reasonably shaped leaf of the same general form as the original target polygon. Both the polygon and the replicas defining the production rules can be more crude than this and still give a strikingly good picture, as in the fern in Fig. 5.18. The collage theorem therefore gives evidence for the stability of the attractor and its insensitivity to the exact form of the shapes used for defining the production rules.

5.8   Lindenmayer Systems

Aristid Lindenmayer (1968a, 1968b) introduced a new approach to "rewriting algorithms" like the Koch curve algorithm. It evolved from what had become

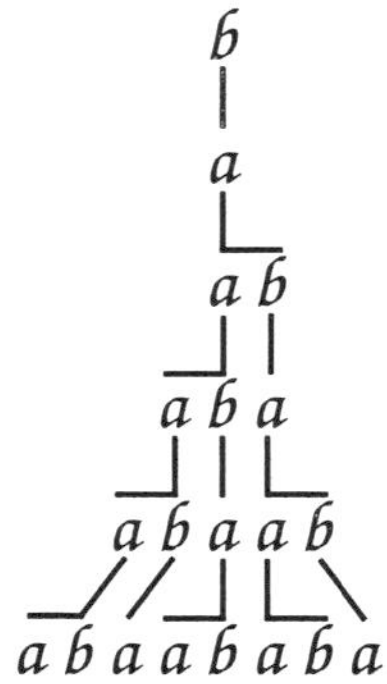

Figure 5.21. String rewriting in a deterministic context-free Lindenmayer system. The replacement rules are: b→a, and a→ab. At each iteration the production rules are applied to each element of the string, so that each is replaced. (From Prusinkiewicz and Hanan, 1989.)

known as Chomsky grammars, rewriting strings or sets of strings that could be computer generated, recognized, or transformed. An example of string rewriting is shown in Fig. 5.21.

For graphical interpretation, the L-system production rules can be implemented using a LOGO-style turtle (Abelson and diSessa, 1982) as proposed by Szilard and Quinton (1979). The "turtle" traces out a line. Its state in a plane is the triplet x, y, α, where x, y is position and α = direction in which the turtle is heading. For repetitive operation, a step size, d, and an angular course shift, δ, are also defined. The words of the language are:

F   Advance one step (length d). The positional change is given by the change of the turtle's "state" from x, y, α to x′, y′, α, where $x' = x + d\cos\alpha$, $y' = y + d\sin\alpha$. The heading is unchanged. A line is drawn along the route (the turtle track).

f   Advance one step without drawing a line.

+   Turn right by δ degrees. (Here δ positive indicates a clockwise turn, but be aware that in the Scale-rotation form of the production rules of IFS the angle changes follow standard mathematical habits, namely that α positive means counterclockwise. Often it won't matter.) The next state of the turtle is x, y, $\alpha + \delta$.

−   Turn left by δ degrees.

ω The description of the initiator.

p   The production rule.

For a simple example, let us revisit the Koch triangle using Lindenmayer iterated string production by applying the rules given in the "turtle" language as follows:

$\delta = 60$ degrees (angle increment)
$d_{n+1} = d_n/3$ (scale contraction factor)
ω: F + + F + + F   giving the plain equilateral triangle. (The initiator.)

p: F → F − F + + F − F   replacing each straight line segment with four. (Rule.)

The result is the Koch snowflake, one side of which is shown in Fig. 5.4 and the whole thing in Fig. 2.1.

Sidedness plays an important role in the final configuration, as seen earlier. The dragon curve given by pattern *m* in Fig. 5.9 and resulting in the Heighway dragon shown in result 9 in Fig. 5.10 can be translated into the following:

$\delta = 90$ degrees
$d_{n+1} = d_n / \sqrt{2}$
ω: F1
p1: $l \rightarrow l + rF +$
p2: $rF \rightarrow -Fl -r$
(p3: F → F)   (not normally stated overtly)

where the symbols r and l for right and left are not interpreted by the turtle but are retained in the string. Each part of the symbol pair is examined and replaced so that the sidedness is iterated upon.

The bush in Fig. 5.7 is likewise simply stated, using brackets as additional notation:

$n = 4$
$\delta = 22.5$ degrees
$d_{n+1} = d_n / 1.58$
ω: F
p: F → FF + [+F−F−F] − [−F+F+F]
[Push the current state of the turtle onto a pushdown stack. (Save this x, y, α + color or line width.)

] Change the current state to the state on the top of the pushdown stack, and proceed with the next instruction.

An example is shown in Fig. 5.22 for a bracketed string and its turtle interpretation. With only a little more complexity, that required to enclose a leaf

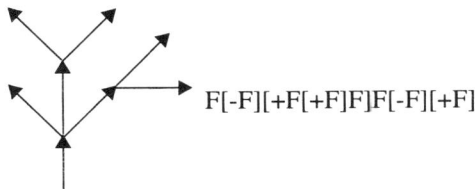

F[-F][+F[+F]F]F[-F][+F]

Figure 5.22. Turtle interpretation of a bracketed string. (From Prusinkiewicz and Hanan, 1989).

Figure 5.23. A three-dimensional Fibonacci bush. The L-system for the bush generates lengths proportional to the elements of the Fibonnaci series. (From Prusinkiewicz and Hanan, 1989, their Figure 3.3.)

shape and to fill with a shade reflecting the light from a source, Prusinkiewicz and Hanan (1989) portray a bush (Fig. 5.23).

These systems can be modified further to include rules for the stem widths, colors, shades, etc. A most beautiful book is that of Prusinkiewicz et al. (1990) where they show in color many wonderful synthetic plants and flowers. This is an advanced stage of the art, and is the result of many years of development.

Some early work was seen in the Star Wars films and the methods are preserved on a videotape from PIXAR (Smith, 1984).

5.9   Cellular Automata

Cellular automata are systems forming and changing under a set of production rules. They are dynamical discrete systems of simple forms in which the form of the system at the next time step is exactly predictable from the present form, but whose form after some time is not necessarily predictable. While they are vehicles for producing signals of fractal-like characteristics, it is probably better to think of them as potentially chaotic systems rather than fractals.

Cellular automata are one-, two-, or higher-dimensional systems that change with time in obedient response to a set of rules imposed at each time step. The rules may be probabilistic or deterministic, but the classic cases are all deterministic. The "cells" are usually displaced as pixels on a square lattice, and in the simple cases the cell is either "dead," value 0, or alive, value 1, a two-state system. The rules are applied all at once, simultaneously to all of the cells; this is "parallel processing," a feature of cellular automata that has led to their use in designing parallel computer architecture. This is a multiple rule recursion machine. In addition to the rules, there must be a starting configuration of the system upon which the rules operate repetitively. The difference between cellular automata and the single-step, multiple-variable recursion machine represented in Fig. 5.12 is that the application of the rules depends upon the state of the neighboring cells as much as it does on the state of the cell to which it is applied.

An example is the "Game of Life" invented by John Conway and popularized in *Scientific American* (Gardner, 1970). It has only two rules that relate a cell's next state to its current state and the state of its eight immediate neighbors on a square lattice: (1) a dead cell becomes live when it has exactly three live neighbors, and (2) a live cell dies (of loneliness or isolation) if it has less than two neighbors, or it dies (of overcrowding or competition?) if it has more than three neighbors. The evolution of the system depends on the form of the initiator, the patterns of live cells at time zero.

An example of a self-sustaining automaton is the "glider." Its initial form consists of five live cells on the square matrix:

The glider changes shape as cells are added or lost, but after four time steps regains its original form, but is shifted down and to the left. Thus it moves across the field in a straight line in this cycling fashion.

More complex initial forms or fields on which there are several separated forms have complex behavior. The behavioral possibilities can be summarized into four types:

1. Rules that produce trivial universes, all dead or all alive.
2. Rules that produce stable repetitive configurations.
3. Rules that produce complex patterns with local organization.
4. Rules that produce chaotic nonperiodic patterns.

These classes have some similarity to the behaviors of the logistic equation at the different rate constants illustrated in Figs. 5.15 and 5.16. So these cellular automata are quite analogous to nonlinear dynamical systems of differential equations that will be examined in detail in later chapters.

The general ideas are outlined in an introductory text by Toffoli and Margolus (1987) and in more mathematical detail in a set of papers introduced by Wolfram (1984). Cellular automata can give rise to fractal networks, as described by Orbach (1986), and which look like the Ising systems discussed by Toffoli and Margolus; both are characteristic of amorphous systems of materials which have not mixed uniformly but have dispersed so that there are complicated interfaces visible at all length scales. The little book and set of DOS programs "Artificial Life" by Prata (1993) is a useful introductory set of cellular automata that can fascinate one for many an evening.

5.10   Cellular Growth Processes

There is a complement to the Mandelbrot set which has the two rules,

$$\{w_1(z) = \lambda z + 1, w_2(z) = \lambda^* z - 1\}, \tag{5.14}$$

with λ^* indicating the complex conjugate of λ. (The number λ is the complex number $\lambda_1 + i\lambda_2$; thus $\lambda^* = \lambda_1 - i\lambda_2$.) An iteration on this map at a particular location gives an attractor that in parts looks like growth at a cellular bud (Fig. 5.24). The cell primordium at the edges appears to show small cells gradually emerging and becoming separable from one another, and then expanding into thick-walled cells that are much larger.

Comparison with Fig. 5.25 shows the similarity to a growth point. The same pattern of growth initiated at a surface and cell layers of increasing size is evident.

5.11   Generating One-Dimensional Fractal Time Series

There are two standard ways of generating artificial signals that have fractal statistical behavior. These are the methods of (1) successive random additions, SRA, and (2) spectral synthesis (SSM, using M for method). Both use random number generators but use them in specific ways to maintain a prechosen level of correlation defined by the chosen Hurst coefficient H. Both have been described in excellent detail by Saupe (Chapter 2 in Peitgen and Saupe, 1988), providing more background and also pseudocode for computing the functions for various values of H.

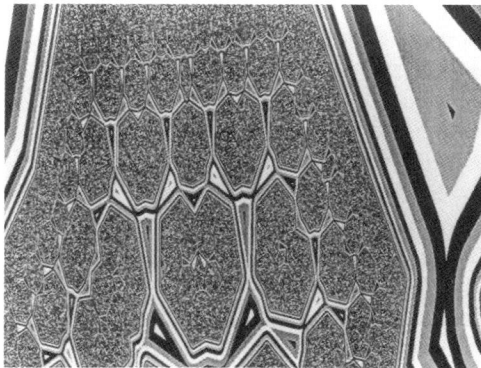

Figure 5.24. An imitation of cellular growth emerging from a dynamical system which is the complement of the Mandelbrot set. (From Barnsley, 1988, Fig. 8.3.2b.)

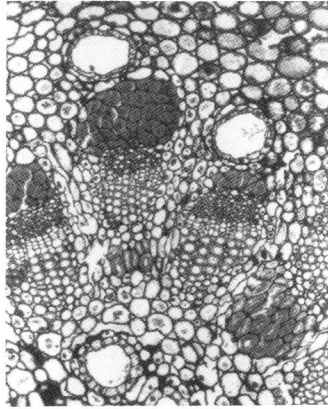

Figure 5.25. Growth of a cell layer. (From Kodak *Techbits* 1, p. 7, 1992, with permission.)

The *successive random additions* method, SRA, is a line replacement method. Displace the midpoint of an "initiator" line by a random distance perpendicular to the line, thereby creating two line segments from the initiator. Displace the two endpoints randomly also, using the same scalar. (This version therefore is a midpoint and endpoint displacement method.) This creates two line segments from the original one segment, but none of the original points remain exactly where they were. On the next iteration the scalar for the random displacements is reduced by a fraction depending on H. Expressed as the variance for a Gaussian random variable:

$$\text{Var}\,(\Delta x) \;=\; \frac{\sigma^2}{(2^n)^{2H}} \cdot (1 - 2^{2H-2})\,, \qquad (5.15)$$

where Δx is the variable displacement, σ^2 is the variance for a Gaussian distribution, and n is the iteration number. For uncorrelated Gaussian noise $H = 0.5$ and the expression reduces to

$$\text{Var}\,(\Delta x) \;=\; \frac{\sigma^2}{2^{n+1}}\,. \qquad (5.16)$$

The method replacing the endpoints as well as the midpoint is better than midpoint displacement alone. The resultant signals give fractional Brownian motion, fBm, with the H defining the correlation. The successive differences of fBm are fractional Brownian noise, fBn. The algorithm gives the Brownian motion, fBm, rather than the noise, fBn, because the successive displacements are added to or subtracted from the midpoints of lines that have already been moved from the original line in the preceding iterations.

The *spectral synthesis method*, SSM, is designed to produce a signal where the amplitude of the Fourier power at each frequency contained in the signal diminishes with frequency proportionally to $1/f^\beta$. When $\beta = 0$ the power spectrum is flat, as is the case for white noise, in which the time series is composed of a sequence of independent random values. Integrating a random white noise signal gives the classical Brownian motion, where the individual steps are random and uncorrelated, but where the sequence of positions of a particle is a continuous function, even if it is infinitely jerky. The Fourier power spectrum of Brownian motion diminishes as $1/f^2$ or $\beta = 2$. The integration gives the correlation and increases β, as expected since $\beta = 1 + 2H$ (Table 4.3). Integration of any stationary signal produces a derived signal with a value of β higher by 2.0 than the original signal.

Fractional Brownian noises, f Bn, have $0 < \beta \leq 2$, ranging from 0 for white noise to just under 2.0 for noises for noise with internal correlation. A definition is that $\beta = 2$ is fractional Brownian *motion*, and values higher than 2.0 are smoother types of f Bm. Fractional Brownian noise signals, f Bn, are composed of the differences between points in a fractal Brownian motion.

The spectral synthesis method, given in detail by Saupe in Peitgen and Saupe (1988), is also called the Fourier filtering method since the power spectrum looks like that of a signal filtered to diminish the high frequency components. The method is to generate the coefficients in the Fourier series,

$$X(t) = \sum_{j=1}^{j=N/2} (A_j \cos jt + B_j \sin jt) , \qquad (5.17)$$

where $A_j + B_j$ are the transform coefficient generated within the *SSM* routine using

$$P_j = (j)^{-\beta/2} \cdot \text{Gauss} ( ) , \qquad (5.18)$$

$$Q_j = 2\pi \, \text{Rand} ( ) , \text{ and} \qquad (5.19)$$

$$A_j = P_j \cos (Q_j) , \qquad (5.20)$$

$$B_j = P_j \sin (Q_j) , \qquad (5.21)$$

and where Gauss( ) is a Gaussian random number generator with mean 0 and variance 1, and Rand( ) is a uniform random number generator over the range from 0 to 1. The power or amplitude terms, the P_j's, contain the Gaussian term that alone is to be scaled by $j^{-1/2}$ and the further scale reduction at each successive frequency for values of $H > 0$. This gives fractal Brownian motion.

The increments between elements of $X(t)$ comprise the fractional Brownian noise. This and the SRA method are described in more detail by Feder (1988).

5.12   Summary

This chapter on geometric fractals has adhered to discussing fractal generation from completely deterministic rules. The only probabilistic influence been the relative probability of applying members of sets of rules using the Iterate Function Systems. Replacement rules for a point were illustrated by the Mandelbrot set itself. This brought out the principle that the Mandelbrot set is the map of points giving rise to Julia sets. It was then only a little stretch of the mapping idea to show that the Logistic Map predicts the general form of the point by point recursion of the logistic equation.

Single rule line replacements, like the Koch curve, gained complexity when sidedness was given as a subsidiary rule. The overt expansion of the algorithmic approach to multiple rules for positioning a point led to very nice examples of complex attractors like the fern or the Sierpinski gasket wherein each step moves the point from one place to another on the attractor building it up, dot by dot, to reveal its form. Are these the mechanisms of cell growth and intraorgan organization, and are our cells rules almost as simple as these?

Notes on the References

Edgar (1990) takes a mathematical view, giving the theorems and proofs needed to provide a solid basis for this type of work. He brings out particularly well the ideas of tiling. Planar tiling has an extensive literature beyond fractals, summarized thoroughly by Grünbaum and Shephard (1987).

Programs allowing an introduction are "Designer Fractal" (van Roy et al., 1988), "Fractal Attraction" (Lee and Cohen, 1990) and "Desktop Fractal Design Systems" (Barnsley, 1990). The former two allow infinite exploration, a means for generating your own fractals. The latter shows Barnsley's beautiful fern and other preprogrammed fractals, but its flexibility is limited to generating Julia sets from the Mandelbrot set, with fair resolution only.

Fractals for the Classroom, by Peitgen et al. (1992), is an excellent introductory work with still a lot of depth. It covers much more than this chapter, and gives exercises and good illustrations; its greatest virtue is its step-by-step approach to each topic, leaving no step untrodden. Another program, available over the Internet, is "Fractint," which includes a large set of fractal functions and good examples of chaotic dynamical systems (Tyler et al., 1991).

Applications to generating patterns and plants are now in several books. Garcia's *The Fractal Explorer* (1991) develops the rationale for learning about fractals but is elementary. More sophisticated details are given by Barnsley's *Fractals Everywhere*. The two books by Prusinkiewicz et al. (1989, 1990) are a good combination, the earlier one taking gentle strides along the path to how to do it, and the latter showing the powerful results that can be obtained.

ɔtic Phenomena

t also in the everyday world of politics and economics, we would
people realized that simple nonlinear systems do not necessarily
al properties.

R.M. May (1976)

6.1   Introduction

Two time series are illustrated in Fig. 6.1. They look similar. They have
approximately the same statistical properties, approximately the same mean, and
approximately the same variance. They both look random, but we now know that
not everything that looks random is random. These two time series are very
different. The time series on the left is random; it was generated by choosing
random numbers. However, the time series on the right is not random at all; it is
completely deterministic. The next value, x_{n+1}, was computed from the previous
value x_n by the relationship $x_{n+1} = 3.95 \, x_n \, (1 - x_n)$. Systems, like the one on the
right, that are deterministic but whose output is so complex that it mimics random
behavior, are now known by the jargon word "chaos." Chaos is perhaps a poor word
to describe this phenomenon. In normal usage chaos means disordered. Here it
means just the opposite, namely a highly ordered and often simple system whose
output is so complex that it mimics random behavior. It is quite important that we
distinguish between chaos and noise in a given experimental time series, because
how we subsequently process the data is determined by this judgment.

Many signals in physiology look random. Examples are the fluctuations in the
rate of the heart and the blood pressure in the arteries. It has always been assumed
that these fluctuations must be produced by random processes. However, if these
fluctuations are really chaotic instead of random, then they would be the result of
deterministic physiological mechanisms. If that is the case, we might then be able to
understand these mechanisms and even learn to control them. This would have
tremendous significance in our understanding of physiological systems. In this
chapter we describe some of the elementary properties of chaos. In the next chapter
we describe the methods of analysis that can be used to determine if a system is
chaotic or just noisy.

It might seem strange that a regular relation such as that depicted between x_{n+1}
and x_n in Fig. 6.1 (right lower) can lead to such erratic behavior in the graph of x_n
versus n depicted in Fig. 6.1 (left lower). The reason is in part that the *order* in

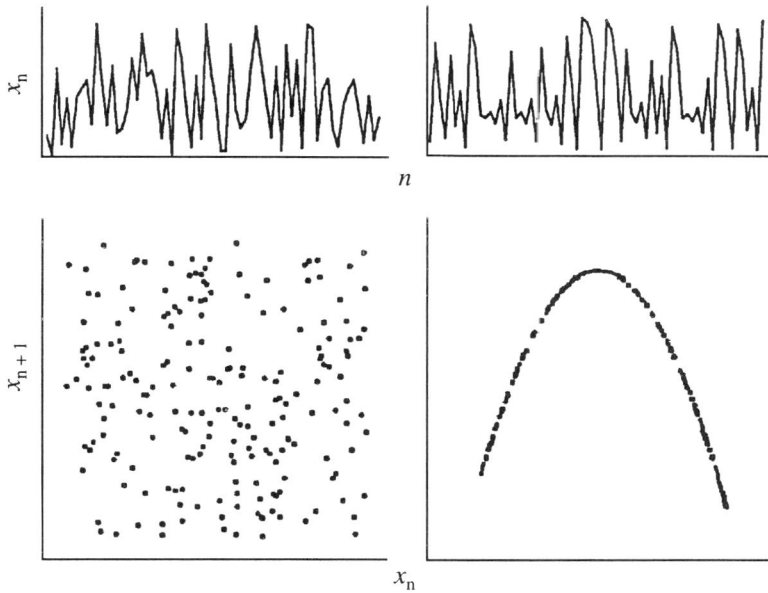

Figure 6.1. *Top:* Two time series that look quite similar. They have approximately the same statistical properties. They both look random. However, they are very different. The time series on the left is random; each new value was chosen by a random number generator. But the time series on the right is not random at all; it is completely deterministic. *Bottom:* The true identity of each time series is revealed in the phase space set where x_{n+1} is plotted versus x_n. The phase space set of the random time series on the left fills this two-dimensional phase space, and would also fill any higher dimensional phase space, indicating that this time series was generated from an infinite set of independent random variables. The phase space set of the chaotic time series on the right is a one-dimensional line indicating that it can be generated by a deterministic relationship having one independent variable.

which the various values of x_n are generated by the parabolic function is not apparent from the function. This is only seen in the "time series," and it is this order which appears random.

A chaotic time series is physiologically interesting for a number of reasons. Conrad (1986) has suggested categories for five functional roles that chaos might play in biological systems. Among these are: i) search, ii) defense, iii) maintenance, and iv) dissipation of disturbance. In the first function, that of search, such phenomena as mutation, recombination, and related genetic operations might be placed. In such cases chaos plays the role of the generator of diversity. Also in the category of search processes is behavior, such as that of microorganisms. The function here is to enhance exploratory activity, whether dealing with macromolecules, membrane excitability, or the dynamics of the cytoskeleton.

The second function is that of defense, wherein the diversity of behavior is used to avoid predators rather than to explore. An organism that moves about in an unpredictable way is certainly more difficult to ensnare than one that moves ballistically.

A third possible function for chaos is the prevention of entrainment, which is to say the maintaining of the process, such as in the beating of the heart. In neurons, for example, either absolutely regular pacemaker activity or highly explosive global neural firing patterns would develop in the absence of neural chaos. The erratic component of the neural signal would act to maintain a functional independence of different parts of the nervous system. It has been argued here and elsewhere that a complex system whose individual elements act more independently are more adaptable than one in which the separate elements are phase locked. This same decoupling effect can be identified in the immune system, and population dynamics.

The possibility of chaos causing dissipation or subduing disturbances in a system has recently received a great deal of attention (see, e.g., Beck, 1990; and Trefán et al., 1992). It is argued that the many degrees of freedom traditionally necessary to produce fluctuations and dissipation in a physical system may be effectively replaced by a low dimensional dynamical system which is chaotic. In the same way a biological system which is disturbed may relax back to regular behavior by the absorption of the perturbation by the chaotic trajectories. As Conrad (1986) said:

> . . . if the system possesses a strange attractor which makes all the trajectories acceptable from the functional point of view, the initial-condition sensitivity provides the most effective mechanism for dissipating disturbances.

6.2   Fractals and Chaos Share Ideas and Methods but They Are *Not* the Same Thing

The studies of fractals and chaos are linked in many ways. They share common ideas and methods of analysis. Each field is used within the other. An object called a phase space set can be formed from the time series data. When the phase space set is fractal the system that generated the time series is chaotic. The fractal dimension of the phase space set tells us the minimum number of independent variables needed to generate the time series. Chaotic systems can be designed that generate a phase space set of a given fractal form.

However, it is important to remember that the objects and processes studied by fractals and chaos are essentially different. *Fractals* are objects or processes whose small pieces resemble the whole. *The goal of fractal analysis* is to determine if experimental data contain self-similar features, and if so, to use fractal methods to characterize the data set. *Chaos* means that the output of a deterministic nonlinear system is so complex that it mimics random behavior. *The goal of chaos analysis* is to determine if experimental data are due to a deterministic process, and if so, to determine the mathematical form of that process.

6.3   The Defining Properties of Chaos

We present an overview of the essential properties of a chaotic system. These systems are only a small subclass of nonlinear systems.

1. ***The chaotic system is a deterministic dynamical system.*** The values of the variables that describe the system in the future are determined by the present values. An example of a chaotic time signal evolving from a third-order, single-variable equation is shown in Fig. 6.2 for a nonlinear damped spring with sinusoidal forcing called the Duffing equation (after the mathematician who first studied its chaotic features): $\ddot{x} + \gamma\dot{x} + \alpha x + \beta x^3 = B\cos t$.

2. ***The chaotic system is described by either difference or differential equations.*** In a difference equation, the values of the variables are computed at discrete steps. The values of the variables at the next time step are a function of their current values. For example, $x_{n+1} = g(x_n)$, where x_n is the value of the variable x at the nth time step. As shown by the Logistic Equation (Chapter 5, Eq. 5.5), a one-variable difference equation can demonstrate chaotic behavior.

   In a differential equation, the values of the variables change continuously in time. The values of the variables in the future depend on current values and the derivatives of their current values. For example, $x(t) = g[x(t), dx(t)/dt]$. (The notation $x = g[y,z]$ means that x is a function of y and z.)

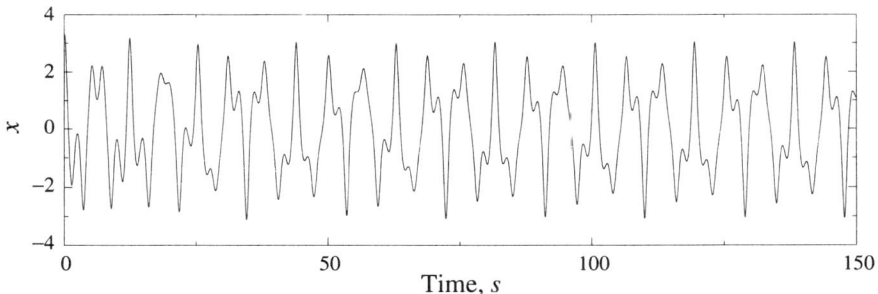

Figure 6.2. Time series of a steady-state chaotic response. The equation is $d^2x/dt^2 + \gamma dx/dt + \alpha x + \beta x^3 = B\cos t$, i.e., the Duffing equation with $\gamma = 0.05$, $\alpha = 1$, $\beta = 1$, $B = 7.5$. At 100 points per cycle of the driving cosine function this series might represent 12,500 points if taken as a discrete time series.

A chaotic system can consist of a single equation with one variable if it is discrete, or a set of coupled equations with more than three variables if it is continuous. Thus a difference equation of one variable can have chaotic behavior. Over a hundred years ago, Poincaré (1878, 1880) proved that a set of *differential* equations must have at least three independent variables (and thus three equations) to have chaotic behavior. The sinusoidal driving term in the Duffing equation mentioned above is viewed as a third variable. The other variables are $y = \dot{x}$ and $\dot{y}$.

3. **The chaotic system has sensitivity to initial conditions.** The values of the variables after a given time depend on their initial values. In a chaotic system, very small changes in these initial values produce very large changes in the later values. That is, if the initial values at time $t = 0$ were $x_1(t = 0)$ and $x_2(t = 0)$, then after a time t, $|x_1(t) - x_2(t)|$ will be proportional to $\exp(\lambda t)$, where λ is called the Liapunov exponent. Thus, the difference in the values diverges exponentially quickly in time. A system that displays this sensitivity to initial conditions is said to be "chaotic."

4. **The values of the variables are not predictable in the long run.** The values of the initial conditions can only be specified with finite accuracy. Because of the sensitivity to initial conditions, the values that we compute as time goes by will diverge ever further from their true values based on their exact initial values. For example, if the values of the initial conditions are specified with 5 digits, as time goes by, the accuracy of their values will fall to 4, then 3, then 2, digits. If we could specify all the infinite digits of the initial conditions then we could continue to fully accurately compute their subsequent values in time. However, because we can only specify the values of the initial conditions to finite precision, we continually lose accuracy, and cannot predict exactly their values in the long run. Thus, although the system is fully deterministic, it is not predictable in the long run, because of the sensitivity to initial conditions. For the system portrayed in Fig. 6.2, the divergence of the trajectories from nearby starting points takes a little while to appear, as shown in Fig. 6.3.

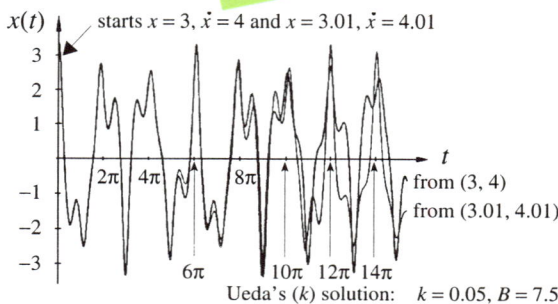

Figure 6.3. Divergence of $x(t)$ for the Duffing equation from nearby starting points. (From Thompson and Stewart, 1986, their Fig. 1.2.)

5. **The values of the variables do not take on all possible values.** In the long run, although the values of the variables seem to fluctuate widely, they do not take on all combinations of values. This restricted set of possible values is called the *attractor*.

6.4   Additional Features of Chaos

1. **Phase space.** The state of the system can be entirely described by the values of its variables. If there are n variables, then the state of the system can be described as a point in an n-dimensional space whose coordinates are the values of the dynamical variables. This space is called a *phase space*. As the values of the variables evolve in time, the coordinates of the point in phase space move. We can thus analyze the time behavior of the system by analyzing the motion of this point that represents the values of the variables.

   The phase space representation was developed by Poincaré. He showed that it was easier to analyze some of the properties of a dynamical system by determining the topological properties of trajectory, which is the line traced out by the point in the phase space, rather than analyzing the time series of the values of the variables directly. That is, we turn the time series of the values of the variables into an object in phase space and study the topological properties of this object. In the phase space plot in the right lower panel of Fig. 6.1, the "object" has a parabolic form, giving a clear and simple representation of the complex time signal shown above it. This was achieved by plotting the value of the $(n + 1)^{th}$ point against the value of the nth point.

2. **Attractors.** As noted above, the values of the variables do not take on all possible combinations of values. In the phase space, this set forms an object called an *attractor*.

   The point in phase space that represents the state of the system moves as the values of the variables evolve in time. If the initial state of the system is not on the attractor, then the phase space point moves exponentially rapidly toward the attractor as time goes by.

   However, because of the sensitivity to initial conditions, two points on the attractor diverge exponentially fast from each other as time goes by, even though they both remain on the attractor.

3. **Strange attractors.** The attractor is typically finite in the phase space. Sensitivity to initial conditions means that any two nearby points on the attractor in the phase space diverge from each other. They cannot, however, diverge forever, because the attractor is finite. Thus, trajectories from nearby initial points on the attractor diverge and are folded back onto the attractor, diverge and are folded back, etc. The structure of the attractor consists of many fine layers, like an exquisite pastry. The closer one looks, the more detail in the adjacent layers of the trajectories is revealed. Thus, the attractor is fractal. An attractor that is fractal is called "strange."

4. **Strange and chaotic.** As noted above, the technical jargon word "chaotic" means that a system has sensitivity to initial conditions and the jargon word "strange" means that the attractor is fractal. The typical chaotic system that we have been describing, such as that pictured in Fig. 6.1, is therefore chaotic and strange. To make life interesting there are also chaotic systems that are not strange (they are exponentially sensitive to initial conditions but do not have a fractal attractor), and nonchaotic systems that are strange (they are not exponentially sensitive to initial conditions but do have a fractal attractor).

5. **Dimension of the attractor.** The fractal dimension of the attractor is related to the number of independent variables needed to generate the time series of the values of the variables. If d is the smallest integer greater than the fractal dimension of the attractor, then the time series can be generated by a set of d differential equations with d independent variables. For example, if the fractal dimension of an attractor is 2.03, then the time series of the values of the variables can be in principle generated by three independent variables in three coupled nonlinear differential equations.

6. **Bifurcations.** Often, the equations of the dynamical system depend on one or more control parameters. A *bifurcation* occurs when the form of the trajectory in phase space changes because a parameter passes through a certain value. In analyzing data it is very often useful to plot how the form of the dynamics depends on the value of a parameter.

7. **Control.** In linear systems, small changes in the parameters or small perturbations added to the values of the variables produce small changes in subsequent values of the variables. However, in nonlinear systems, small changes in the parameters or small perturbations added to the values of the variables can produce enormous changes in subsequent values of the variables because of the sensitivity to initial conditions. Thus, in principle, a chaotic system can be controlled faster and finer and requires smaller amounts of energy for such control than a linear system. In the last few years it has been shown how to determine the input perturbations needed to drive a chaotic system into a desired state.

The phase space for a dynamic system consists of coordinate axes defined by the variables in the system. A simple harmonic oscillator is described by its displacement $x(t)$ and velocity dx/dt, so its phase space is two-dimensional with axes x and dx/dt or $\dot{x}$. Each point in the phase space corresponds to a particular duo (x,v) that uniquely define the state of the oscillator. This point leaves a trail as it moves in time, and this trail is referred to as the orbit or trajectory of the oscillator. In general a phase space orbit completely describes the evolution of a system through time. Each choice of an initial state produces a different trajectory. If, however, there is a limiting set in phase space to which all trajectories are drawn as time tends to infinity, we say that the system dynamics are described by an attractor. The attractor is the geometric limiting set on which all the trajectories eventually find themselves, i.e., the set of points in phase space to which all the trajectories are attracted. Attractors come in many shapes and sizes, but they all have the property of occupying a finite volume of phase space. As a system evolves it sweeps through the attractor, going through some regions rather rapidly and through others quite slowly, but always staying on the attractor. Whether or not the system is chaotic is determined by how two initially adjacent trajectories cover the attractor over time. As Poincaré stated, a small change in the initial separation of any two trajectories will produce an enormous change in their final separation. The question is how this separation is accomplished on an attractor of finite size. The answer to this question has to do with the layered structure necessary for an attractor to be chaotic.

Rössler (1976) described chaos as resulting from the geometric operations of stretching and folding often called the Baker's transformation. The baker takes some dough and rolls it out on a floured bread board. When thin enough he folds the dough back onto itself and rolls it out again. This process of rolling and folding is repeated again and again. Arnold gave a memorable image of this process using the image of the head of a cat (Arnold and Avez, 1968). In Fig. 6.4 a cross section of the square of the square of dough is shown with the head of a cat inscribed. After the first rolling operation the head is flattened and stretched, i.e., it becomes half its height and twice its length (arrow 1). It is then cut in the center and the segment of

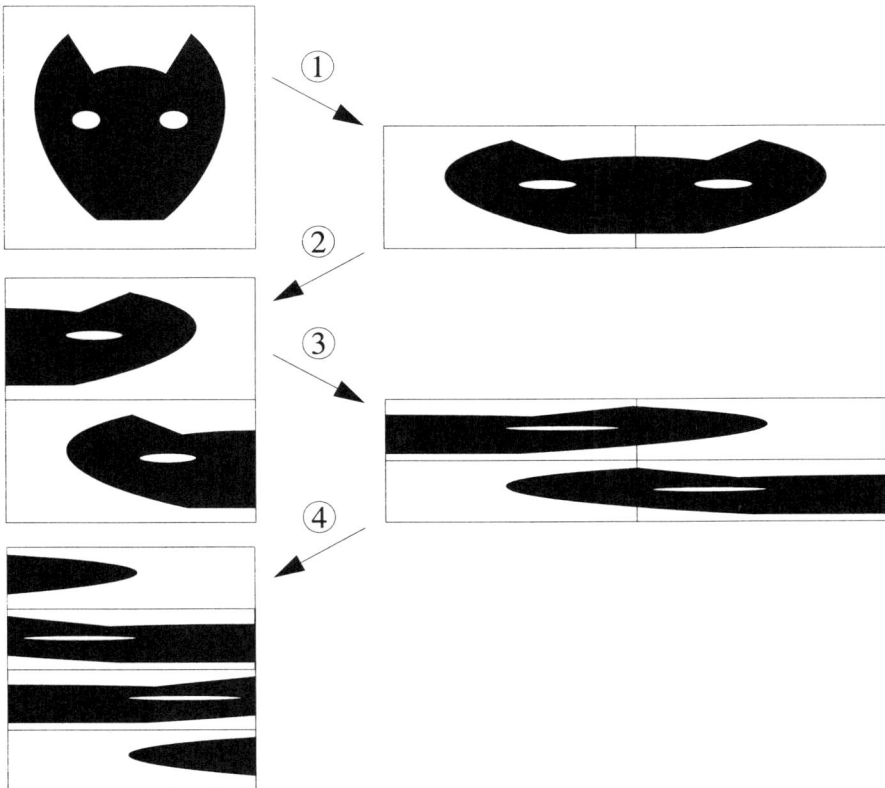

Figure 6.4. Arnold's cat undergoing the Baker's transformation. Meow.

dough to the right is set above the one on the left to reform the initial square (arrow 2). The operation is repeated again (arrows 3 and 4) and we see that at the bottom the cat's head is now embedded in four layers of dough. After these two transformations the cat's head is obscured. After twenty transformations the head will be distributed across 10^6 layers of dough—not easy to identify. As so charmingly put by Ekeland (1988), "Arnold's cat has melted into the square, gradually disappearing from sight like the Cheshire Cat in Wonderland."

Two initially nearby orbits cannot rapidly separate forever on a finite attractor; therefore, the attractor must eventually fold over onto itself. Once folded, the attractor is again stretched and folded again. This process is repeated over and over, yielding an attractor structure with an infinite number of layers to be traversed by various trajectories. The infinite richness of the attractor structure affords ample opportunity for trajectories to diverge and follow increasingly different paths. The finite size of the attractor ensures that these diverging trajectories will eventually pass close to one another again, albeit on different layers of the attractor. One can visualize these orbits on a chaotic attractor as being shuffled by this process, much as a deck of cards is shuffled by a dealer. Thus, the randomness of the chaotic orbits is a consequence of this shuffling process. This process of stretching and folding creates folds within folds ad infinitum, resulting in the attractor having a fractal structure in phase space. The essential fractal feature of interest here is that *the greater the magnification of a region of the attractor, the greater the degree of detail that is revealed*, i.e., one observes that there is no characteristic scale for the attractor and therefore none in the resulting time series given by a trajectory traversing the attractor.

6.5   A Change in Perspective

Mathematical models of physiological systems have traditionally relied on the paradigm of classical physics. A paradigm is a shared set of assumptions at any given time in a particular scientific community that molds and orients its area of investigation. The potency of the physics paradigm lies in its ability to relate cause and effect and thereby enable scientists to make predictions. For discussion of this, see Crutchfield et al. (1986). Pierre Simon de Laplace (1776) clearly stated the paradigm:

> The present state of the system of nature is evidently a consequence of what it was in the preceding moment, and if we conceive of an intelligence which at a given instant comprehends all the relations of the entities of this universe, it could state the respective positions, motions, and general effects of all these entities at any times in the past or future.

Uncertainty for him was a consequence of imprecise knowledge, so probability theory is necessitated by incomplete and imperfect observations.

In 1887 the King of Sweden, Oscar II, offered a prize of 2,500 crowns for the answer to the question "Is the universe stable?" The King wanted to know if the planets in their present orbits would remain in those orbits for eternity, or whether the moon would crash into the earth, or the earth would hurl itself into the sun. One of the respondents and the eventual winner of this contest was the French mathematician Henri Poincaré. The 270-page monograph that he submitted to the King in 1890 established a new branch of mathematics, *topology*. The topological ideas were used to determine the general properties of dynamical equations without actually solving them. Poincaré was able to show that if the universe consisted of only the sun and a single planet, then the planetary orbit would periodically return to the same point in space. He was also able to show that if a third body was added to the above universe, and if that body was much lighter than the other two, then its orbit would have a very complex structure that we would now call chaotic. Thus, the orbits of the planets of the solar system are chaotic and the fate of the solar system is not predictable.

Poincaré's conclusion was a sharp break with the views of the past:

> A very small cause which escapes our notice determines a considerable effect that we cannot fail to see, and then we say that the effect is due to chance. If we knew exactly the laws of nature and the situation of the universe at the initial moment, we could predict exactly the situation of that same universe at a succeeding moment. But even if it were the case that the natural laws had no longer any secret for us, we could still only know the initial situation *approximately*. If that enables us to predict the succeeding situation with the same approximation, that is all we require, and we should say that the phenomenon had been predicted, that it is governed by laws. But it is not always so: it may happen that small differences in the initial conditions produce very great ones in the final phenomena. A small error in the former will produce an enormous error in the latter. Prediction becomes impossible, and we have the fortuitous phenomenon.

There is no contrast greater than this in today's science. Laplace championed the intrinsic knowability of the universe, whereas Poincaré focused on the implicit limits of our ability to make predictions.

6.6   Summary

We have seen that not everything that looks random is random. Some nonlinear deterministic systems produce signals that mimic random behavior. This is now called *chaos*. Chaotic systems have five essential characteristics: 1) They are deterministic dynamic systems; the values of the variables in the future are a function of their past values. 2) They can be described mathematically by difference equations. 3) They have sensitivity to initial conditions and thus 4) the ariables cannot be predicted in the long run. 5) The values of the take on all possible values but are restricted to a set called a strange

some of the physiological signals that we now think are random turn out instead to be examples of chaos, we may be able to understand and perhaps even control much more of the workings of the human body than we ever thought possible.

Background references

We have focused on the properties of chaotic systems, which are only a small part of nonlinear dynamics. There are many excellent books that describe the analysis tools and the properties of nonlinear dynamic systems at different levels of mathematical detail. At the introductory nonmathematical level is the book by Gleick (1987). At the intermediate calculus level are the books by Thompson and Stewart (1986), Moon (1992), and Parker and Chua (1989). At a more advanced mathematical level are the books by Guckenheimer (1983) and Hao (1984, 1988).

7

From Time to Topology: Is a Process Driven by Chance or Necessity?

> Human beings . . . [who have seen only] the shadows . . . of all sorts of artifacts . . . cast by the fire on the side of the cave.
>
> Plato, *The Republic*, Book VII (translated by Bloom, 1968)

7.1   Introduction

The data from our senses and the measurements recorded by our instruments are only a shadow of reality. We analyze the data and the measurements to reconstruct the Platonic "artifacts" that produced those shadows. For example, we use the spots on an X-ray diffraction pattern to infer structural information about the three-dimensional shape of a protein. The physical spots on the photograph, that we can touch with our fingers, are further removed from the reality of the protein than the "cartoon" of protein structure that we form in our mind's eye and display on our computer screen.

Similarly, a time series of measurements is only a shadow of the process that produced it. We recover information about the process by analyzing the time series. The analysis techniques used over the last four hundred years have concentrated on linear properties such as the mean and variance. Linear properties are adequate for describing systems with few, weakly interacting parts. Since no other tools were available until recently, all systems were treated as if they were linear. Yet, many systems, including most biological ones, have numerous strongly interacting parts and thus are highly nonlinear. To reveal the nonlinear properties of the data, we need to use a different set of analysis tools. Some of these were first explored by Poincaré (1879) about a hundred years ago and have been extensively developed over the last twenty years.

In this chapter we use these new analysis tools to transform the time series into a geometric object called a set in phase space and then analyze the topological properties of this set. Hence, we shall replace studies of time behavior with those of spatial behavior. In the previous chapters we used fractals to analyze the values of the time series itself. Now we use fractals to analyze a representation of the time series in a phase space. The fractal dimension of the phase space set is different than the fractal dimension of the values of the time series itself, and it conveys different information to us about the nature of the process that produced the data.

147

In the preceding chapter we showed that deterministic systems, which can be described by a few independent variables, can produce behavior that is so complex that it mimics that of random systems, with many independent variables. The analysis tools described in this chapter allow us to determine if random-looking data have been produced by truly random processes or by deterministic chaos.

The reasons for wanting to make the distinction between a chaotic signal and a merely noisy signal are both theoretical and practical: for description, for insight and prediction, or for providing a basis for intervention. If the signal has a deterministic component, then it can be described mathematically and this can provide some predictive capability that could not be reached for a noisy system. As we shall see below, even an incomplete and insecure level of description and understanding can be sufficient to allow intervention and control. In the case of fatal cardiac arrhythmias, improving the capability for therapeutic intervention is a driving incentive for investigation, whether or not the signal is truly chaotic (Jalife, 1990). Also raised with respect to the cardiac rhythm is the question of the complexity of the variation in the signal: Goldberger makes the case that a healthy young person shows greater complexity than an older person, and that great loss of complexity (reduction to an almost constant rhythm) may be premonitory to a fatal arrhythmia (Lipsitz et al., 1990; Kaplan et al., 1990, 1991).

7.2   Distinguishing Chaos from Randomness

How to make the distinction between determinism and randomness is a deep and interesting question. The methods developed to analyze physical systems with low noise and large data sets are sometimes difficult to apply to measurements from biological systems, which often have significant noise and small data sets. This is an area of active research. There are conditions where a random signal can masquerade as a chaotic signal (Osborne and Provenzale, 1989). Low-pass filtering of a random signal can produce short-range correlation in the signal that can look like a deterministic relationship. Considering that most natural signals are augmented with a combination of externally imposed noise and measurement error, it is not easy to distinguish correlated noise from a signal from a dynamical system even when the order of the system is low. A list of potential methods for distinguishing a chaotic from a noisy signal can be divided into methods that suggest deeper analysis is useful and more definitive methods that produce measures of the chaotic nature of the signal (see Table 7.1).

A general principle to keep in mind when trying to make the distinction between chaos and noise with these methods is that a very high order system such as that described by a dozen nonlinear differential equations, even though smooth and coordinated in a twelve-dimensional domain, will not be distinguishable from noise unless the system is so dominated by a few variables that it behaves as a low-dimensional attractor. This is not to recommend giving up, since most physiological and biochemical systems in block diagrams appear this complex, but rather to say

Table 7.1. Methods for determining if signals are chaotic

A. Methods suggestive of underlying chaos:

  1. Visual inspection for irregularity and bifurcations in periodicities

  2. The power spectral density

  3. Autocorrelation function

B. Special methods for characterizing low dimensional chaotic signals:

  1. Visual analysis of phase plane plots

  2. Capacity, correlation, and information dimension

  3. Estimation of Lyapunov exponents

  4. Calculations of entropy

C. Special situations:

  1. Observations of changing states exhibiting bifurcations or changes of periodicities

  2. Interventions resulting in period doublings or changes in apparent chaotic state

that low-dimensional systems are much easier to identify and characterize. But all is not lost for high-dimensional systems; for example, the chain of reactions in glycolysis is composed of sets of high velocity reversible reactions lying between relatively low velocity control reactions, and so the overall system behaves as a low order system. Such systems, even though inherently complex and seemingly naturally chaotic, often are locally more or less linear, and consequently stable. (See Glass and Malta, 1990.)

7.3   Methods Suggestive of Underlying Chaos

1. *"Visual" methods* depend on the appearance of the signal to give an impression of irregularity in rhythm or amplitude. If the variability is subtle, measurement is better. Examples are the signals in Fig. 6.1, right panel, and Fig. 6.2. The variability in peak heights and wave forms, and the absence of periodic or repeating segments of the same form, *suggest* chaos.

2. *Power spectral analysis*, plotting amplitudes at each frequency determined from the Fast Fourier Transform, gives useful hints. A chaotic signal commonly shows broadly spread peaks in the power spectrum, Fig. 7.1.

3. The *autocorrelation function* was defined in Section 4.6. Measuring the fractal correlation is valuable for two reasons. The first is that positive correlation over even a short range distinguishes the chaotic signal from random noise even though it will not necessarily distinguish chaos from

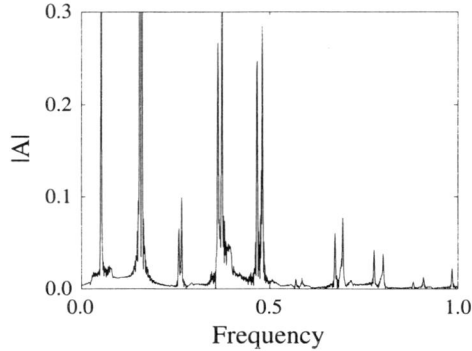

Figure 7.1. Fourier power spectrum of the time series from a nonlinear spring where the restoring force is not proportional to the displacement. The parameters are those used in Fig. 6.2.

filtered or smoothed noise. The autocorrelation function is shown in Fig. 7.2 for a random noise (left panel) and for the driven spring function (middle panel) whose time course was shown in Fig. 6.2. The noise function shows no correlation even over short times. The spring function shows only short range correlation, and then some periodicity in the correlation function. The second reason for measuring the fractal correlation is that it provides an estimate of the time lag to be used in the embedding to create a pseudo-phase space reconstruction of the signal to see if it has the features of a chaotic attractor (right panel).

7.4   Phase Space and Pseudo-Phase Space

The Phase Space Set

The phase space set reveals important information about the physiological process that produced the time series data. First, we use the time series to construct a phase space set. If the fractal dimension of the phase space set is high, then the time series may have been produced by a large number of independent random variables. If the fractal dimension of the phase space set is low, then the time series may be generated by a deterministic process having a small number of independent variables and the phase space set displays a form of the relationship between those variables. We are then faced with the interesting and challenging task of discovering the physiological mechanism that produced the observed relationship.

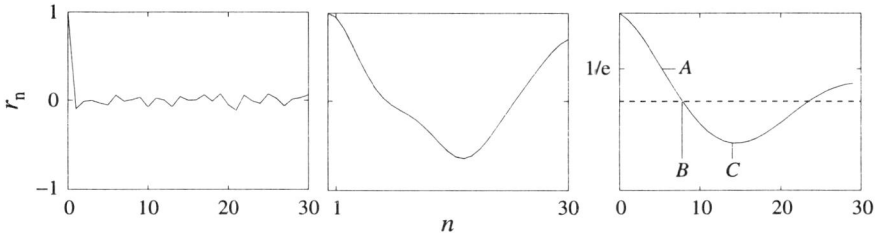

Figure 7.2. Autocorrelation functions. *Left panel:* For Gaussian random white noise. *Middle panel:* For the equation for a nonlinear driven spring position versus time function shown in Fig. 6.2. Sampled at 0.2 s. *Right panel:* Choices for determining the lag for pseudo-phase space reconstruction of an attractor. A is positioned where r_n falls to $1/e$, B at the first zero crossing and C at the first nadir.

How to Analyze Experimental Data

Embedding: Turning the time series into a set in pseudo-phase space.
If a system is entirely determined by the values of a set of variables, then knowing those values completely specifies the state of the system. We can represent one set of those values as a point in a space with coordinates corresponding to those variables. Such a space is called a *phase space*. The set of states of the system can thus be represented by a set of points in the phase space. Analysis of the topological properties of these phase space sets gives us insight into the nature of the system.

Thus, we would like to analyze the phase space set of experimental data. Sometimes, we can experimentally measure all of the variables that specify the system and thus directly construct the phase space set. On the other hand, it is often not practical or possible to measure all the relevant variables. However, it may be possible to construct a pseudo-phase space from experimental data where only one of the variables is measured. This is the method that must be used when measurements are made of only one of the variables, or when the number of variables is not known.

The attractor is the phase space set generated by a dynamical system consisting of a set of difference or differential equations. A phase space consists of *independent coordinates*. For example, consider the dynamical system of three equations having the independent variables X, Y, and Z. Each of these variables is a function of the other variables and of time. The phase space set could be the values of the variables at each time plotted on three mutually perpendicular axes. Thus, a point (x, y, z) in the phase space uniquely specifies the values of the three variables and hence the state of the system at each instant. Note that we have used uppercase letters such as X, Y, Z to represent the values of variables measured as a function of time and lowercase letters such as x, y, z to represent the values of coordinates in phase space.

Another choice of independent coordinates for this same system is to plot one of the variables and its derivatives such as X, dX/dt, and d^2X/dt^2 on three mutually perpendicular axes (x, y, z). The derivatives are also three independent variables. In this case, we have reconstructed the entire phase space set from the time series of only one of the three variables, namely from the derivatives of X. This can be done because the Y and Z variables are intimately coupled with X through the nonlinear equations.

Given the time series for the variable $X(t)$ obtained from measurements of $X(t)$ recorded at time intervals Δt,

$$X(0), X(\Delta t), X(2\Delta t), X(3\Delta t), X(4\Delta t) \ldots , \qquad (7.1)$$

we could in principle numerically differentiate these values to determine dX/dt, d^2X/dt^2, etc., but since these derivatives depend sensitively on the measurements, even small experimental errors may introduce large errors in evaluating them. This means that we cannot reconstruct a real phase space set using the derivatives. To substitute for this there is a widely used method of constructing a *pseudo-phase space set* by plotting points whose coordinates are the measurements of the time series separated by time delays Lt, as shown in Fig. 6.1. That is, for a E_m-dimensional phase space, the coordinates of each point in the phase space have the values:

$$\{x(t), x(t + L\Delta t), x(t + 2L\Delta t), x(t + 3L\Delta t) \ldots x(t + (E_m - 1)L\Delta t)\} . \quad (7.2)$$

When the measurements of the time series are recorded at equal time intervals Δt, then each point will have the coordinates

$$\{x(m\Delta t), x(m\Delta t + L\Delta t), x(m\Delta t + 2L\Delta t) \ldots x(m\Delta t + (E_m - 1)L\Delta t)\} , \quad (7.3)$$

where $L\Delta t$ is the time lag. The appropriate time lag $L\Delta t$ for analyzing experimental data is determined to some extent by mathematical principles, and to an equal extent by empirical criteria based on previous experience using the autocorrelation function (as suggested by Fig. 7.2) or the lag taken from the highest frequency peak in the power spectrum (as in Fig. 7.1). The phase space points can be constructed using every measurement $X(t)$, or some can be excluded, using only measurements separated by a time shift $S\Delta t$. That is, instead of choosing

$$m = 1, 2, 3 \ldots , \qquad (7.4)$$

we choose

$$m = 1, 1 + S, 1 + 2S, 1 + 3S, \ldots , \qquad (7.5)$$

where $S\Delta t$ is called the time shift. Often, we will use as many of the measurements as possible and thus S will be set equal to 1.

The coordinates of the phase space constructed from these time delays are linear combinations of the derivatives. For example,

$$dX/dt \approx ([X(t + \Delta t) - X(t)] / (\Delta t)) . \tag{7.6}$$

The plots in Fig. 7.3 show that the pseudo-phase space set constructed by using the time-lagged measurements is a linear transformation of that constructed from the derivatives of the original dynamical equations. The topological properties of the phase space set that reveal the nonlinear properties of the data are invariant under such a linear transformation. One of these topological invariants is the fractal dimension of the phase space set.

It is remarkable that the entire phase space set determined by all the independent variables can be generated from the measurements of only one of these variables (Takens, 1981; Packard et al., 1980; Eckmann and Ruelle, 1985; Gershenfeld, 1988). The construction of the phase space set is technically called an *embedding*. These methods are presently an active area of mathematical research and many important issues are not yet fully understood. For example, it is not known exactly which procedures are required to produce valid embeddings for different types of time series. Takens (1981) proved that the embedding method using time lags, described above, is valid if certain assumptions are met. Thus, only if these assumptions are true are we guaranteed that this method is valid. For example, it is assumed that the time series is "smooth," that is, twice differentiable. When this is not true, for example, when the time series is fractal, then the fractal dimension determined from the embedding may not equal the true fractal dimension of the phase space set (Osborne and Provenzale 1989). Moreover, we also do not know if the assumptions of Takens's theorem are unnecessarily restrictive. For example, Takens's theorem requires that to reliably reconstruct a phase space set of dimension D, it must be embedded in a space of dimension $2D + 1$ to provide enough dimensions so that the E_m-dimensional orbits do not falsely intersect themselves. In

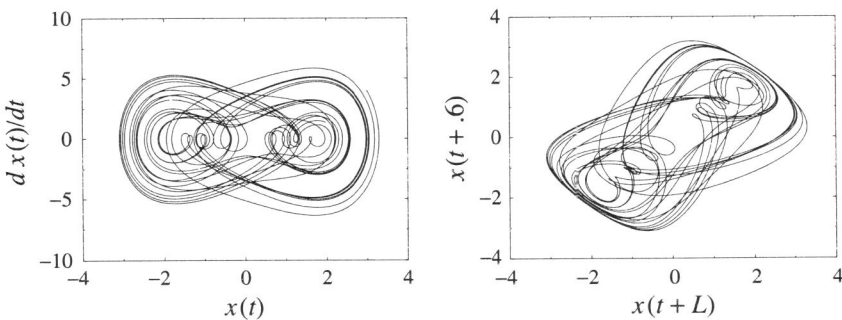

Figure 7.3. Phase space (*left panel*) and pseudo-phase space (*right panel*) reconstructions for the nonlinear spring (Duffing-type) equation with sinusoidal forcing. The equation, given in the legend for Fig. 6.2, here represents a damped nonlinear oscillator with $\gamma = 0.05$, $\alpha = 1$, $\beta = 1$, and B = 7.5. *Left panel:* $x(t)$ versus its derivative. *Right panel:* $x(t)$ versus $x(t + L)$, where L is the lag time. This delay plot or pseudo-phase space plot has the same general characteristics as the phase space of the left panel.

practice, however, phase space sets are often reliably reconstructed if the embedding is done in a space of dimension greater than or equal to D.

The most difficult part of the embedding method described above is to determine the appropriate value of the time lag Lt that should be used in Eqs. 7.2 and 7.3. There are two cases: 1) where the choice of the appropriate time lag may be (relatively) easy to guess, and 2) where it is not easy to guess. Once we have used the embedding to construct the phase space set we must then evaluate the properties of that set. When the phase space set is a simple figure, then its shape reveals a form of the equation responsible for the observed time series and no further analysis is necessary. When the phase space plot is difficult to interpret, then additional numerical analysis is required.

Guessing the Time Lag

Sometimes the appropriate time lag is (relatively) easy to guess. The appropriate time lag to use to construct the phase space set may be suggested by the system itself. This happens when either the physiological process or the experimental protocol has some apparent natural time scale. For example, the density of biological populations with nonoverlapping generations, such as temperate zone arthropods, varies in a seasonal yearly cycle and thus a natural choice for the time lag is one year (May and Oster, 1976). Glass et al. (1984, 1986) measured the transmembrane electrical potential of aggregates of embryonic chick heart cells which were periodically stimulated. They measured the phase, namely, the time between each stimulus and the following beat of the heart cells. A natural choice of the lag is simply the interbeat interval, e.g., the previous phase. Thus, the coordinates of each point in the E_m-dimensional phase space are the E_m consecutive values of the phase.

The logistic growth equation, $X_{n+1} = X_n + r \cdot X_n (1 - X_n)$, described in Section 5.5, is a discrete equation whose natural period is 1. In such a case using $L = 1$ in Eq. 7.2 is the natural analog to the first derivative, and the phase plane plot reveals the parabolic structure of the attractor, as in Fig. 7.4. Using a two-interval lag produces a more complex curve, no longer a simple parabola, and so in this case it is clear that taking too long a lag ($L = 2$) is disadvantageous, but it does have utility in analyzing the system for stability.

When the Appropriate Time Lag Is Not Easy to Guess

Sometimes there will not be an obvious choice for the appropriate time lag. Physiological measurements where this is the case include time series of the electroencephalogram (Rapp et al., 1985; Babloyantz, 1989; Mayer-Kress and Layne, 1987; Skarda and Freeman, 1987; Xu and Xu, 1988), the electrocardiogram (Kaplan, 1989), and the number of children with measles in New York City (Schaffer and Kot, 1986). In such cases we try to choose a time lag that corresponds to the best natural time scale of the system that we can find. For example, it is

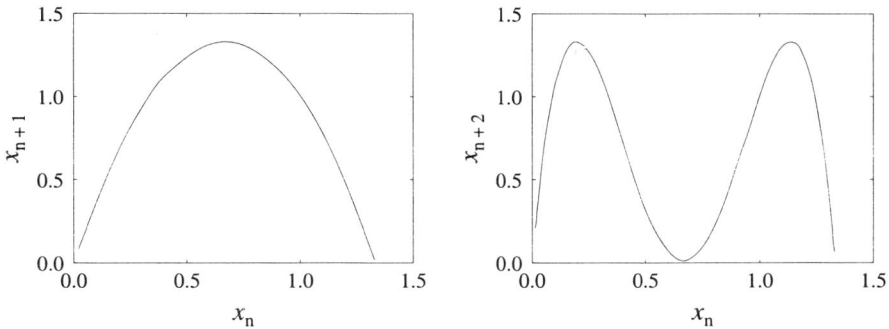

Figure 7.4. Delay plots for the logistic growth equation, Eq. 5.5. *Left panel:* Using a single lag reveals that the first iteration function is the parabolic attractor. *Right panel:* Using a lag of two steps ($L = 2$) gives the second iteration function, which, if it were the only lag used, would give the impression of a more complex structure in the attractor. For both panels $r = 2.99$.

common to choose the time span over which different measurements become uncorrelated. One method is to use the mutual information, based on information entropy, to determine the time span over which the value of one measurement is no longer predicted from the value of a previous measurement (Gershenfeld, 1988).

Another commonly used method to choose a natural time is based on the autocorrelation function, r_n, described in Section 7.3 and illustrated in Fig. 7.2. The autocorrelation has its maximum value of 1.0, at $t = 0$, because a measurement correlates with itself, and it diminishes as the distance, $n\Delta t$, between the points increases, since the correlation between subsequent measurements fades with time. Hence, the time it takes for the autocorrelation function to change "significantly" defines an approximate natural time scale. There is no consensus as to the exact value of this "significant" change. For example, different choices for the lag are shown in the right panel of Fig. 7.2. Recommended choices for the lag range from the number of intervals required for the autocorrelation to fall to $1/e$ (arrow A), to the first zero crossing (arrow B), or to the first nadir in r_n (arrow C) (Albano et al., 1988; Abarbanel et al., 1989; Theiler, 1990). Moreover, some people choose the time lag as a fraction of this time (for example, one twentieth), equal to this time, or a multiple of this time (Abarbanel et al., 1989; Schuster, 1988; Theiler, 1990). To construct the phase space set some use a time lag equal to one twentieth of the time it takes for the autocorrelation function to reach its first zero. This is often wasteful, since it oversamples the data. On some test problems, our most accurate results were achieved when we chose a time lag equal to the time interval during which the autocorrelation function decreases to $1/e = 0.37$. We do not guarantee that this is the best choice for all cases.

The relationship between the dimensionality of an embedding space and the real phase space that one is trying to identify for a chaotic signal depends on the length of lag chosen. Using a very short lag means that many more successive embeddings will have to be used than if the lag is chosen to give the closest representation to the derivative of the true but unknown function. Choosing too large a lag will merely create noise in the embedding, and one will miss observing the form of the attractor. The three choices shown in Fig. 7.2 represent good guesses as to the most efficient embedding. The choice of $1/e$ errs on the conservative side. The first zero crossing represents a choice which requires fewer embeddings without much risk of losing sight of the attractor. The time to the nadir is often too long.

The other method is to use the power spectrum, taking the lag in time to be the reciprocal of the frequency centered within the band of the highest frequency peak of importance. Sometimes this does not work well even though to the eye there are some apparently obvious higher-frequency waves. Instead, as is shown by the double scroll equation shown in Fig. 7.5, there is a broad-band Fourier spectrum without an "obviously important high frequency peak." For such functions the visual impression of the best lag comes from making repeated trials, then picking out the lag which seems to be the most revealing from the pseudo-phase plane plots.

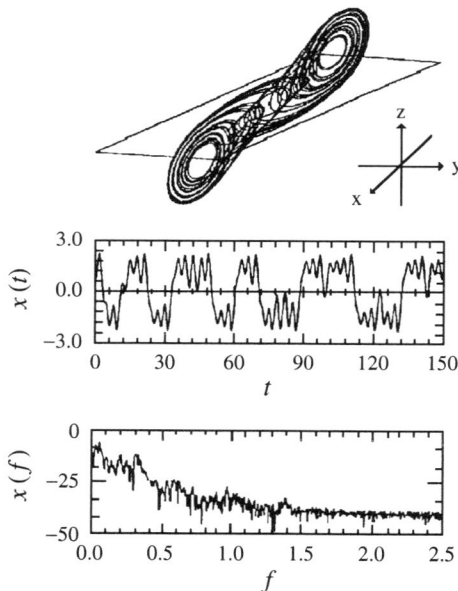

Figure 7.5.   The double scroll equation in the chaotic domain. The equation for the double scroll is three-dimensional: $dx/dt = \alpha \ (y - h(x))$; $dy/dt = x - y + z$; $dz/dt = -\beta y$. The source of the nonlinearity is $h(x)$, which is piecewise linear: $h(x) = m_1 x + (m_0 - m_1)$ for $x > 1$; $h(x) = m_0 x$ for $|x| <$ or $= 1$; and $h(x) = m_1 x - (m_0 - m_1)$ for $x < -1$. The parameters are $\alpha = 9$, $\beta = 100/7$, $m_0 = -1/7$, and $m_1 = 2/7$. Upper panel: Trajectory in three-space; Middle panel: The time course for $x(t)$; Lower panel: The Fourier power spectrum for the signal in the middle panel. (From Parker and Chua, 1989, their Fig. 1.9, with permission.)

This approach is reasonable when different lags give very different structures to the reconstructed attractor, as is the case for the double scroll equation; as seen for four different lags in Fig. 7.6, lags of 0.05 and 0.25 are too short, that of 6.25 are too long, and that of 1.25 are just right.

This is in effect using a visual impression that the attractor is "revealed" best by a particular lag. A very tedious but mathematically sensible way of achieving the desired result when one has very large amounts of data (as one can get for mathematical functions, but not often in real life) is to find by numerical experimentation the lag that most efficiently reveals the N-dimensional shape of the attractor. The method is the same as that used for estimating the capacity or box dimension of the attractor, using lags recognized to be too short only after the answer is found. See Parker and Chua (1989) for a good discussion of the problem.

The importance of the time lag in constructing the phase space set can be understood from Eq. 7.6. The time lag Δt is like a step size used in the numerical evaluation of the derivatives which are the independent coordinates of the phase space. If the time lag is too small, then there is little change so that $x(t + \Delta t) - x(t)$ and thus the derivatives cannot be accurately determined. If the time lag is too large, then $x(t + \Delta t)$ is completely uncorrelated and independent of $x(t)$, and again the derivatives cannot be accurately determined. Thus, the appropriate time lag should approximately correspond to the time scale at which the system changes, for example, the time scale determined by the autocorrelation function. The properties of the phase space set, such as its fractal dimension, remain approximately constant

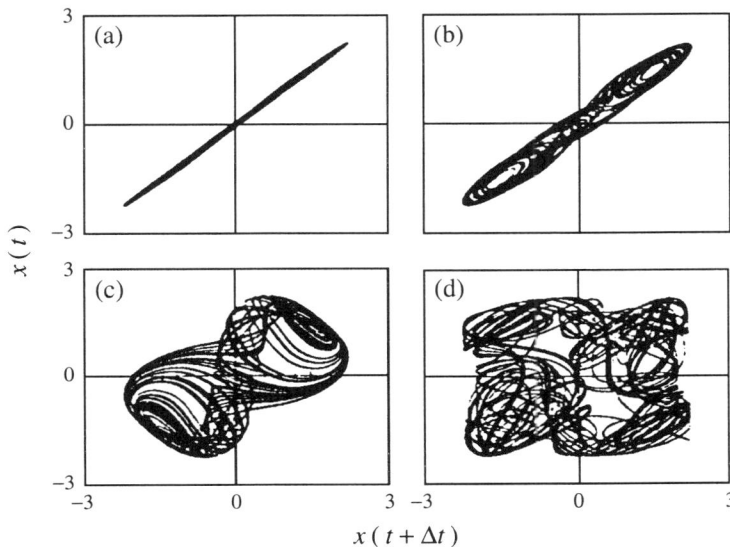

Figure 7.6. Reconstructions of the double scroll equation given in the legend for Fig. 7.5. The lags used were: (a) 0.05, (b) 0.25, (c) 1.25, and (d) 6.25. (From Parker and Chua, 1989, their Figure 7.10, with permission.)

when the time lag varies near this appropriate value. "Approximately" means that we can indeed determine the fractal dimension of the phase space set even though the value of the most appropriate time lag is only approximately known. "Approximately" also means that for some data sets this embedding procedure may be difficult in practice. For example, if there are multiple time scales of the autocorrelation function, which one should we use? If the time lags based on these different time scales yield phase space sets with different dimensions, then which is the real one?

7.5   Additional Types of Deterministic Relationships

The self-similarity of fractals means that there is a relationship connecting structures or processes at different spatial or temporal scales. See Chapter 4 also. Thus, an analysis of the fractal properties of a time series can also reveal deterministic relationships in the data. These relationships may *not* be expressible by the simple difference or differential equations that characterize chaos.

Methods for Determining If Signals Are Fractal

1. Power spectral analysis.
2. Autocorrelation function.
3. Dispersion analysis.
4. Rescaled range (Hurst) analysis.
5. Standard statistical measures.

1. *Power spectral analysis.* It is uncommon to find a clear fractal signal where the logarithm of the amplitude falls off with the logarithm of the frequency as a straight line, the $1/f^\beta$ relationship. However, finding a power spectrum that diminishes smoothly with frequency, f, as $1/f^\beta$ where the exponent β is between zero and 2 is an indicator that the time series is fractal. However, this fractal behavior can be the result of colored noise, which is a signal with Gaussian statistics but an inverse power-law spectrum. A random white noise process has $\beta = 0$; Brownian motion, the integral of white noise, has $\beta = 2.0$. Other values of β could be a consequence of fractional Brownian motion, or from a process having a phase space attractor that is fractal. The power spectrum cannot distinguish between chaos and colored noise.

2. *The autocorrelation function*, defined in Chapter 4, Section 4.4, is zero for white noise, and extends only over a short range with filtered noise. For a fractally correlated function r_n shows a slow fall-off and an extended range of positive correlation. (This statement applies to functions of time and also of space where the spatial range is very large; it has to be modified to account for negative correlation when there is a fractal relationship within an organ of finite size.)

Because chaotic signals are themselves not usually fractal, one does not expect to find that the autocorrelation function, r_n, is continuously positive over a long range as shown by the analytic function illustrated in Fig. 4.23.

3. *Dispersional analysis* is what we first used for looking at regional flow distributions (Bassingthwaighte, 1988; Bassingthwaighte et al., 1989). The basis of the method is to determine the spread of the probability density function (variance, V, or standard deviation, SD, or coefficient of variation, CV, relative dispersion) at the highest level of resolution, then to lump nearest-neighbor elements of the signal together to obtain the local mean of the pair of elements and recalculate the dispersion at the reduced level of resolution, and to repeat this with successively larger groupings of near neighbors. The slope of the log-log relationship between dispersion and level of resolution gives the fractal D or H: a slope of 0.5 and an $H = 0.5$ is obtained for an uncorrelated random signal. A smaller slope and a higher H indicates positive correlation that allows us to call it "fractal" and to determine the degree of correlation between neighbors over a wide range of separations.

4. *The rescaled range or Hurst method* likewise shows the correlation structure by an analysis of the range of differences of the integral of the signal from the expected mean, making this estimate over subsets of the data set and using subsets of widely differing size (Hurst, 1951; Mandelbrot and Wallis, 1968; Feder, 1988), but it requires more points than does dispersional analysis (Bassingthwaighte et al., 1994).

5. *Standard statistical methods.* Fractals are incomplete statistical measures. The fractal measures we have described provide a limited description of the statistical properties of a system. Voss (1988) makes the point that when a set of observations of lengths of coastlines or of the length of a Koch snowflake boundary are fractal, all of the moments are fractal, with the same fractal dimension. The dispersional analysis and the rescaled range analysis have the virtues of giving scale independent measures of the change in variance of a signal. But neither provides a measure of the mean, variance, skewness, kurtosis, or other of the shaping parameters of the distribution of values for the set of observations. Some of this information is in the intercept, RD_0, the relative dispersion at a particular reference size. With the rescaled range, R/S, both the numerator and the denominator relate to different aspects of the signal, the range of cumulative deviations from the mean and the dispersion of the values, but do not tell one the mean or the variance.

Thus, fractal measures are not a replacement for standard statistical measures, but are a supplement to them. They provide additional information of two sorts: (1) that there are scale independent features of the data over some range of unit sizes or resolution, and (2) that the fractal dimension defines the correlation within the data set. *This is new information that the standard statistical measures did not provide or stimulate one to look for.* Consequently, we gently remind the reader not to abandon the standard measures but rather to calculate in addition the new fractal measures that create new insight or at least the provocation to figure out why there is correlation over space or time.

The probability density function (*pdf*) may or may not be Gaussian, and one can be more suspicious about a highly skewed or markedly platykurtic (flat peaked) or leptokurtic (high peaked) function than one that is normally distributed. A variant on this is not only to look at the *pdf* of the whole data set, but to look at the *pdf* for subsets such as tenths or quarters of the whole set. This allows one to look for shape changes as well as changes in the means, and to gain an assessment as to the stationarity of the signal. (A system is defined to be "stationary" if it behaves the same at one time as at another time, for example, giving the same response to a pulse input.)

Probability density functions are shown in Fig. 7.7 for a Gaussian random signal (left panel) and for the nonlinear spring (middle panel); these *pdf*s correspond to the autocorrelation functions shown in Fig. 7.2. The Gaussian function is the familiar bell-shaped curve. The *pdf* for the spring function has peculiar peaks and is sharply bounded. The shape of the *pdf* is different for each set of parameters in the chaotic range. For contrast the shape of the pdf for a sine function is shown in the right panel. (It is stimulating to note that a very small fraction of the observations of sinusoidally varying function are near the mean.)

7.6   Capacity, Correlation, and Information Dimensions

Analyzing Complex Phase Space by Determining Its Fractal Dimension

When the phase space set is complex further analysis is required to reveal the information that it contains. Perhaps the most useful analysis technique is to determine the fractal dimension of the phase space set. *This has been used by investigators to determine if the process that generated the time series is a deterministic one based on a small number of independent variables or one based on random chance having a large number of independent variables.* The phase

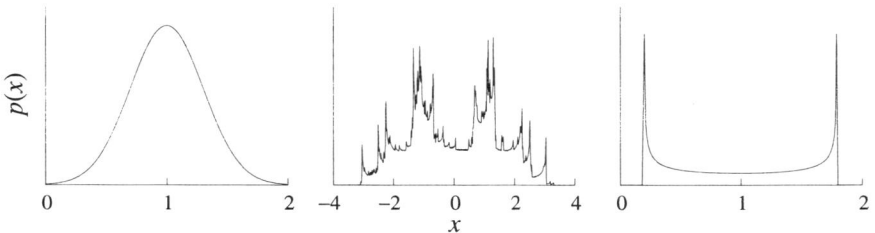

Figure 7.7. Probability density functions. *Left panel:* Gaussian function. *Middle panel:* Nonlinear spring with same parameters as those used for the phase plane plots in Fig. 7.3. *Right panel:* Sine wave.

space set of a deterministic process has low fractal dimension and that of an uncorrelated random process has high (infinite) fractal dimension. *If the phase space set has dimension* E_m, *then the number of independent variables needed to generate the time series is the smallest integer that is greater than or equal to* E_m. When the phase space set consists of one or more lines we can immediately conclude that it has a fractal dimension of approximately one. When the form of the phase space set is more complex or it is embedded in dimensions greater than 3, a numerical analysis is required to determine its fractal dimension. We will describe in detail a method that we have found useful, and list some of the many other techniques, to determine the dimension of such phase space sets. (A more rigorous treatment of the definitions of the dimensions described below can be found in the Aside in Section 3.3.) Additional information on these methods can be found in the excellent reviews by Eckmann and Ruelle (1985), Gershenfeld (1988), Holzfuss and Mayer-Kress (1986), Moon (1987), Schuster (1988), Theiler (1990), Casdagli et al. (1992), Gibson et al. (1992), and Grassberger et al. (1991). Note that colored noise can also have a low fractal dimension and is often confused with chaos.

Determining the fractal dimension of a phase space set. Capacity dimension and box counting. As shown in Chapter 2, the fractal dimension describes how many new pieces come into view when we examine a set at finer resolution. If the phase space has dimension E_m, we can evaluate the capacity dimension D_{cap} of the phase space set using E_m-dimensional spheres. For spheres of a given radius r, we count the minimum number of spheres $N(r)$ that are needed to contain every point in the phase space set. We repeat this count for spheres of different radii. The capacity dimension D_{cap} of the phase space set is then given by

$$D_{cap} = d\log N(r) / d\log(1/r) , \qquad (7.7)$$

in the limit as r approaches zero. Eq. 7.7 is usually valid over a scaling regime that extends over a range of values of r, and thus the capacity dimension D_{cap} can be found from the slope of a plot of $\log N(r)$ versus $\log(1/r)$ over that range of r. This procedure is similar, but not identical, to the Hausdorff-Besicovitch dimension (Barnsley, 1988). Barnsley (1988, Fig. 5.3.1 and Tables 5.3.2 and 5.3.3, pp. 190–191) illustrates this method by counting the number of ever larger circles needed to contain the dots that make up the clouds on a fine Japanese woodcut.

    A variant of this method, called *box counting*, is to divide the space into a grid of E_m-dimensional boxes and count the number of boxes $N(r)$ of edge size r that contain at least one point of the phase space set. This is shown in Fig. 7.8. We recommend *box counting as a fast, reliable method to determine the fractal dimension* (Liebovitch and Tóth, 1989). Box counting was ignored for some years because of reported limitations (Greenside et al., 1982), but several groups have recently circumvented these limitations Liebovitch and Tóth, 1989; Block et al., 1990; Hou et al., 1990). Note, however, that the box-counting algorithm 1) requires at least 10^D points to determine the fractal dimension of a set of dimension D, and 2) averages the fractal dimension over different scales so that it is not appropriate where the fractal dimension varies with scale, as it does for a multifractal.

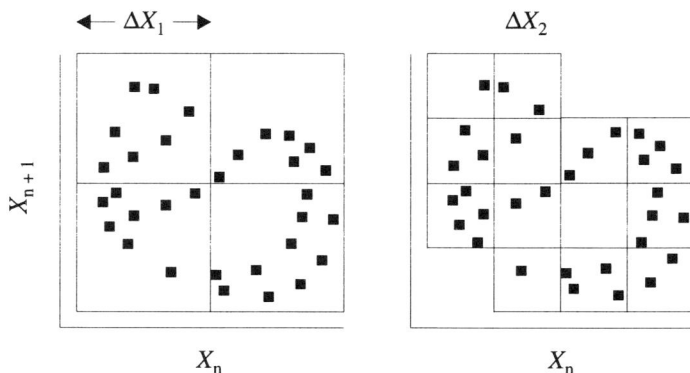

Figure 7.8. Box counting to determine the fractal dimension is illustrated for a set in a two-dimensional space. The points in the set are shown at top left. The number of boxes $N(r)$ occupied by at least one point of the set is determined for boxes of ever smaller size r. The fractal dimension is the slope of $\log N(r)$ versus $\log(1/r)$.

Generalized dimensions. In determining the capacity dimension every box that contains a point of the set is counted once and every box that does not is not counted at all. However, the contribution of each box can be weighted in different ways (Hentschel and Procaccia, 1983), leading to the generalized dimensions. The two most used of these are the correlation and information dimensions. The *correlation dimension* D_{corr} (Grassberger and Procaccia, 1983) is based on correlations between pairs of points. It is determined from the number of pairs of points $C(r)$ of the set that lie within a radius r from each point:

$$D_{corr} = d\log C(r) / d\log r. \tag{7.8}$$

This has been the most popular method to analyze phase space sets because it can evaluate high fractal dimensions and determine the variation of the dimension with spatial scales. However, since D_{corr} often varies as a function of r, the estimate of the best value of the fractal dimension is often ambiguous. (We prefer the box-counting method, which is simpler to program, faster to calculate, less ambiguous, and less influenced by errors due to points near the edge of the set.) The *information dimension* is based on a weighing of the points of the set within a box that measures the rate at which information changes, called the Kolmogorov entropy, as one moves through the phase space set.

Pointwise mass dimension. The generalized dimensions are based on boxes that cover the set. We can also determine the dimension from the properties of the points of the set directly. For example, D_{mass} is determined from the number of points $M(r)$ of the set that lie within a radius r from each point:

$$D_{mass} = d\log M(r) / d\log r. \tag{7.9}$$

In practice, it is sometimes more accurate to determine radii that contain a given number of points of the set (Holzfuss and Mayer-Kress, 1986).

Lyapunov exponents. The rate of the change in the volume of phase space along independent directions is characterized by the Lyapunov exponents, which can be determined from the phase space set (Wolf et al., 1985). Kaplan and Yorke (1979) proposed a way to calculate the fractal dimension from the Lyapunov exponents. Their conjecture has been verified numerically for many cases, although it has not yet been proved. West (1990) gives a good review of their basis, as do Parker and Chua (1989).

Singular value decomposition. The number of independent variables of a set can be determined by singular value decomposition (Albano et al., 1988; Pulskamp, 1988). This can be used to estimate the smallest integer larger than the fractal dimension. This method was introduced in an attempt to determine the dimensionality of chaotic time series contaminated by random uncorrelated noise. The procedure involves the calculation of eigenvalues and it was hoped that the eigenvalues for a low-dimensional attractor would be well separated from those of the noise. In principle, this method of determining the dimension can be orders of magnitude faster than those described above. However, a clear separation may not arise, making the value of the dimension difficult to determine.

Embedding the Phase Space Set in Spaces of Increasing Dimension

The fractal dimension of the phase space set can only be lower than or equal to the dimension of the phase space in which it is embedded. Thus, to determine the true fractal dimension of the phase space set we construct it in spaces of increasing dimension. If the time series is due to *an uncorrelated random process* having a large number of independent variables, the *phase space set will always fill up the space in which it is embedded.* Thus, the fractal dimension D of the phase space set will always equal the dimension E_m of the phase space, as shown in Fig. 7.9. However, if the time series is due to a *deterministic process* having only a few independent variables, then the fractal dimension D of the phase space set will reach a plateau equal to its true fractal dimension D when the space it is embedded in has dimension $E_m > D$, as shown in Fig. 7.10.

The method devised by Grassberger and Procaccia (1983) has the strongest heritage and the longest usage, but should nevertheless be regarded as "under investigation." Its limitations and its accuracy are not yet well defined, nor is it clear how noise affects the estimates of the characterizing factor, the so-called correlation dimension, known as D_{corr} or D_{gp}, after the originators.

The basic idea of quantitative measures of the dimensionality of a chaotic signal is similar to that using fractal characterization, namely, how much of the next dimension in N-space does it fill. There is a difficulty that is not so apparent for fractals: when one asks how much of a plane does an iterated line fractal fill, or how much of three-space does a rough surface fill, one starts with a signal of known dimensionality, e.g., one for a line or two for the surface, and seeks a measure of the complexity of form by asking how much of the next dimension up is used. (So an

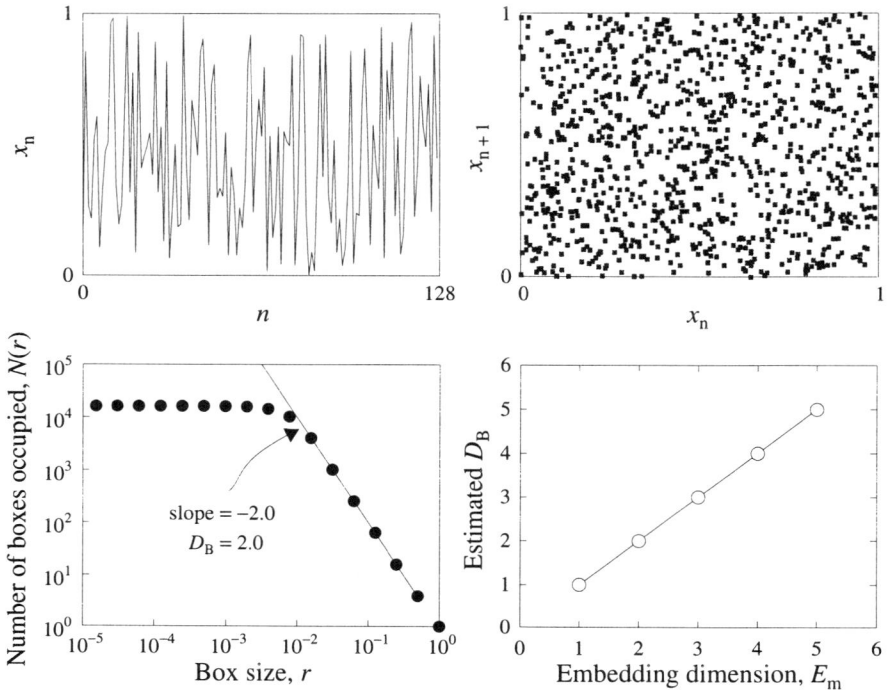

Figure 7.9. Analysis of a time series of randomly generated numbers. *Upper left:* The time series. *Upper right:* The two-dimensional pseudo-phase space set constructed with a time lag $L = 1$. *Lower left:* The estimate of the fractal dimension $D_B = -\log N(r)/\log r$ of the two-dimensional phase space set determined by counting the number $N(r)$ of two-dimensional boxes of edge size r that contain at least one point of the phase space set; here $D_B = 2$. *Lower right:* The fractal dimension D_B of the phase space set as a function of the dimension E_m of the phase space. This phase space set always fills up the space it is embedded in, that is, $D_B = E_m$. Hence, we accurately conclude that this time series was generated by an uncorrelated *random process* having a large number of independent variables.

infinitely recursive line-segment algorithm gives a result with fractal D between 1 and 2, and a mountainous surface has for example a dimension 2.4 compared to hilly country with a dimension 2.2.) A signal from a chaotic attractor has clear dimensionality that is revealed when the trajectory in projected in a space where there are no intersections with itself: a ball of wool is fully expressed in three-space but a two-space silhouette will show intersections of the strand with itself. But we do not know from the signal or from a delay plot what this dimensionality might be. The delay or pseudo-phase plots (Fig. 7.6) are attempts to display the signal as if it were the ball of wool in the appropriate dimension so that the next question, "How

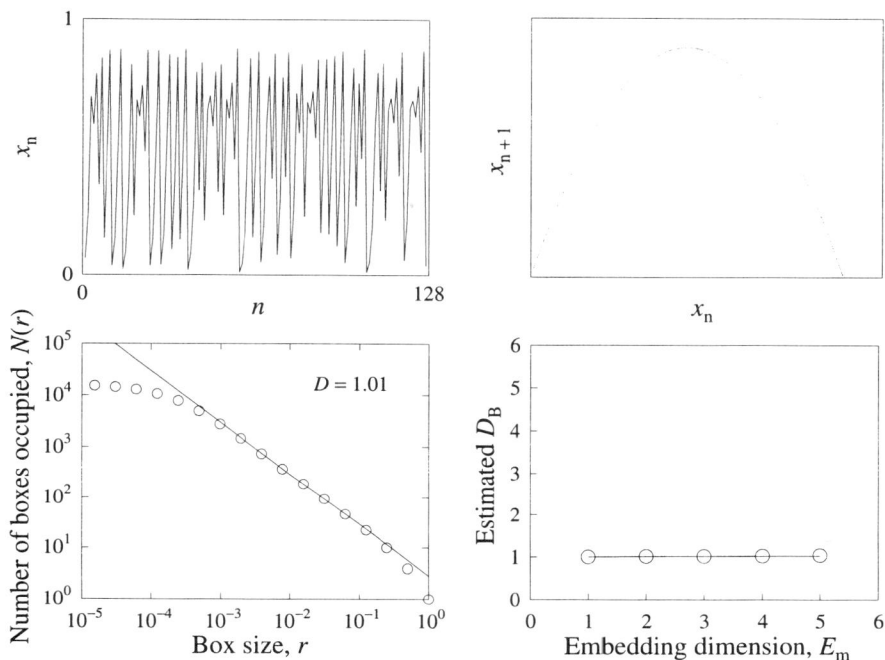

Figure 7.10. Analysis of a time series generated by the deterministic process $x_{n+1} = 3.95\,x_n\,(1 - x_n)$, a variant on the logistic function; the rate constant 3.95 puts it in the chaotic realm. *Upper left:* The time series. *Upper right:* The two-dimensional phase space set constructed with a time lag $L = 1$. *Lower left:* The fractal dimension $D_B = -\log N(r)/\log r$ of the two-dimensional phase space set determined by counting the number $N(r)$ of two-dimensional boxes of edge size r that contain at least one point of the phase space set. *Lower right:* The fractal dimension D_B of the phase space set as a function of the dimension E_m of the pseudo-phase space. Unlike the random process shown in Fig. 7.9, here the estimated fractal dimension of the phase space set D_B *reaches a plateau equal to 1.* Thus, we accurately conclude that the time series can be generated by a *deterministic process* having one independent variable.

much of space does it fill?", can be asked. The problem is that the delayed signal is not necessarily orthogonal to the original, as the derivative would be. If the chosen delay is too short, then one or even a few delays will not unravel or prevent intersections. The embedding dimension is unrelated to the actual dimensions of the object, the complex signal. So a succession of embeddings must inevitably be used.

For a true three-dimensional signal, embedding in four or higher real dimensions reveals no more details of its structure, just as a sphere or cube is completely revealed in three-space. However, if the successive embeddings are partially interdependent, as must occur with a lag less than one equivalent to giving the derivative, one often has to go to embedding spaces of 10 or more. The sum of the distances between all pairs of points is maximal when the embedding dimension (of the plotting space) is high enough to reveal all the true dimensionality of the signal. This is the basis for calculating the Grassberger-Procaccia (1983) correlation dimension. The box dimension used by Liebovitch and Tóth (1989) is based on the same idea except that instead of measuring all of the individual lengths between point pairs to see how spread out the signal is, one puts N-dimensional boxes around the phase plot signal to determine the volume or number of boxes required to contain it. As one decreases the size of the boxes, the number of boxes required to enclose the signal increases while the total area or volume of the boxes decreases and tightens around the signal. An improved algorithm is available for workstations from Hou et al. (1990).

[Aside]

Determining the box dimension, D_B . The capacity or box fractal dimension is the logarithmic slope of the number of boxes required as a function of the box size. The strategy is to calculate the box fractal dimension, D_B, at successively higher embedding dimensions until D_B increases no further. This is a similar strategy to that used by Grassberger and Procaccia (1983) for the correlation dimension while measuring lengths. The steps in calculating the box fractal dimension take advantage of the discretization of the signal, since the signal has lost its continuous nature to form a set of successive points in the embedding space. The steps are:

1. From the original continuous signal take samples uniformly spaced in time at interval durations Δt. This gives the time series $X(t_i)$, i = 0, 1, 2, 3, 4, . . .N.
2. Plot the two-dimensional pseudo-phase plane representation using an appropriate lag. For embedding in two-dimensional space, use $X(t_i + \tau)$ versus $X(t_i)$ for each time point, t_i, in the original signal. Pick a length scale ΔX_1, and an arbitrary starting location for laying out a rectangular grid on the plane. Count N_1, the number of squares in which there are one or more points $X(t_i + \tau)$, $X(t_i)$. This provides one point N_1, versus ΔX_1, in two-space, E_2. See Fig. 7.8 where, for the example, $N_1 = 4$ for ΔX_1.
3. Repeat with a smaller length ΔX_2 and count the number N_2 of corresponding boxes. This gives a second point N_2, ΔX_2. Continue with successively smaller ΔX_i, to the limit of the smallest ΔX_i at which there is not more than one point per box. Going to this limit is not necessary for the next step. Now one has the set, N_i versus ΔX_i for embedding in two-space, $E_m = 2$.
4. Plot $\log N_i$ versus $\log \Delta X_i$, and determine the slope of the line. The slope is $-d(\log N_i)/d(\log \Delta X_i) = D_B(2)$, the estimated box dimension for embedding in two-space. As the box size approaches the limit of one point per box, the curve reaches saturation so the slope should be estimated below this level as suggested by Liebovitch and Tóth (1989), for example, between 25% and 75% of the maximum.

5. Repeat the steps 2 to 4 using three-dimensional and successively higher-dimensional embedding. For E_3, calculate the location of each point from the original signal and two signals delayed by τ and 2τ. For each of the embedding dimensions E_m, plot log N_i versus $\log \Delta X_i$. The slope gives D_B at each E_m.

6. The next step involves interpreting how D_B varies with embedding dimension, E_m. Plot the *fractal dimension* D *of the phase space set versus the embedding dimension* E_m. If enough data points are available, D_B should reach a plateau as E_m is increased. This plateau level is taken to be the fractal dimension. If there are too few data points, demonstration of a plateau may not be possible.

The steps 1, 2, 5, and 6 are illustrated in Figs. 7.9 and 7.10 for a time series that consists of uncorrelated random variables which was created by a pseudorandom number generator and for a time series generated by the deterministic chaotic process $x_{n+1} = 3.95\, x_n(1 - x_n)$, the logistic equation.

Determining the correlation dimension, D_{corr}. The method defined by Grassberger and Procaccia (1983) is similar in intent, namely, to find a measure of the attractor in successive embedding spaces and to find a region where the rate of increase of the measure is a power function of the measure. This method uses the Euclidean distances between all point pairs in the data set. The idea of "correlation" is that points on the attractor may be arbitrarily close to each other in the embedding space even though their occurrence may be widely separated in time as the attractor is being traced out. Thus, steps 2 to 4 of the box analysis are replaced by

2. In the embedding space or pseudo-phase plane of dimension E_m, calculate the distance r between all point pairs. Order the set of distances in ascending order of lengths r.

3. Construct the cumulative probability $C(r)$ is a fraction of unity: $C(r)$ is the fraction of point-to-point distances that are equal to or less than r.

4. Plot the logarithmic slope to find an estimate of D_{corr} for $E_m = 2$:

$$D_{corr} = \frac{d\log C(r)}{d\log r}. \tag{7.10}$$

Since $\log C(r)$ versus $\log r$ is usually curved, one may plot $\Delta\log C(r)/\Delta\log r$ versus $\log r$ and plot the results as estimates of D_{corr}. One looks for a plateau in $d\log C(r)/d\log r$ over a range of r. This plateau gives the estimate of D_{corr} for the particular E_m. The noise tends to give high spurious slopes at small r, which are to be ignored.

5. Repeat steps 2 to 4 for $E_m = 3$, 4, etc.

6. Plot the estimates of D_{corr} against E_m; one expects D_{corr} to increase from low values at small E_m to a plateau at higher level embeddings. This plateau gives the fractal dimension D_{corr} of the attractor.

An alternative but equivalent view is to plot all the curves of $\Delta\log C(r)/\Delta\log r$ versus $\log r$ on the same plot for all embedding dimensions, and then to look for the plateau over a range of r that is the same for the higher embedding spaces. A high enough E_m has been reached when the plateau region is the same on a few successively higher embeddings.

Lyapunov Exponent and Lyapunov Dimension

The Lyapunov exponent measures the divergence of two trajectories originating from neighboring points. In general, chaotic signals from systems of continuous differential equations have at least three Lyapunov exponents, representing the eigenvalues of the system. For a chaotic system with three independent variables one exponent must be positive, representing a tendency toward expansion, one must be zero for there to be a stable limit cycle, and a third must be negative in order that the sum be negative, for this is the requirement that the signal tend toward the attractor and remain in the basin of attraction. From this it can be seen that a first- or second-order continuous system cannot be chaotic. Observation of a positive exponent in a signal on a limit cycle is diagnostic of chaos.

The Lyapunov exponent is notoriously difficult to calculate. Most methods involve picking points that are close, and then following their trajectories to see how fast they separate. But picking the points is tricky, and repetitions are required. The methods of Wolf et al. (1985) paved the way. Parker and Chua (1989) give a broad picture of the problems. Abarbanel et al. (1991) provide a method which requires less than 1000 data points and so is practical to apply.

Entropy and Information Dimensions

Entropy calculations are based on the information content of the observed system. For example, while the box dimension gives a geometric measure of the space occupied by the signal on the attractor, there is an information dimension, D_I, which in addition accounts for the number of times a particular box is visited. D_I is slightly smaller than D_B; the entropy, H, is the amount of information needed to specify the system to an accuracy ε when the signal is on the attractor, $H(\varepsilon) = k\varepsilon^{-D_I}$, just as the number of boxes in the box-counting method is $N(\varepsilon) = k\varepsilon^{-D_B}$. The usual problem is, as before, that large amounts of data are needed to make the full calculations with any assurance of accuracy in estimating the dimension. Pincus (1991) is developing a simplified variant on the entropy calculation that can really only be termed empirical, but it may prove useful because of its simplicity and applicability to short data sets.

[Aside]

Approximate Entropy: Pincus's method (1991) is called *ApEn*, approximate entropy. It is used somewhat in the style of the standard entropy calculations and its estimation has some similarity to that for D_{corr}, but it uses only low embedding dimensions, typically $E_m = 2$ and 3. The steps for $E_m = 2$ are

1. Calculate the mean and SD of the data set, X_i, $i = 1$, N_x. Define a length $r = 0.2$ SD; this will be used as a tolerance limit to reduce the effects of noise.

2. Define a set of points in a two-dimensional space from successive pairs of values in the one-dimensional signal. The pair X_i, X_{i+1} define a point in two-space. For an X array of 1000 points, there are 999 points in two dimensions.

3. Obtain a measure of the distances between (X_i, X_{i+1}) and all other points (X_j, X_{j+1}). The distance may be the proper Euclidean distance, or may be simplified to be the $Max[(X_j - X_i), (X_{j+1} - X_{i+1})]$, which makes the calculations faster and shortens the distances somewhat. (If Euclidean distance were calculated, one might use an $r = 0.3$ instead, for example.) There are N_{x-2} distances for each X_i in two-space.

4. For each of the pairs of data points, X_i, find the fraction of the distances to the other points which are less than r, and let this fraction be called $C_i(r)$. This fraction is a measure of the regularity or the frequency of similar patterns between successive points in the two-dimensional (phase plane) plot.

5. Find the average of the $C_i(r)$s for the N_{x-1} fractions, and take the natural logarithm, so defining $\Phi(m = 2, r = 0.2)$ as this average.

6. Repeat steps 2 through 5 using $E_m = 3$, which means using triples (X_i, X_{i+1}, X_{i+2}) to define the point positions in three-space. Again calculate the average fraction of distances less than r for N_{x-2} triples. This gives $\Phi(m = 3, r = 0.2)$.

7. The measure, $ApEn$ $(m = 2, r = 0.2, N = 1000) = \Phi(m = 2, r = 0.2) - \Phi(m = 3, r = 0.2)$. Thus, $ApEn$ is a very specific example of a set of statistical measures that might be taken from the signal. Because it uses only small m and a particular r, its value does not approximate the formal entropy measures. (This is not a particular problem, since the formal measures are not useful on mildly noisy signals.) Values of $ApEn$ increase with increasing N_x. Therefore, $ApEn$ is used as a comparative statistic. with a fixed E_m, r, and N_x.

* * *

For heart-rate variability, $ApEn$ $(m = 3, r = 0.2, N = 1000)$ is around 1.0 for young adults and diminishes with age to around 0.7 at age 80 years (Pincus and coworkers, unpublished, with permission: see also Fig. 13.8). The suspicion is that a diminution in $ApEn$ may indicate an unhealthy degree of constancy of the heart rate, and research by Goldberger and others (e.g., Goldberger, 1992) should determine whether or not $ApEn$ can be used as a measure of risk of sudden death by arrhythmia.

Special Situations

Even low-order processes of generation and decay (chemical reaction networks, predator-prey relationships, etc.) can exhibit dependence on a rate coefficient. A cycling or periodic system in which an increase of the rate results in a bifurcation or period doubling is diagnosed as potentially, if not actually, chaotic. A bifurcation is a doubling of the number of stable settings for the observed variable, which is equivalent to doubling the length of the period required to repeat the pattern. In an iterated system in which a rate increase causes a steady signal to take 2 alternating values, then 4, 8, and 16, and thereafter appears unpredictable at a higher value of the rate, the system is said to follow period-doubling bifurcations on the route to chaos. In continuously observable systems the same kind of thing can occur, a stable variable shifts to a single cycle (circular or elliptical in the phase plane), then to a two-cycle (two linked but distorted ellipses in the phase plane) and so on; at higher rates the signal may never be exactly repetitive and is called chaotic. Even though the trajectories are nicely bounded and smooth they are not exactly predictable.

These kinds of shifting in periodicity can occur spontaneously in living systems or may sometimes be induced by temperature changes or another stimulus. Observations of jumps in behavior patterns should therefore raise one's suspicions that the system operates on a low-dimension attractor in either a periodic or chaotic mode. The attractor, the region of phase space covered by the trajectories, is commonly fractal; like taking a slice through Saturn's rings, the distances between neighboring slices tend to show fractal variation in the intervals, which means the spacings show correlation structure, near-neighbor spacings being more alike than spacings between N^{th} neighbors. Thus, the spacings are neither random nor uniform.

7.7   Good News and Bad News About This Analysis

The *good* news is that the analysis techniques described above yield important information about physiological processes that cannot be obtained by other methods. Namely, we can use them to tell if a time series is generated by a random or a deterministic process and to find its mathematical form if it is deterministic. The *bad* news is that the practical application of these techniques is often difficult and frustrating. We now describe some these difficulties.

Amount of Data Needed

It is not clear how many values in a time series are needed in order to accurately construct a phase space set and determine its dimension. Estimates from different authors based on different analytical arguments, each supported by numerical simulations, are given in Table 7.2. To illustrate the disagreement between these estimates we have used each one to compute the number of values of the time series needed to analyze a process with a six-dimensional attractor. These results are complicated by the fact that each estimate was based on a different range of the dimension D. Nonetheless, this crude comparison shows that the estimates of how many values of a time series are needed to analyze a six-dimensional attractor range from 200,000 to 5,000,000,000. This makes it very difficult to design experiments.

If the higher estimates are correct (Wolf et al., 1985; Gershenfeld, 1988; Liebovitch and Tóth, 1989; Theiler, 1990), great care must be exercised to maintain the physiological system under constant conditions and in the same state during the

Table 7.2. Estimates of the number of measurements in a time series needed to accurately construct and determine the dimension of a phase space set of dimension D

	N Data points needed for an attractor of dimension D	N when $D = 6$
Smith, Jr., 1988 *Phys. Lett. A* 133:283	42^D	5,000,000,000
Wolf et al., 1985 *Physica D* 16:285	30^D	700,000,000
Wolf et al., 1985 *Physica D* 16:285	10^D	1,000,000
Nerenberg and Essex, 1990 *Phys. Rev.* A42:7065	$\dfrac{2^{1/2} \left[\Gamma (d/2 + 1) \right]^{1/2}}{\left[A \ln (k) \right]^{(D+2)/2}} \dfrac{D + 2}{2}$ $\times \left[\dfrac{2 (k - 1) \Gamma ((d + 4)/2)}{\left[\Gamma (1/2) \right]^2 \Gamma ((d + 3)/2)} \right]^{D/2}$	200,000

long time that the data are collected. This may mean that reports of the six-dimensional attractors found in EEG and ECG data based on time series of 1,000 to 10,000 measurements may be unreliable. On the other hand, test data of phase space sets with fractal dimension 5 (Holzfuss and Mayer-Kress, 1986) have been accurately analyzed with as few as 20,000 measurements.

Another way to state this problem is to say that the data need to be spread around the phase space set in some approximately uniform way. For example, 10^D points for each circuit of the set and 10^D such circuits (Wolf et al., 1985). But to do this requires that we know the properties of the phase space set. Thus, we are caught in the paradox that only if we already know the phase space set can we really be sure how to construct it.

Surrogate Data Sets for Evaluating the Presence of Nonlinear Dynamics

A surrogate data set is one produced from the data set under test by manipulating the data so that critical characteristics are lost. Randomizing the order of the elements of the set preserves the probability density function of values but obliterates any point-to-point correlation. Then both the original signal and its surrogate are analyzed using identical methods. If they do not differ, for example, in the estimated correlation dimension, D_{corr}, then the conclusion is that the original signal did *not* evolve from a low-dimensional chaotic system. This is too strong a measure to be considered a subtle test.

A better approach is to form a new data set from the experimental data with the same correlation structure but without the deterministic relationship implied by the existence of an attractor. This can be done by determining the Fourier series of the time series, and then randomizing the phases of the Fourier components, and transforming the signal back into the time domain to produce the surrogate signal, as described by Theiler et al. (1992). This generates a null hypothesis test signal with the same correlation structure of the original data set, but with its deterministic relationship (if any is present) randomized away. An even milder treatment is to shuffle the phases in the Fourier series before transforming back into the time domain, as used by Slaaf et al. (1993). The difference between randomizing and shuffling may be too subtle to worry about, since the difference in the resultant form of the time domain signal is not great, and not enough tests have yet been done to know the statistical difference. In any case, if the attractor is present in the data but not present in this null hypothesis test data set, then our confidence in the existence of the attractor is greatly increased. If the fractal dimension remains unchanged the time series is probably colored noise.

The development of surrogate data sets is important because long data sets on experimental data are not generally available. There are a few ways of doing this, which are presented by Theiler et al. (1992), and more work is being done by them and by us.

Noise

All experimental measurements are contaminated by noise from the system itself or from the measurement instruments. Some preliminary studies have been done to suggest how the fractal dimension can be determined so that it is not influenced by such noise (Grassberger, 1986). More recent analyses by Casdagli et al. (1991, 1992) illustrate that the effects of noise severely handicap the estimate of the fractal dimension of known attractors. This is potentially discouraging for biologists whose signals are always noisy.

Fractal Dimension of the Phase Space Set

The fractal dimension is usually determined by how a property $P(r)$ scales with spatial scale r. In principle, $P(r) \propto r^{\pm D}$. In practice, this scaling is only approximately true. Thus, D varies as a function of r. One tries to choose the range of r for which the scaling relationship is best satisfied. This is often difficult and subjective. See, for example, Figs. 1, 3, 5–10 of Albano et al. (1988), Figs. 3.4–3.10 of Kaplan (1989), or Figs. 2 and 3 of Liebovitch and Tóth (1989). We recommend the box-counting algorithm (Liebovitch and Tóth, 1989), because it requires fewer decisions than the more often used correlation dimension (Grassberger and Procaccia, 1983). This point is debated, and Theiler et al. (1992) point out their preference for the correlation dimension. Since computation time, though always an issue for this kind of work, is lessened with modern computers, one can make an argument for using the information dimension; its virtue is that it puts weighting on the *number* of visits by the attractor to particular regions of phase space, not just that a box is occupied or not, and so reduces the impact of wild or noisy points on the calculation of D. See, for example, Parker and Chua (1989).

Meaning of a Low-Dimensional Attractor

It is important to remember that if we find a low-dimensional attractor in the phase space set, this means that the time series *could have been generated* by a deterministic relationship; it does *not* mean that *it was generated* by a deterministic relationship. At the other end of the spectrum of complexity, it is safe to say that if the biological system is really a high-dimensional attractor we will have great difficulty in distinguishing even a noise-free deterministic signal from noise, and that this method of analysis will contribute little to the investigators' search for either mechanism or description. The optimistic note is that finding a low-dimensional attractor does help in designing the next set of experiments: at least it suggests that studies might be done to change rate constants to see if the system behavior changes, and at best may suggest specific variables to add to the list of measured variables in the study.

7.8   Summary

In this chapter we gave a brief overview of a most active field of research, methods for determining a characterizing dimension of a function. The methods help to distinguish the chaos of a nonlinear dynamical system from noise, but cannot prove the distinction in all cases.

Part III

Physiological Applications

8

Ion Channel Kinetics: A Fractal Time Sequence of Conformational States

> These equations can be given a physical basis if we assume that potassium can only cross the membrane when four similar particles occupy a certain region of the membrane ... and if the sodium conductance is assumed to be proportional to the number of sites on the inside of the membrane which are occupied simultaneously by three activating molecules but are not blocked by an inactivating molecule.
>
> Hodgkin and Huxley (1952)

8.1 Introduction

Cell Membrane Transport through Protein Pumps, Carriers, and Channels

Ions such as sodium, potassium, and chloride can move freely through water but cannot cross the hydrophobic lipids that form the cell membrane. However, these ions can interact with proteins in the cell membrane that transport them across the cell membrane. Three types of proteins are involved in ion transport: 1) *Pumps*, such as the sodium-potassium ATPase, bind tightly to a few ions at a time and use energy from ATP to move these ions against their electrochemical gradient. 2) *Carriers*, such as the sodium-potassium-chloride cotransporter, bind tightly to a few ions at a time and help them move down their electrochemical gradient. 3) *Channels*, such as the sodium channel, bind weakly to many ions at a time and allow them to move down their electrochemical gradient.

A typical epithelial cell has approximately 10^6 pump, 10^4 carrier, and 10^2 channel molecules in its cell membrane. The number of ions transported per second through each type of protein is approximately 10^2 per pump, 10^4 per carrier, and 10^6 per channel. The current through each route is the product of the number of proteins and the number of ions transported per second through that type of protein. Thus, approximately equal ionic currents pass though each of these three routes.

Ion Channels

Ion channels of different types are ubiquitous in cells (Hille, 1984; Sakmann and Neher, 1983; Hille and Fambrough, 1987). In nerve, for example, they generate and propagate the action potential; in muscle they control calcium levels that initiate contraction; in sensory cells they trigger electrical changes that result in signal transduction; in epithelial cells they are part of the pathway of net ionic movement across cell layers, and they control the ion concentration in intracellular organelles such as mitochondria.

Each channel consists of approximately one thousand amino acid residues and one hundred carbohydrate residues, arranged in several subunits. As shown in Fig. 8.1, this entire structure spans the cell membrane (Stroud, 1987). The channel protein can have different shapes called conformational states. Sometimes, the channel has a conformational state with a central hole. Electrochemical gradients can force ions through this hole to enter or exit the cell. Pieces of the channel can also rearrange to block the hole and thus block the flow of ions. These open and closed conformational structures differ in energy. The ambient temperature is high enough to provide enough energy to spontaneously switch the channel between conformations that are open, and those that are closed, to the flow of ions through the channel. In this chapter we analyze the kinetics of ion channels, that is, how the channel protein changes between its different conformational states.

8.2   The Patch Clamp

Opening and Closing of an Individual Ion Channel

The spontaneous fluctuation between open and closed conformational states of an individual channel can be detected using the *patch clamp* technique developed by Neher, Sakmann, Sigworth, Hamill, and others (Hille, 1984; Sakmann and Neher, 1983). In contrast to the standard voltage clamp technique in which the currents are the sum of the currents through many individual channels, often of several types at once, the patch clamp technique provides a current through a single channel. Patches less than 1 square micron usually have only one or very few channels. As shown in Fig. 8.2, a small piece of cell membrane is sucked up and sealed inside the micron-wide tip of a glass pipette. Because the cell membrane forms a high electrical resistance seal against the glass, the current measured across the patch is due to ions moving through the channels in the patch. Recordings can be made with the patch attached to the cell, or, as shown, the patch can be entirely removed from the cell. How the patch is pulled off the cell determines if the outside face of the cell membrane faces the solution inside or outside the pipette.

The electronic amplifier maintains a fixed voltage ("clamp") between the electrode in the pipette and a ground electrode in the solution outside the pipette. The amplifier also converts the current through the patch into a voltage, which is amplified, low pass filtered, stored on FM tape or on videotape using a digital-VCR interface, and subsequently digitized by a computer. When the channel is closed no current is recorded, and when it is open a measurable picoamp current is recorded. To determine the open and closed durations, a threshold is set equal to the average

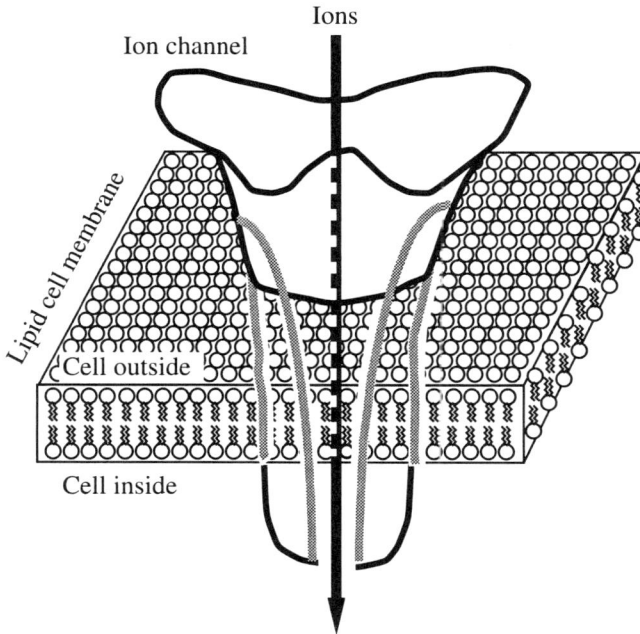

Figure 8.1. Ion channels are proteins that span the lipid bilayer that forms the cell membrane. Ions, such as sodium, potassium, and chloride, cannot cross the lipid bilayer. However, when the channel is in an open conformational state, ions can pass through the inside of the channel protein and thus enter or exit the cell.

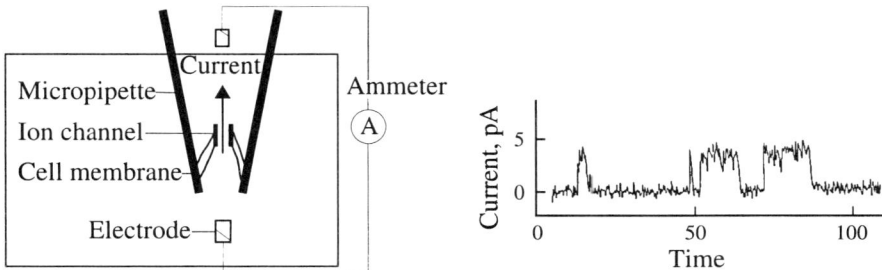

Figure 8.2. The patch clamp technique can record the sequence of ion channel conformations of an individual protein molecule that are open and closed to the flow of ions. *Left:* Schematic view of a patch clamp experiment. A small patch of cell membrane is sealed in a glass micropipette. The current measured through the patch reveals the sequence of the open and closed conformations of the channel. *Right:* Patch clamp recording of a potassium channel in the apical membrane of a corneal endothelial cell (Liebovitch and Tóth, 1990a). (*Left:* redrawing of Liebovitch and Tóth, 1990a, Fig. 1. *Right:* Liebovitch and Tóth, 1990a, Fig. 2.)

179

of the minimum and maximum value of the current. When the current crosses this threshold in the upward direction an opening has occurred and when the current crosses this threshold in the downward direction a closing has occurred.

This technique permits the kinetics, the sequence of closed and open conformations of an individual protein molecule, to be followed. This is a much more sensitive probe of protein kinetics than other common biochemical or biophysical techniques where the experimental signal is averaged over a very large number of molecules, each of which may be in a different conformational state. The patch clamp is able to do this because 1) there is approximately one channel per square micron of cell membrane so that there are only a few channels, and sometimes only one channel, in each patch. 2) Each channel acts like a gate to control the flow of ions. When it is closed, no current is flowing through the channel. When it is open, many ions are flowing through the channel, resulting in a measurable picoamp current that reports if the channel is closed or open. Very approximately, a typical channel opening may last 1 ms during which a current of 5 picoamps passes through the channel. A current of 5 picoamps over 1 ms corresponds to the movement of approximately 30,000 ions.

Dwell Time Histograms

It is useful to analyze how often open (or closed) durations of a given time are observed, which are called the *dwell time histograms*. That is, given a bin size, Δt, we count how many times the channel was open (or closed) for durations less than or equal to Δt, for durations greater than Δt and less than or equal to $2\Delta t$, for durations greater than $2\Delta t$ and less than or equal to $3\Delta t$, etc. The dwell time histogram consists of $N(i\Delta t)$, the number of open (or closed) times longer than $(i - 1)\Delta t$ and shorter than or equal to $i\Delta t$, plotted versus the open (or closed) time at the center of each bin, which is equal to $(i - 1/2)\Delta t$.

The *probability density* f(t), evaluated at the center of the bin, $t = (i - 1/2)\Delta t$, that the channel is open (or closed) for a duration greater than t and less than or equal to $t + dt$, is equal to $N(i\Delta t)/(\Delta t N_o)$, where N_o is the total number of open (or closed) times. The *cumulative probability* P(t) that the channel is open (or closed) for a duration greater than t, is equal to the integral of $f(t)$ evaluated from durations equal to t, to durations equal to ∞.

Transformations of the dwell time histogram can help in analyzing the data. To cover a wider range of open or closed times, it is useful to use a logarithmic transformation. Liebovitch et al. (1987a) described how this can be done by constructing a series of dwell time histograms with different bin sizes. The bin size of each histogram in the series is twice that of the previous histogram. The probability density, $f(t)$, determined from each histogram can then be plotted together as log$f(t)$ versus t, or log$f(t)$ versus log$t$. McManus et al. (1987) described how this can be done by logarithmic binning. Sigworth and Sine (1987) showed that the square root of the logarithmically binned $f(t)$ is useful in finding the time constants of exponential terms in the data.

Fitting Models of the Dwell Time Histograms

We would like to use the experimental data to determine the physicochemical mechanism that opens and closes the channel. Typically, a hypothetical model of such a mechanism predicts a mathematical form with adjustable parameters for the open and closed time histograms. To determine the values of those parameters we vary their values until the mathematical form of the model, $f_{model}(t)$, has the *best goodness of fit* with respect to the experimental data, $f_{data}(t)$.

Different criteria can be used to define the goodness of fit of the model to the data (Colquhoun and Sigworth, 1983). For example, the goodness of fit can be defined as the sum of the square of the difference between the data and the model values for all the bins $i = 1, 2, 3, \ldots n$ in the histogram, namely $\Sigma_{i=1,n} (f_{data}(t_i) - f_{model}(t_i))^2$, where t_i is the time at the center of each bin. Since longer dwell times occur less frequently, the error in their values is proportionately larger and thus we may choose to weight them less in the fit. This can be done by using χ^2 as the goodness of fit, namely, $\Sigma_{i=1,n} (f_{data}(t_i) - f_{model}(t_i))^2 / f_{model}(t_i)$, which weights the values in the bins inversely proportional to their estimated variance, $f_{model}(t_i)$.

The methods of minimizing the square of the deviations and χ^2 assume that the differences between the model and data have a normal distribution. That assumption is not required if we use the method of maximum likelihood that determines the values of the parameters that have the greatest probability of producing the observed data. If we assume that the probability density of the dwell times has the form $f_{model}(t)$, then the probability that we observe a dwell time of duration t_1 is proportional to $f_{model}(t_1)$. The joint probability that we observe the dwell times t_1, t_2, t_3, $\ldots$, t_n, actually found in the data, is the product of the probability of observing each one, namely, $f_{model}(t_1) f_{model}(t_2) f_{model}(t_3) \ldots f_{model}(t_n)$. This product is called the likelihood function. The values of the parameters that maximize the value of the likelihood function thus correspond to the greatest probability of producing the observed dwell time data.

The model that best fits the data is determined by finding the values of the parameters that minimize the sums of the squares of the errors or χ^2, or maximize the likelihood function. These minima or maxima can be found numerically by search algorithms that explore the values of the goodness-of-fit function, and by faster gradient algorithms that use its derivatives. Our experience is that the search algorithms, such as the simplex AMOEBA (Press et al., 1986), are more reliable. Numerical errors in evaluating the derivatives cause gradient algorithms to erroneously identify, as minima or maxima, regions where the goodness of fit changes slowly in one direction and rapidly in another direction. Finding the minima or maxima of the these highly nonlinear goodness of fit functions in the multidimensional parameter space is a difficult problem with many computational pitfalls that are not adequately described in patch clamp articles (Acton, 1970; Press et al., 1986).

In principle, we can compare the fit of different models to find the model that best fits the data and is therefore most likely to be the one that is correct. In practice, this is *not* possible because the models that have been proposed to describe ion channel kinetics can have an arbitrary number of adjustable parameters. As the number of adjustable parameters is increased, many models, even ones that are *invalid*, will provide a better fit to the data. Thus, we cannot tell if the model fits the data because it is valid or because it has a large number of adjustable parameters. Although procedures have been used to compare the fit of models with different numbers of parameters (Colquhoun and Sigworth, 1983; Horn, 1987), it has been shown that these procedures are not valid (Liebovitch and Tóth, 1990c).

Hence, in order to analyze the open and closed time histograms requires that we already have a model in mind that specifies the mathematical form to be fit to the experimentally measured open and closed time histograms. Thus, we find ourselves in the *very* uncomfortable position of being forced to *choose a mechanism in order to* analyze the data.

8.3   Models of Ion Channel Kinetics

We need a mathematical form in order to analyze the experimental data, such as the open and closed time histograms. This mathematical form is derived from our ideas about how a channel opens and closes. First, we describe the Hodgkin-Huxley model of ion channel kinetics and the Markov mathematical form derived from it. Then we describe how fractal concepts can be used to analyze the experimental data and the new insights this reveals about the physicochemical structure and dynamics of ion channel proteins.

Hodgkin-Huxley model

In 1952, Hodgkin and Huxley chose mathematical functions to represent the voltage dependence of the potassium and sodium currents they measured across the cell membrane of the giant squid axon. As quoted at the beginning of this chapter, "these equations can be given a physical basis if we assume" that each ion channel consists of a *few* (three to four) *pieces* that assemble together to form an open channel, or disassemble to close the channel. Hence, each channel has a *few substates* that have less than the full number of pieces. In this model, the switching between substates is an inherently random process. The probability per second to switch between substates is a constant that depends only on the present substate and not on how long the channel has been in that substate, nor on the history of previous substates.

It is now known that ion channels function as integral units rather than transitory assemblies. A few pieces assemble to form an open channel, and the mathematical basis of this model, called a *Markov process*, has provided the mathematical form used to analyze many different types of experiments involving currents through ion channels over the last forty years (DeFelice, 1981; Sakmann and Neher, 1983; Hille, 1984). Andrei Andreevich Markov was a Russian mathematician who died in 1922. He did work on the theory of differential and integral equations, but he is best known for his work on probability theory. The mathematical model he published in 1906 has been the basis of many analyses in biology, chemistry, and physics. However, the only real-world application of this mathematical model by Markov himself was in studying the alternations between vowels and consonants in Pushkin's novel *Eugene Onegin* (Youschkevitch, 1974).

A *Markov model of ion channel kinetics* consists of a *set of open and closed states*, the *connections between those states*, and the *probabilities of switching from one state* to another. Typically, such models have a relatively small number of states (2 to 10). The probability per second of switching from one state to another is called the *kinetic rate constant*.

The open (or closed) time histograms of this model have the mathematical form of the sum of exponentials (Colquhoun and Hawkes, 1983). That is, $f(t)$, the probability of having an open (or closed) time of duration greater than t and less than $t + dt$, is given by

$$f(t) = a_1 e^{-r_1 t} + a_2 e^{-r_2 t} + a_3 e^{-r_3 t} \ldots a_n e^{-r_n t}, \tag{8.1}$$

where there are n open (or closed) states and $a_1, a_2, a_3, \ldots a_n$ and $r_1, r_2, r_3, \ldots r_n$ are the adjustable parameters of the model. As described above, the values of these adjustable parameters are determined by the best fit of Eq. 8.1 to the measured open (or closed) time histogram data.

The parameters of Eq. 8.1 are related to the number of open and closed states and the kinetic rate constants by the Chapman-Kolmogoroff equations (Papoulis, 1984; Colquhoun and Hawkes, 1983). Thus, we can use the parameters determined from the open and closed time histograms to determine the set of open and closed states and the kinetic rate constants. However, except for the simplest kinetic schemes, there are more kinetic rate constants than independent equations, and thus many different kinetic schemes with different sets of kinetic rate constants can fit the same experimental data (Bauer et al., 1987; Kienker, 1989; Kilgren, 1989).

Physical Interpretation of the Markov Model

The Markov model assumes that the ion channel protein and its switching between open and closed states have certain physical properties. These properties are rarely described in patch clamp articles that use Markov models to analyze ion channel kinetics. These physical properties were consistent with the knowledge of chemical reactions at the time that the Hodgkin-Huxley model was proposed.

The physical structure of a protein can be characterized by a potential energy function which is the sum of all the energies in the bonds and the interactions between the atoms in the protein (McCammon and Harvey, 1987). Stable conformational shapes of the protein correspond to local minima in the potential energy function. The kinetic rate constants of the transitions from one conformational state to another depend on the height of the potential energy barrier that separates those two states (Barrow, 1979).

The mathematical properties of the Markov model assume the ion channel protein has certain physical properties: 1) Since Markov models typically have only a few conformational states, there must be only a *few, deep local minima* in the potential energy function that are well separated from each other. 2) Since the probability distribution of the open and closed times has the form of Eq. 8.1, the *energy barriers* separating the conformational states must each have a *unique value*, rather than a continuous distribution of energies (Liebovitch and Tóth, 1991). 3) Since the kinetic rates that connect different states are determined as independent parameters from the data, the physical processes that cause the transitions between different conformational states must be due to *independent physical processes* that *do not interact with each other*. 4) Since the probability to switch from one state to another is constant in time, the *potential energy structure must remain constant in time*.

Fractal Model

At one session of the Biophysical Society meeting in San Francisco in 1986, each speaker reported that the rate of activity of channel openings and closings fluctuates in time, changing suddenly from periods of great activity to periods of little activity. Each speaker interpreted these changes as due to physical changes in the channel protein. However, one of us suddenly realized that this pattern can also be produced by the type of fractal process with infinite variance that was described in Chapter 2. That is, if there are bursts within bursts within bursts, of openings and closings, then data collected within the upper hierarchies of bursts would show very high activity, data collected between the bursts would show very low activity, and data collected at the borders of these hierarchies would show sudden changes in the level of activity, even though there was no physical change in the ion channel protein. This fractal description was supported by the qualitative self-similar appearance of the current records, as illustrated in Fig. 8.3.

This suggested a new way to analyze the open and closed time histograms. The switching probabilities between states at one time scale are related to those at other time scales. Instead of using the mathematical form of Eq. 8.1 derived from Markov models, we could analyze the data by asking the fractal-oriented question: "How are the probabilities to switch between the open and closed states at one time scale related to those at other time scales?"

The effective kinetic rate constant is the probability per second of switching between the open and closed states when the current record is analyzed at a given time scale. To determine how the probability to switch between the open and closed states depends on the time scale at which it is evaluated, Liebovitch et al. (1987a, 1987b) proposed that the open and closed time histograms be analyzed by determining the *effective kinetic rate constant*. The *kinetic rate constant* is the probability per second that the channel changes state. However, the channel must remain in a state long enough for us to be able to detect it in that state. That sufficient time for detection defines the effective time scale, t_{eff}. The *effective kinetic rate constant*, k_{eff}, is the conditional probability per second, that the channel changes

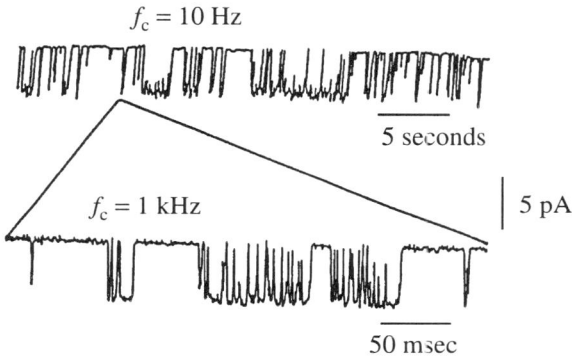

$f_c = 10$ Hz

5 seconds

$f_c = 1$ kHz

5 pA

50 msec

Figure 8.3. The opening and closing of an ATP-sensitive potassium channel is self-similar in time. These recordings of the current through the channel vs. time were made by Gillis, Falke, and Misler. As seen when the data are viewed on a 100 times expanded time scale at the bottom, each opening actually consists of a sequence of openings and closing. That is, there are bursts within bursts within bursts of openings and closings. (This figure by Gillis, Falke, and Misler was published in Liebovitch and Tóth 1990b, Fig. 1.)

state, *given* that it has already remained in a state for a sufficient time, t_{eff}, to be detected in that state. This conditional probability has been given different names in different fields. In life insurance it is called the risk function, in epidemiology it is called the survival rate (Mausner and Bahn, 1974), in renewal theory it is called the age-specific failure rate (Cox 1962), and in tracer washout studies it is called the fractional escape rate (Bassingthwaighte and Goresky, 1984). In all these cases, it is the probability per second of the occurrence of an event (such as dying), given that the system has already survived a time t without that event occurring. There are two effective kinetic rate constants, one for the transitions from the open to closed states, and one for the transitions from the closed to the open states.

The dependence of the effective kinetic rate constant, k_{eff}, on the time scale, t_{eff}, characterizes a *fractal scaling*. That is, like the other fractal scalings described in Chapter 2, it tells us how the kinetics depends on the time resolution used to measure it. Sometimes this scaling has a *simple* power law form. At other times, the scaling characterized by the function $k_{eff}(t_{eff})$ has a more *complex* form.

There are many different ways of analyzing data to determine the scaling function $k_{eff}(t_{eff})$. We describe in detail two methods that are simple and reliable. We illustrate these methods by using the closed times. The scaling for the open times is determined in a similar way. In the first method, as shown in Fig. 8.4, the same set of closed times is used to construct a series of closed time histograms each with a different bin size, Δt. For example, each histogram in the series can have a bin size twice that of the previous histogram. Each histogram consists of $N(i\Delta t)$, the number of closed times longer than $(i - 1)\Delta t$ and shorter than or equal to $i\Delta t$, plotted versus the closed time at the center of each bin, which is equal to $(i - 1/2)\Delta t$. The bin size

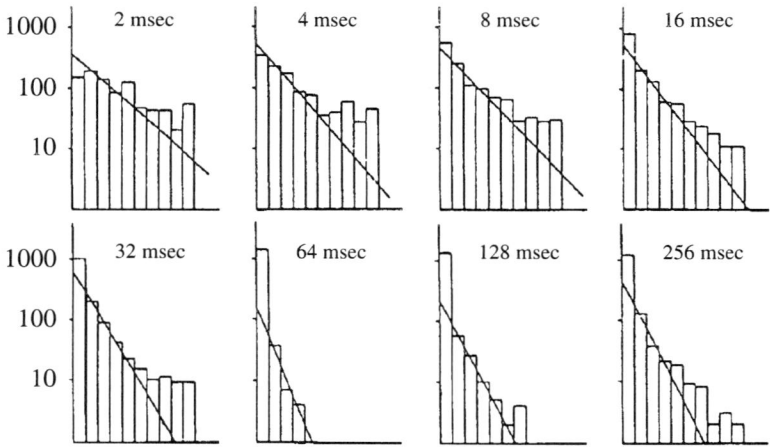

Figure 8.4. Closed time histograms on semi-log plots from a potassium channel in the apical membrane of a corneal endothelial cell (Liebovitch et al., 1987a). The same data of closed time durations were used to construct histograms of bin sizes 2, 4, 8, 16, 32, 64, 128, and 256 ms. The abscissa is the bin number, where the number of closed times ≤ the binning interval for the panel are in the first bin, the number with closed times between 1 and 2 binning interval durations are in the second bin, and so on. This set of histograms can be used to determine the scaling function that is the dependence of the effective kinetic rate constant, k_{eff}, on the effective time scale, t_{eff}.   Each histogram determines the value of k_{eff} at one value of t_{eff}. The effective time scale t_{eff} is equal to the bin size of each histogram. The effective kinetic rate constant, k_{eff}, is the kinetic rate constant determined over a small range of closed time durations, that is, bins. The first bin is excluded because it lumps together all the very short durations. Then the next three bins are used to fit a function proportional to e^{-kt}, where t are the closed times, which is a straight line on these semilog plots. The effective kinetic rate constant is the value of k, which is the negative of the slope of the lines shown. The values of k_{eff} versus t_{eff} determined from these histograms appear in Fig. 8.7. (From Liebovitch et al., 1987a, Fig. 11.)

determines an effective time scale. Thus, from each histogram we determine one value of k_{eff} corresponding to $t_{eff} = \Delta t$. The effective kinetic rate constant, k_{eff}, is the kinetic rate constant determined over a small range of closed time durations, that is, bins. The first bin, $i = 1$, is excluded because it lumps together all the very short durations. The function $N(i\Delta t) = Ae^{-k(i - 1/2)\Delta t}$ is then fit to the data over the next three bins $i = 2, 3, 4$. As can be seen from the plots of log frequency versus bin number in Fig. 8.4, these bins contain most of the data and are least subject to random variation. This can be done by a least squares fit of the straight line Log $N(i\Delta t)$ versus $(i - 1/2)\,t$ on a semilogarithmic plot of the histogram. The effective kinetic rate constant, $k_{eff} = k$, is the negative of the slope of this line. One effective kinetic rate constant, k_{eff}, is thus determined from each histogram, and that set of values is used to construct a plot of $\log k_{eff}$ versus $\log t_{eff}$.

[Aside]

Another good method, developed by French and Stockbridge (1989), uses the maximum likelihood estimator of a kinetic rate constant determined over a range of closed times t, $t_1 \le t \le t_2$. One value of the effective kinetic rate constant is determined from each range of closed times. This is repeated for a series of different ranges of closed times. We can choose the t_1 of each new range to be twice that of the previous one, and let $t_2 = 5t_1$. For each range, we find the average $T(t_1, t_2)$ of all the closed times longer than t_1 and shorter than t_2. The shortest closed time determines the effective time scale and thus $t_{eff} = t_1$. The maximum likelihood estimator, k, of the kinetic rate constant over this range of closed times is given by the solution to the equation

$$1/k = T(t_1, t_2) - (t_1 e^{-kt_1} - t_2 e^{-kt_2}) / (e^{-kt_1} - e^{-kt_2}) . \tag{8.2}$$

This equation can be solved by setting $k(i = 0) = 1/T$ and iterating

$$k(i+1) = 1 / \left[T(t_1, t_2) - \frac{\left(t_1 e^{-k(i)t_1} - t_2 e^{-k(i)t_2} \right)}{\left(e^{-k(i)t_1} - e^{-k(i)t_2} \right)} \right] . \tag{8.3}$$

Stockbridge and French (1988) also show how the likelihood function can be used to find the confidence limits of each value of the effective kinetic rate constant.

✳ ✳ ✳

We briefly review other methods of determining the scaling function $k_{eff}(t_{eff})$:

1. The effective kinetic rate constant is related to $P(t)$, the cumulative probability that the duration of the open (or closed) state is greater than t, and thus can be determined from numerical differentiation of the relationship

$$k_{eff}(t_{eff}) = -\left[\frac{d}{dt} \ln P(t) \right]_{t = t_{eff}} . \tag{8.4}$$

2. A series of closed time histograms can be constructed from the same current versus time data sampled at different analog-to-digital (A/D) conversion rates. The effective time scale, t_{eff}, of each histogram is proportional to the A/D sampling rate. The effective kinetic rate constant at each time scale is determined by fitting a kinetic rate constant to the first few bins of each histogram, as shown in Fig. 8.5.

3. Liebovitch et al. (1986) showed that the kinetic rate constant can be found from the autocorrelation function of the fluctuations, and the square of the fluctuations, of the current. Thus, it is *not* necessary to measure the open and

closed times to determine the kinetic rate constants. The time delay at which these correlation functions are evaluated determines the effective time scale t_{eff}. Thus, the effective kinetic rate constant, k_{eff}, can be determined at different time delays corresponding to different effective time scales, t_{eff}.

Using the effective kinetic rate constant, k_{eff}, to analyze the open and closed time histograms. Looking at experimental data from a new perspective may lead to new ways of analyzing the data that may lead to a better understanding of the mechanisms that produced the data. The sensitivity to fractal concepts, where self-similarity can be revealed by properties measured at different scales, was the motivation for the development of the effective kinetic rate constant. The analysis using the effective kinetic rate constant has led to important new insights from channel data. This analysis can be successfully used on data that are *fractal* as well as data that are *not fractal*. Thus, an understanding of fractal concepts has led us to develop a new and useful tool that has broad applicability to extract information from the data. Without the insights provided from an understanding of fractals, such a tool would not have been developed.

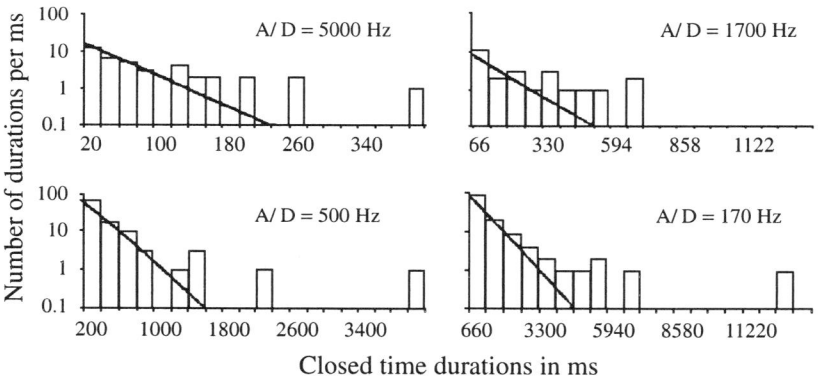

Figure 8.5. Closed time histograms from a potassium channel in the apical membrane of a corneal endothelial cell (Liebovitch et al., 1987a). Each histogram was measured from the same data sampled at a different temporal resolution determined by the analog to digital (A/D) conversion rate. Short closed durations are found when the data are sampled at fine resolution for short times, and long closed durations are found when the data are sampled at coarse resolution for longer times. Thus, as noted by the horizontal time scale, the numerical values of the histograms are very different. However, all the shapes of the histograms are similar. These channel data are statistically self-similar in time. The short and long durations are linked together by a fractal scaling. The information about the channel is not contained in any one histogram, but in the scaling function that characterizes how the histograms change when they are measured at different time scales. That scaling is characterized by the effective kinetic rate constant, k_{eff}, which is the negative of the value of the initial slopes indicated by the straight lines.

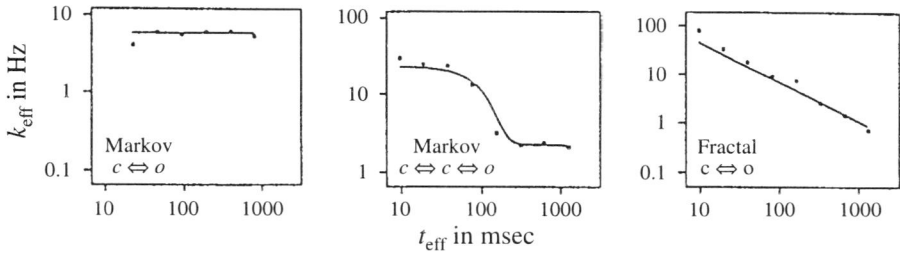

Figure 8.6. Different processes have clearly different forms when the logarithm of the effective kinetic rate constant, k_{eff}, is plotted versus the logarithm of the effective time scale, t_{eff}. Shown above are plots of the effective kinetic rate constant for leaving the closed state calculated from analytic solutions (lines) and numerical simulations (squares) of three models: (a) closed $\Leftrightarrow$ open Markov process, (b) closed $\Leftrightarrow$ closed $\Leftrightarrow$ open Markov process, and (c) a simple fractal process. The Markov models with discrete states have plateaus equal to the number of closed states, while the simple fractal model has a continuous power law scaling that appears as a straight line on these log-log plots, suggesting that there are a large number of "discrete" states differing only a little between states. (Adapted from Liebovitch et al., 1987a, Fig. 6.)

The experimental data are analyzed by plotting the *logarithm of the effective kinetic rate constant, k_{eff}, versus the logarithm of the effective time scale, t_{eff}*. These plots are very useful because different processes have clearly different shapes on such plots, as shown in Fig. 8.6. Previously, ion channel data had been analyzed by choosing models *a priori*, fitting the mathematical forms of those models to the data, and then comparing the goodness of fit of those models. However, we can evaluate $k_{eff}(t_{eff})$ from the data *without making any assumptions dictated by a model*. Thus, by constructing these plots and seeing what they look like, *we may discover functional forms for $k_{eff}(t_{eff})$ that we might not have anticipated*, and these may suggest new types of models to consider.

Scalings with different functional forms for $k_{eff}(t_{eff})$ can be defined with adjustable parameters, and then the methods described above can be used to fit these models to the data and to determine the values of the adjustable parameters. However, as explained above, the power of this fractal analysis is that it aids us in finding new and unanticipated functional forms in the data. Thus, rather than fitting assumed functional forms by the quantitative determination of their adjustable parameters, we emphasize the qualitative interpretation of the different forms that are revealed on the plots of $\log k_{eff}$ versus $\log t_{eff}$.

The kinetics of some channels has a simple fractal form. If the plot of $\log k_{eff}$ versus $\log t_{eff}$ is a *straight line*, as in the right panel of Fig. 8.6, then the data can be described by a *simple fractal scaling*. As seen in Fig. 8.7, Liebovitch et al. (1987a) found that the data from a potassium channel in corneal endothelial cells have this form, namely

$$k_{eff}(t_{eff}) = A \, t_{eff}^{1-D} . \tag{8.5}$$

The values of D and A can be determined from the slope and intercept of a straight line fit to the data on a plot of $\log k_{eff}$ versus $\log t_{eff}$. The fractal dimension, D, is equal to one minus the slope of that line, and $\log A$ is equal to the intercept. The best fit can be determined by the least squares method of minimizing the square of the deviations between the line and $\log k_{eff}$ of the data. This method assumes that the deviations between the line and the data have a normal distribution. Another useful method to fit a straight line to data is to use the Hodges-Lehmann estimators, which are the median slope and the median intercept of all pairs of the logarithms of the experimental points (Hollander and Wolfe, 1973). This nonparametric method does not assume that the deviations have a normal distribution and therefore the values of the slope and intercept determined by this method are less biased by outliers (data points with unusually large deviations).

The justification for identifying D as a fractal dimension can be understood by considering the measurement of the total length, L, of the trace of current as a function of time as it might appear on the face of an oscilloscope, or as on a recording over time. The effective time scale t_{eff} corresponds to the resolution at which the length of the current trace is measured. If we use a ruler of length r to measure the current trace, then $r = t_{eff}$. Each opening or closing increases the length of the trace. Thus, the length is proportional to the frequency of open to closed transitions, which is proportional to k_{eff}. If a graphical recorder is run at a slow speed, or has a thick pen, then multiple openings and closings over very short intervals will not be distinguished from a single opening and closing, which is to say

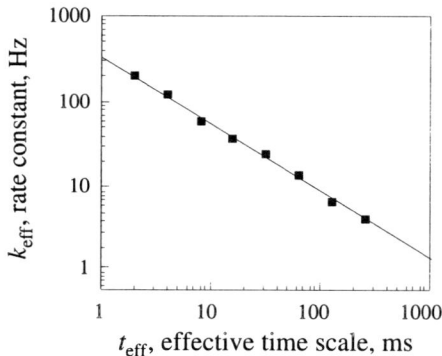

Figure 8.7. The logarithm of the effective kinetic rate constant versus the logarithm of the effective time scale determined from the closed time histogram of a potassium channel in the apical membrane of a corneal endothelial cell (Liebovitch et al., 1987a). The straight line on this log-log plot indicates that these data have a simple fractal scaling that can be described by $k_{eff}(t_{eff}) = A \, t_{eff}^{1-D}$.

that t_{eff} is a function of the recorder resolution. Since, k_{eff} is proportional to t_{eff}^{1-D}, the total length of the current trace when measured at resolution r is given by

$$L(r) = B\, r^{1-D},\tag{8.6}$$

where B is a constant. Eq. 8.6 is analogous to the measurement of the length L of a coastline at scale r, where D is the fractal dimension, as described in Chapter 2.

When the channel data have this fractal form, Eqs. 8.4 and 8.5 can be used to show that $P(t)$, the cumulative probability that the duration of the open (or closed) state is greater than t, is given by

$$P(t) = exp\{[-A/(2-D)]\, t^{2-D}\},\tag{8.7}$$

and the probability density $f(t) = -dP(t)/dt$ that the duration of the open (or closed) state is greater than t and less than $t + dt$, which is used to fit the open (and closed) time histograms, is given by

$$f(t) = A\, t^{1-D}\, exp\{[-A/(2-D)]\, t^{2-D}\}.\tag{8.8}$$

The component of Eq. 8.7 in the curly braces is a form that occurs in many fractal objects and processes, has been rediscovered many times, and is known by many different names, including *stretched exponential*, Weibull distribution, and Williams-Watts law (Klafter and Shlesinger, 1986). As shown in Fig. 8.8,

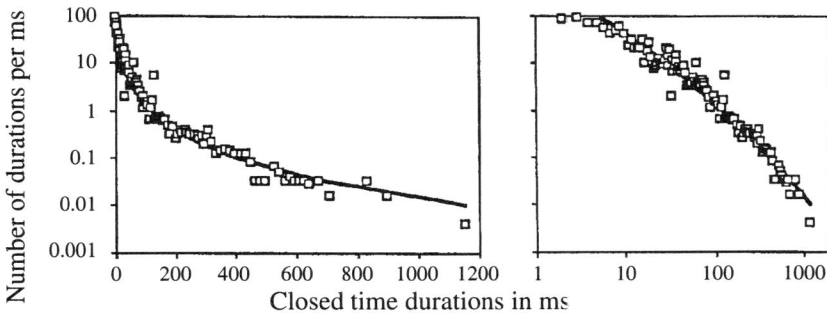

Figure 8.8. Closed time histograms on a semilog plot (*left*) and log-log plot (*right*) from a potassium channel in the apical membrane of a corneal endothelial cell (Liebovitch et al., 1987a). The line shown is based on the fit of a probability density function $f(t)$ that has the form $f(t) = At^{1-D}exp\{[-A/(2-D)]t^{2-D}\}$ characteristic of a simple fractal scaling.

Liebovitch et al. (1987a) found that the data from a potassium channel in corneal endothelial cells have this form. When the fractal dimension D is approximately 2, then Eqs. 8.7 and 8.8 have the power law form

$$P(t) \propto t^{-A}, \text{ and} \tag{8.9}$$

$$f(t) \propto t^{-A-1}, \tag{8.10}$$

which has been observed for some channels, as shown in Fig. 8.9.

The kinetics of some channels does not have a simple fractal form.

If the plot of $\log k_{eff}$ versus $\log t_{eff}$ consists of a set of a few *well-separated plateaus*, as shown in Fig. 8.6(b), then the channel has *a few, well-resolved, discrete states*. None of the approximately dozen channels that have been analyzed in this way has shown such a set of clearly defined, well-separated plateaus.

Even though well-separated plateaus are not found, the plot of $\log k_{eff}$ versus $\log t_{eff}$ for some channels is a *not straight line*, and thus cannot be characterized by a *simple* fractal scaling. However, if this plot still has a relatively simple functional form, then it can be characterized by a *complex fractal scaling*. Such complex scalings can be represented by models with additional substates (Starace, 1991) or multiple fractal dimensions (Dewey and Bann, 1992).

The plot of $\log k_{eff}$ versus $\log t_{eff}$ for some channels, as illustrated in Fig. 8.10, has a straight line at short time scales and a plateau at longer time scales (French and Stockbridge, 1988; Stockbridge and French, 1989; Liebovitch, 1989a). The corresponding form in the open (and closed) time histograms of a power law form at

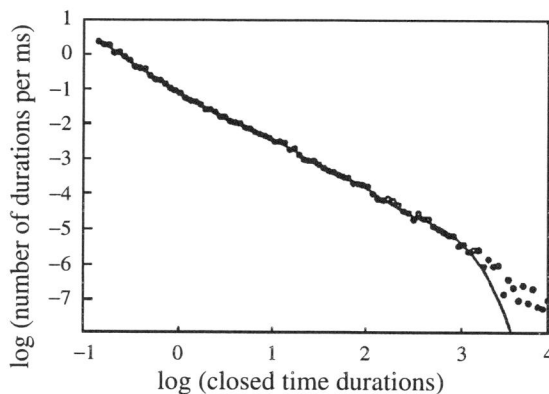

Figure 8.9. Closed time histogram on a log-log plot from a potassium channel in cultured neuroblastoma x glioma hybrid cells (McGee, Jr., et al., 1988). The probability density is a straight line that has the power law form $f(t) \approx t^{-\alpha}$ that is characteristic of a simple fractal scaling. (From McGee, Jr., et al., 1988, Fig. 9.) The continuous line is a model composed of the sum of six exponentials.

short times and an exponential form at long times has also been found (Teich, unpublished). Thus, the kinetics of these channels can be described by a complex scaling consisting of a *mixed model that has a simple fractal scaling at short time scales and one discrete state at long time scales*. This form was not anticipated by any theoretical model. As described in above, the kinetic analysis done until now has relied on determining the adjustable constants of the mathematical forms of *known* models and thus could not have recognized this new form in the data. This demonstrates the ability of the fractal analysis to reveal new forms in the data.

French and Stockbridge (1988) and Stockbridge and French (1989) found that the potassium channels in fibroblasts have this form of a simple fractal scaling at short time scales and a discrete state at long time scales. The open and closed times of this channel depend on the voltage applied and the calcium concentration in the solution, which they found can be characterized by a change *only* in the value of the plateau of the discrete state. On the other hand, to fit the same data exclusively with a Markov model required three closed states and two open states, and the many kinetic rate constants of that model showed no consistent trends with voltage or calcium concentration. The ability of the $\log k_{eff}$ versus $\log t_{eff}$ plot derived from a

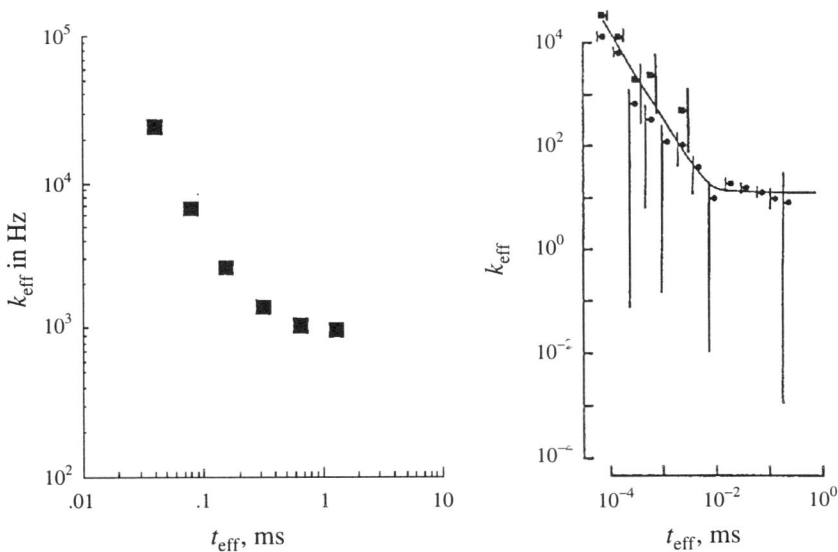

Figure 8.10. The logarithm of the effective kinetic rate constant versus the logarithm of the effective time scale determined from the open time histograms from two channels. The effective kinetic rate constant of these channels has a simple fractal scaling at short time scales (slanted straight line) and a discrete state (plateau) at long time scales. *Left:* Data from a potassium channel in cultured pituitary cells recorded by Korn and Horn (1988) and analyzed by Liebovitch (1989a). *Right:* Data from a cation channel in cultured chick fibroblasts recorded and analyzed by French and Stockbridge (1988). (*Left:* adapted from Liebovitch, 1989a, Fig. 1; *right:* from French and Stockbridge, 1988, Fig. 2.)

fractal viewpoint to characterize these data in such a simple way suggests that this approach, rather than the Markov approach, may be more likely to lead to an understanding of the mechanism of the voltage and calcium sensitivity.

The plot of $\log k_{eff}$ versus $\log t_{eff}$ for some channels is a *complex fractal scaling* consisting of segments of *approximate simple fractal scalings* and *approximate discrete components* with different intensities at different time scales. The corresponding segments of the open (and closed) time histograms are *approximately power laws* of the form $t^{-\alpha}$, or *approximately exponentials* of the form $e^{-\alpha t}$, where t is the open (or closed) times. There is a continuum of different forms of these histograms, from 1) those with nearly exact single power law forms, such as the potassium channel from cultured nerve cells shown in Fig. 8.9, 2) those with overall approximate power law scalings weakly perturbed at some time scales by approximate exponential forms, such as the chloride channel from cultured muscle cells shown in Fig. 8.11, to 3) those with weak power law scalings dominated by multiple approximate exponential forms, such as the acetylcholine-activated sodium channel from cells cultured from a mouse brain tumor shown in Fig. 8.12. The latter figure also illustrates the observation by Oswald et al. (1991) that the histograms of ligand gated channels are dominated by exponential forms at low ligand concentrations and a single power law form at high ligand concentrations.

Physical Interpretation of the Fractal Model

The simple and complex fractal scalings found from the open and closed times of some channels suggest that the ion channel protein has a different set of physical properties than those assumed by the Markov model.

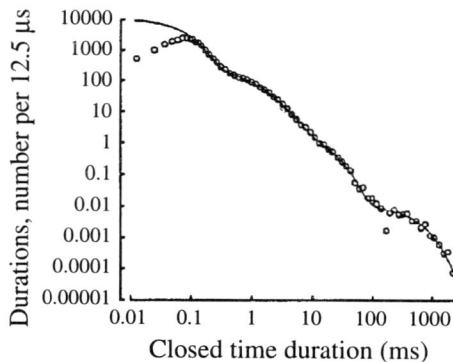

Figure 8.11. Closed time histogram on a log-log plot from a chloride channel in rat skeletal muscle (Blatz and Magleby, 1986). Over the range of closed times t, an overall approximate power law scaling $t^{-\alpha}$ (straight line) is perturbed at some time scales by approximate exponential components $e^{-\alpha t}$ (bumps). This might be seen as a modulated power law relationship. (From Blatz and Magleby, 1986 Fig. 6a.)

1. The effective kinetic rate constant describes how the open and closing probabilities scale with the time scale at which they are evaluated. In many channels this scaling is a smooth, slowly varying function. This implies that the channel protein has a continuous sequence of nearly identical stable conformational shapes that are separated by small activation energy barriers. That is, the channel protein has a *very large number of similar conformational states* corresponding to *shallow, local minima* in the potential energy function.

2. In some channels the scaling described by the effective kinetic rate constant has a simple power law form. In other channels this scaling may be

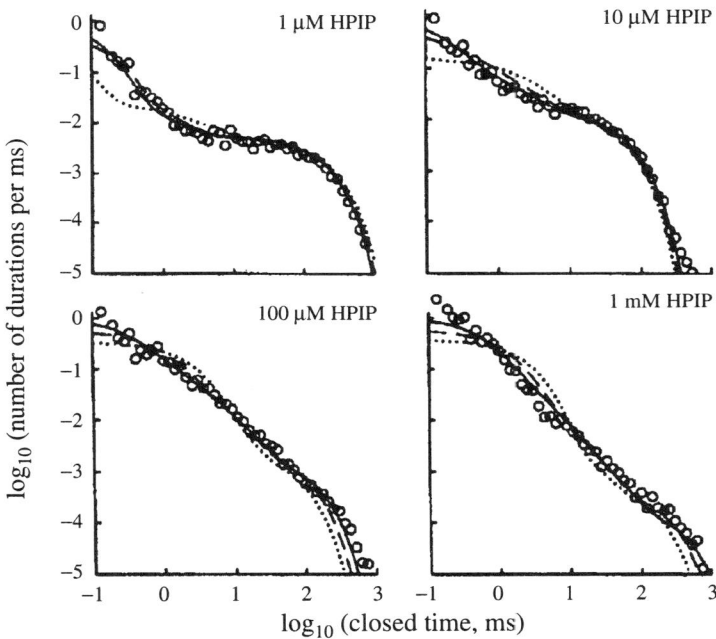

Figure 8.12. Closed time histograms on log-log plots from an acetylcholine-activated sodium channel, at different concentrations of an acetylcholine agonist (HPIP), in cells cultured from a mouse brain tumor (Oswald et al. 1991). There are both approximate power law components $t^{-\alpha}$ (straight lines) and approximate exponential components $e^{-\alpha t}$ (bumps) at different time scales t. The relative strength of the power law and exponential components depends on the agonist concentration. As the concentration of the agonist increases, the histogram approaches the straight line form of a power law of a simple fractal scaling. The dotted, dashed, and continuous lines represent increasing numbers of available states (4, 10, and 25, respectively) in a multistate "expanded diffusion" model fitted to the data. The model with the highest number of states gives the best fits. (From Oswald et al., 1991, Fig. 1.)

approximate or it may be more complex. The existence of these scalings indicates there is a relationship between dwell times of different durations. That is, there is a relationship between short and long times spent in the open (and closed) states. Short dwell times arise primarily when the transition from one conformational state to another passes over a small energy barrier, while long dwell times arise primarily when the transition passes over a large energy barrier. Since the times spent in each conformational state are related, the energy barriers separating these states are related.

3. Since there is a relationship between the energy barriers, there is a relationship between the physical processes that correspond to those energy barriers. Thus, the fractal model implies that *different physical processes* in the channel act together as a *cooperative, dynamic whole* to open or close the channel. This contrasts with the Markov model, which assumes that physical mechanisms in different parts of the protein do not influence each other and thus their kinetic rate constants are independent.

The relationship between these energy barriers can be interpreted in two different ways, which Dewey and Spencer (1991) call "dynamic" and "structural":

1. In the *dynamic* interpretation only one closed and one open conformational state, separated by an energy barrier, exist. The height of this energy barrier varies in time. The time variation of the energy barrier determines the scaling described by the effective kinetic rate constant (Liebovitch et al. 1987a).

2. In the *structural* interpretation there is a landscape of local energy minima and energy barriers between them. Each conformational state consists of many similar, but not identical conformational substates corresponding to a set of local minima in the energy landscape. The distribution of energy barriers, that is, the number of barriers of each energy, between the open conformational substates and the closed conformational substates, determines the scaling described by the effective kinetic rate constant (Liebovitch and Tóth, 1991).

These two interpretations are related to each other. We can think of the fixed energy surface of the structural model as corresponding to the set of still frames of the motion picture of the time-varying energy surface of the dynamical model.

Thus, the fractal description implies that channel proteins have *energy barriers that vary in time and/or many states separated by activation energy barriers that are related to each other.*

The open and closed histograms always consist of certain forms. The entire histogram may consist of only one of these forms, or it may consist of several of these forms, each at a different time scale. The observed forms are single exponentials $\exp(-at)$, stretched exponentials $\exp(at^b)$, and power laws $t^{-\alpha}$. Liebovitch and Tóth (1991a) showed that all these forms can be described by a progression that depends on a single parameter.

1. In the *dynamic* interpretation, these forms represent a progression in how the activation energy barrier between the open and closed stares changes in time

(Liebovitch et al., 1987a; Liebovitch and Sullivan. 1987). In all cases (except for the single exponential), the longer the channel remains in a state, the less the probability per second that it exits that state. It is as if the channel structure becomes increasingly stable with time and it becomes ever more difficult for it to acquire the energy needed to exit that state. When the energy barrier between states is constant in time, then the single exponential form results. When the energy barrier increases slowly with time, then the stretched exponential form results. When the energy barrier increases fastest in time, which is inversely proportional to the time already spent in the state, then the power law form results.

2. In the *structural* interpretation, these forms represent a progression of different types of distributions of energy barriers between the open conformational substates and the closed conformational substates (Liebovitch and Tóth, 1991). When there is a single conformational substate and a unique energy barrier to exit it, then the single exponential form results. When there are a number of conformational substates and a distribution of energy barriers to exit them, then the stretched exponential form results. When there is a very large number of conformational substates and a very broad distribution of energy barriers to exit them, then the power law form results. Thus, the progression from single exponential to stretched exponential to power law forms corresponds to a progression from a very narrow, to an intermediate, to a very broad distribution of energy barriers.

The Markov model assumes that the energy barriers between states are *constant* in time and occur *only* at discrete energies rather than being spread over a distribution of energies. Thus, in the dynamic interpretation, the less the time variation of the energy barriers, the more the kinetics is like a discrete Markov state. In the structural interpretation, the narrower the distribution of energy barriers, the more kinetics is like a discrete Markov state. As the time variation or the spread of energy barriers increases, the kinetics become less like a discrete Markov state and increasingly fractal in character.

In the dynamic interpretation physical processes in the channel can occur over a narrow or a broad range of time scales. In the structural interpretation physical processes can occur over a narrow or a broad range of energy barriers. A similar pattern occurs in many systems, such as the clearance curves of the decay of concentration of a chemical in the blood (Wise and Borsboom, 1989) and the contours of geological regions where "certain processes do operate only at certain discrete scales (e.g., glaciation, the formation of volcanoes) whereas others (e.g., wind and water erosion) operate at a wide range of scales" (Burrough, 1989).

8.4   Comparison of Markov and Fractal Models

The Markov model with a small number of states has been used to interpret data from ion channels for the last forty years. The fractal approach is new and its development has challenged many of the notions of how to analyze and interpret ion channel data. At this time many issues are not yet resolved, and so the comparison between the Markov and fractal approaches presented here will change as new discoveries are made. In this section we first review how each model interprets the experimental data in a different way. Then we describe the statistical tests used to compare each model to the patch clamp data. Lastly, we show how the physical properties of proteins determined by other biophysical techniques compare to the physical properties of the ion channel protein assumed by each model.

The Physical Interpretation of the Markov and Fractal Models

A comparison of the Markov and fractal interpretations is shown in Fig. 8.13. The physical interpretation of the Markov model is that the ion channel protein has 1) a *few, discrete conformational states* corresponding to the well-defined, *deep, local minima* in the potential energy function, 2) *sharply defined energy barriers* separating these states, 3) transitions between states that are due to *independent physical processes* that *do not interact with each other*, and 4) a *potential energy structure that is constant in time*.

  The physical interpretation of the fractal model is that the ion channel protein has 1) *a broad continuous distribution of many, similar conformational states* corresponding to *shallow, local minima* in the potential energy function, 2) *broad distributions of energy barriers* separating these states, 3) transitions between states that are *linked* together, and 4) a *potential energy structure that varies in time*.

Markov and Fractal Models Give Different Interpretations of the Same Data

The Markov and fractal models can fit the same experimental data. However, each model interprets the data in a different way. For example, consider the histogram of closed durations shown in Fig. 8.8. *Few-state Markov interpretation:* These data can be fit by Eq. 8.1 with three exponential terms (Korn and Horn, 1988). Each exponential term appears as an approximate straight line segment on this semilog plot when each rate constant is several-fold different from the others. Hence, to fit Eq. 8.1 to the data requires one such approximate straight line segment to fit the shortest times, one for the middle times, and one for the longest times. Thus, there must be three stable, discrete closed conformational states of this channel. *Fractal interpretation:* As shown in Fig. 8.7, the plot of the logarithm of the effective kinetic rate constant versus the logarithm of the effective time scale for these data is a straight line, indicating that it can be described by a simple fractal scaling (Liebovitch et al., 1987a), and hence the histogram of closed durations is well fit by Eq. 8.8. Thus, there must be many conformational states, due to time varying or parallel activation energy barriers, and the dynamic processes connecting these states are not independent, but are linked together in a way that produces a simple fractal scaling.

Few-state Markov

few, sharp, static, independent

Continuum fractal

many, broad, variable, linked

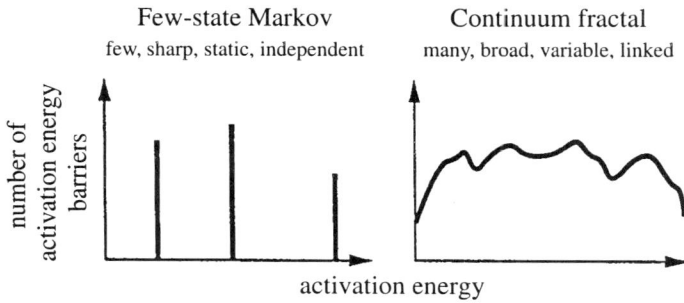

number of
activation energy
barriers

activation energy

Figure 8.13. The Markov and fractal models imply different physical properties about the ion channel protein. Some of the differences between those models are summarized here in a qualitative plot of the distribution of the number of energy barriers of a given energy that separate different conformational states. *Left:* The Markov model implies that there are a small number of such energy barriers, that each barrier is very sharp (a delta function), that these barriers remain constant in time, and that the parameters that describe them are independent. *Right:* The fractal model implies that there are many such energy barriers, that each barrier has a width, that these barriers may vary in time, and that the many barriers are linked together by a scaling.

Statistical Comparison of Markov and Fractal Models

We try to determine the true nature of a system through statistical comparisons of theories and data. The true nature of the system is most likely to have the properties of the theoretical model that most closely matches the experimental data.

Since the Markov and fractal models imply different physical properties for the ion channel protein it seems reasonable to determine the true nature of the channel by finding which theoretical model most closely matches the experimental data. Unfortunately, this *cannot* be done because the Markov and fractal models have an unusual property that makes this procedure fail. The problem is that both these models can have an arbitrary number of adjustable parameters. Thus, each model can fit the experimental data to unlimited accuracy by simply increasing the number of adjustable parameters. For example, the fit of Markov models to the data can always be improved by adding additional adjustable parameters corresponding to additional states. The first fractal model had only a simple fractal scaling with two adjustable parameters (Liebovitch et al., 1987a, 1987b). However, for fractal models with more complex scalings, the fit to the data can always be improved by adding additional adjustable parameters corresponding to additional states (Starace, 1991) or multiple fractal dimensions (Dewey and Bann, 1992) or more complex time dependencies of the energy barriers (for the dynamic interpretation) or more complex distributions of energy barriers (for the structural interpretation). Thus, the goodness of fit of these models to the data depends on the number of adjustable parameters used rather than on the intrinsic validity of the model. Hence, we cannot use the goodness of fit to determine which model is the most valid.

Since both the Markov and fractal models with enough adjustable parameters can be fit to any data, they can also be related to each other as well. For example, Millhauser et al. (1988a, 1988b), Condat and Jäckle (1989), and Liebovitch (1989b), have shown how simple fractal models can be approximated by Markov models with a large number of states.

Several papers have compared the simplest fractal model with that of Markov models having an unlimited number of states (McManus et al., 1988; Korn and Horn, 1988; Horn and Korn, 1989; McManus et al., 1989; Sansom et al., 1989). It was found that Markov models with an unlimited number of adjustable parameters are a closer fit to the data from some channels than a fractal model with two adjustable parameters. However, that result does not justify the conclusion that "by such statistical analysis we have provided support for the Markov model of channel gating" (Sansom et al., 1989), because almost any model with an unlimited number of adjustable parameters, valid or invalid, may be fitted to experimental data more closely than a two-parameter model. To overcome this problem, these papers used methods to compare the goodness of fit of models with different numbers of parameters. However, those methods have been shown to be invalid (Liebovitch and Tóth, 1990c). In two reports that compare Markov and fractal models that have the same number of adjustable parameters, the fractal model was a closer fit to the data than the Markov model (Liebovitch et al., 1987b; Liebovitch, 1989a).

The Markov model implies that the ion channel protein has only a few stable conformational states while the fractal model implies that it has a relatively large number of conformational states and/or that the energy barriers between these states vary in time. It has been shown that the goodness of fit of Markov models to the open and closed time histograms does not improve when there are more than a few discrete states (McManus et al., 1988; Korn and Horn, 1988; Horn and Korn, 1989; McManus et al., 1989; Sansom et al., 1989). This has been used to support the idea that the channel protein has *only* a few stable conformational states. However, finding that more than n states does not improve the goodness of fit does not imply that there are only n states; rather, it implies that there are n or more states. Since models with few states and models with many states fit the data equally well, one cannot conclude that channels have either few states or many states (Liebovitch and Tóth, 1990c).

There are differences of opinion as to whether physical insight is better achieved by fitting the salient features or the small-scale details of the data. For most scientists the nearly exact power law form in Fig. 8.9, and the approximate power law form in Fig. 8.11, suggest that we search for the physical mechanism responsible for such a striking and unexpected relationship. On the other hand, those who prefer to emphasize the few-state Markovian description of the processes look upon the data of Fig. 8.11 as an opportunity to identify the many parameters required to fit the several "bumps" on the curve rather than to regard the data as giving approximately a straight line on this plot. This particular channel thus provides support for both points of view, potentially satisfying both "lumpers" (preferring the broader, simpler fractal view) and the "splitters" (preferring the particular detailed several-state or Markovian view), or perhaps satisfying neither.

What Criteria Should We Use to Decide Which of Two Competing Interpretations Is Valid?

In the previous section it was assumed that progress in science is helped by finding the model that best fits the data. This may not be true, as stated by Machacek (1989):

> The standard model of *planetary motions—the epicycle theory—*has enjoyed an enormous amount of success. Indeed, it appears to be consistent with all established *astronomical observations.* This being the case, the first question one should ask is why one should even be looking for something better. Most criticisms of the standard model are based on the fact that it requires a number of arbitrary choices and fine-tuning adjustments of parameters. These features do not prove that it is wrong or even incomplete. However, given the history of successes in *mathematical astronomy,* it is natural to seek a deeper underlying theory that can account for many of the arbitrary choices of parameters. These include the *choice of equans, the number of epicycle generations,* and the specific values of various parameters.
>
> Ptolemy's epicycle theory could, within the accuracy then obtainable, correctly reproduce and predict the results of observations. What is more important, it could even be improved by adding another generation of epicycles now and then and fine-tuning the parameters. From a positivist's point of view the theory was quite a success.
>
> And yet it was fundamentally in error—not because of its aesthetic shortcomings, but because it could not serve as the first link in the chain that led to Newton's laws. Copernicus's theory, on the other hand, could, despite all its residual epicycles. Without the Copernican world picture, Kepler would not have been able to calculate the trajectory of Mars and formulate his laws, even if he had used today's observational data instead of Brahe's. Note that Copernicus himself did not need any new data in proposing his theory; he was even somewhat hampered by too accurate observations, which were showing deviations from uniform circular movement about the sun.
>
> . . . More experiments with a lot of accurate data do not necessarily mean progress in physics. They could, instead, perpetuate old theories, just refining their free parameters and adding new generations of something now and then. This danger is even greater today, when the data are screened by computers; these are looking for what we want to find and tend to mask the rest . . .
>
> There is, unfortunately, no *a priori* way of telling Copernicus-type theories from Ptolemy-type ones. As physical theories cannot be

proved, the only evidence we can have for any of them is in events compatible with it. The strength of such evidence varies and there is no objectivity in assessing it.

The Markov model of ion channel kinetics has been widely used. It can be made to fit the patch clamp data by adding as many adjustable parameters as needed. The fractal interpretation is new. It will have been proven worthwhile if it leads to a better understanding of the physics and chemistry of how an ion channel protein switches from one conformational state to another. *The chief new insight of the fractal approach is to interpret the open and closed time histograms in terms of the time variation of the energy barrier between the open and closed conformational states of the channel protein, or in terms of the distribution of energy barriers between the open and closed conformational substates.*

The Physical Properties of Proteins Implied by the Fractal Model and by the Markov Model

The difficulties in using the patch clamp data to differentiate between the physical properties implied by the Markov and fractal models suggest that we should use other methods in order to determine the nature of the ion channel protein.

First we consider other types of electrophysiological data, such as the net current recorded from a single cell. These measurements can be well described by the Hodgkin-Huxley model. Some of the assumptions of the Hodgkin-Huxley model are essential in reproducing these phenomena, while other assumptions are not essential. An essential assumption is that the ionic conductances of potassium and sodium are nonlinear with voltage. That is, once the voltage changes, the changes in these conductances force the voltage to change even more. This underlies the generation and propagation of an action potential. On the other hand, it is not essential to assume that the kinetics is a Markov process. The nonlinearity assumption alone can produce the action potential phenomena. This was demonstrated by Fitzhugh (1961) and Nagumo et al. (1962), who showed that mathematical models with *continuous functions* have the same properties as models with *discrete Markov* functions. Thus, the existence of a few discrete *Markov* states *cannot be justified* on the *electrophysiological* data alone. We must look to other experimental techniques to determine the nature of the ion channel protein.

Ion channels are membrane-bound proteins. Because technical problems make it more difficult to study membrane proteins than globular proteins, much less is known about the physical properties of membrane proteins than of globular proteins. For example, globular proteins can be isolated and crystallized, which makes possible X-ray diffraction that can be analyzed to solve the three-dimensional spatial structure of the protein. Membrane proteins that have some regions surrounded by hydrophobic lipids and other regions surrounded by hydrophilic salt solutions are much more difficult to extract and crystallize. Nuclear magnetic resonance (NMR) can be used to determine the spatial structure of small globular proteins in solution, but is not adequate to determine the spatial structure of the much larger membrane proteins. Lacking this structural information also limits simulation techniques, such as molecular dynamics, which require such structural information as a starting point. Hence, the physical properties of ion channel proteins are not yet known. However, it is likely that membrane proteins share many properties in common with globular proteins. Thus, we now review the properties of globular proteins and compare them to the properties implied by the Markov and fractal models.

The potential energy function of globular proteins has a *very large number of shallow, local minima* (Karplus and McCammon, 1981; McCammon and Harvey, 1987; Welch, 1986). This is *consistent* with the *fractal* model and *inconsistent* with the few, deep minima of the *Markov* model.

The time course of ligand rebinding to enzymes can be used to determine the distribution of energy barriers that the ligand must cross as it passes through the protein. Many such experiments, such as the rebinding of CO to myoglobin, demonstrate that these ligands pass through a *broad, continuous distribution of energy barriers* (Austin et al., 1975). The time course of fluorescence depends on the surrounding distribution of energy barriers. For example, fluorescence decay can be used to determine the distribution of energy barriers through which a tryptophan ring flips (Alcala et al., 1987). These experiments also demonstrate that there is a *broad, continuous distribution of energy barriers*. These distributions are *consistent* with the *fractal* model and *inconsistent* with the set of a few, discrete energies predicted by the *Markov* model.

A variety of techniques measure the fluctuations of structure within proteins (Careri et al., 1975; Karplus and McCammon, 1981). For example, the amount of blur of X-ray diffraction spots can be used to measure the fluctuations of conformational shape. The rate of exchange of tracers, such as deuterium, between the interior of the protein and the solution, and the interaction of fluorophores in the interior of the protein with fluorescent quenchers in the solution, depend on the fluctuations in structure that create transitory openings which allows these small tracers and quenchers to move between the interior and exterior of proteins. These experimental studies demonstrate that there are considerable and important *fluctuations in protein structure*. Molecular dynamic simulations also show that the *structure varies in time*. For example, the energy barrier determined from the static crystallographic structure of myoglobin is so large that oxygen should not be able to reach its binding site. However, small fluctuations in the structure open up a passageway for oxygen to reach the binding site (Karplus and McCammon, 1981), and the data on the rate of reoxygenation of hemoglobin after a sudden dissociation produced by a burst of laser light show a broad range of rate constants, consistent with there being a multitude of conformational states of the protein (Antonini and Brunori, 1971). This time dependence of the structure is *consistent* with the *fractal* model and is *inconsistent* with the static-energy structure of the *Markov* model.

Thus, these studies demonstrate that proteins have 1) a very large number of conformational states corresponding to many, shallow, local minima in the potential energy function; 2) broad continuous distributions of energy barriers; and 3) structure and energy barriers that vary in time. These properties are consistent with the physical properties of ion channels implied from the fractal model and are inconsistent with the physical properties assumed by the Markov model.

An important difference between the Markov and fractal models has not yet been tested. The potential energy function of channel proteins must consist of both deep minima and shallow minima. The Markov model implies that one can think of the shallow minima as small perturbations on the deep minima. In the fractal model, the existence of a fractal scaling implies that the shallow minima and deep minima are related and thus cannot be thought of independently. This means that the processes that correspond to the shallow minima are cooperatively linked to the processes that correspond to the deep minima. The distinction between these viewpoints was summarized by Frauenfelder (unpublished): "Do the little wheels have to move before the big wheels can move?" That is, the little and the big wheels correspond to the small and the large structural domains in the channel protein. If the small and large domains are functionally independent, that is, the small domains do *not* have to move before the large domains can move, then the Markov model is correct. On the other hand, if the small and large domains are functionally dependent, that is, the small domains *do* have to move before the large domains can move, then the fractal model is more appropriate. Of course, because a protein is of finite size, its behavior can be fractal over only a moderate range of time scales, and it cannot mimic a mathematical fractal ranging from zero to infinity.

8.5   Uncovering Mechanisms Giving Fractal Channel Kinetics

Mechanisms

The use of a fractal perspective to analyze the channel data is an important advance, because it has revealed *scalings* that were not previously appreciated. However, the fractal model is only a phenomenological description of the data. The next step is to use these scalings as clues to uncover the *mechanisms* responsible for them. Three types of models have been proposed as the cause of these scalings: 1) inherently random transitions between conformational states that are linked together in either space or time, 2) deterministic chaotic transitions between conformational states described by a small number of independent variables, and 3) deterministic transitions between conformational states described by a large number of independent variables.

Many-State Markov Models

Inherently random transitions between conformational states are linked together. The fractal description suggests that there are many kinetic processes and that these processes are linked together to generate the fractal scaling. This linkage can be "dynamical," resulting from the relationship that describes the time-dependent changes in the structure of the ion channel protein, or it can be "structural," resulting from the relationship between the conformational states or the transitions between them (Dewey and Spencer, 1991).

Dynamical models have been proposed that interpret the fractal scaling in terms of time-dependent changes in the physical structure of the channel. Examples of such models are: (1) Croxton (1988) proposed that the activation energy barrier between the open and closed states fluctuates slowly. The scaling depends on the time average of the activation energy barrier distribution, which he chose to be Gaussian. (2) Rubinson (1986) proposed that diffusion of charged components in the channel protein results in a broad distribution of energy barriers, which determines the scaling. (3) Agmon and Hopfield (1983) had modeled a globular protein as having a slow diffusion within nearly identical substates and rapid transitions between more distant major states. Levitt (1989) proposed a similar channel model with a slow diffusion within open and closed substates and rapid transitions between the set of open and closed states. The scaling is determined by the shape of the continuous potential energy functions that define the open and closed substates and the dynamics of the diffusion along these potentials, which is driven by the voltage applied across the channel.

Structural models have been proposed that interpret the fractal scaling in terms of the set of physical structures that the channel can have. They are Markov models that have many states, unlike the Markov models previously used to model channel data, which have only a few states. Examples of such models are: (1) Millhauser et al. (1988b) and Millhauser (1990) proposed that the helices of a channel protein twist back and forth in a random walk until they move far enough to suddenly knock the channel open or closed (Millhauser, 1990). This can be represented by an open state and a long sequence of many closed states. The scaling arises from the linear set of approximately equal kinetic rate constants that link the closed states. (2) Läuger (1988) and Condat and Jäckle (1989) proposed that a piece of the channel protein moves into and blocks the pore that the ions pass through. The probability that this piece removes itself from the pore depends on the constantly changing shape of the rest of the protein, which they modeled as pieces randomly moving between different locations in a grid. The many different possible locations of these pieces mean that the channel has many different states. The geometry of the grid and the transition probabilities of the moving pieces determine the scaling. (3) Liebovitch (1989b) and Liebovitch and Tóth (1991) proposed that there are many parallel pathways connecting many open and closed substates. The scaling is determined from the distribution of energy barriers between these open and closed substates. They derived the distributions of energy barriers that lead to fractal scalings with different fractal dimensions.

Chaotic Models

Models of channel gating had always assumed that the switching between conformational states was an inherently random event. However, as we showed in

Chapter 7, a chaotic deterministic system with a few independent variables can have such complex behavior that it mimics a random process. The motions in the channel protein could be chaotic if the fluctuations in the channel structure were due to the motion of a few large pieces of the channel. Two chaotic models of channel kinetics have been developed: (1) Liebovitch and Tóth (1991) proposed a chaotic model that predicted distributions of open and closed durations similar to that found in the channel experiments. This model was based on a mathematical model of an iterative map, which they were able to give a physical interpretation, namely, that the channel functions as a nonlinear oscillator that amplifies its own motions, perhaps through organizing its unperiodic thermal fluctuations into coherent motion. As soon as the channel enters a conformational state, its internal motions continue to increase, until it drives itself into another conformational state. The scaling is determined by the mathematical form of the iterative map. (2) Liebovitch and Czegledy (1991, 1992) proposed a model based on physical principles by using a potential energy function with two potential wells, corresponding to the fully open and fully closed states, to characterize the channel protein. The average acceleration of the motion within the channel is proportional to the sum of three forces: a velocity-dependent friction due to a viscosity caused by the relative motions within the protein, a force due to the potential energy function, and a driving force from the environment that induces chaotic motion in the system. The scaling depends on the forces and the potential energy function.

The open and closed time histograms of the simplest chaotic models have an approximately single exponential form rather than a power law fractal scaling. Fractal scalings are found only in chaotic models with additional states or time-dependent energy barriers. The increased structural or dynamic complexity required to produce fractal scalings suggests that the fluctuations in channel structure may be due to many, small interacting units, rather than the few, large units of a simple chaotic model.

Self-Organizing Critical System Models

Chaotic systems are deterministic and can have as few as three independent variables. Recently, there has been considerable interest in a different type of deterministic system, called a self-organizing critical system (Bak et al., 1987; Bak and Chen, 1991). These systems can produce fractal patterns in space and in time. They have many interacting pieces. Energy can be added randomly or deterministically to a location in the system. This energy accumulates locally, until it exceeds a critical value, and then it is dispersed to the neighboring locations. The rule that determines when and how this excess energy is relieved is a deterministic one. Because the energy is relieved only when it is over the critical value, the local strain is always just under that critical value everywhere in the system. Thus, this system is poised between stability and instability in a way that is very similar to a system at a phase transition poised between two different phases, such as a liquid and a gas. The properties of systems near such phase transitions scale as power laws with fractal characteristics.

A channel protein may be a self-organizing critical system. The channel protein consists of many pieces that interact with their neighbors. The energy added to the protein from the environment cause local strains that are spread throughout the structure. If these distortions spread faster than the time it takes for the structure to thermally relax, then the channel protein may be a self-organizing critical system. If that is the case, then the fluctuations in the channel structure will be due to a global organization of the local interactions between many small interacting pieces of the channel protein. The fractal scaling would then be due to the fact that the channel structure is poised at a phase transition between its open and closed conformational shapes. Models of ion channel proteins based on self-organizing critical systems are now being explored (Liebovitch and Czegledy, 1992).

8.6   Summary

Ions such as sodium, potassium, and chloride can cross the lipid bilayer that forms the cell membrane through the interior of protein ion channels that span the cell membrane. The ionic current through an individual channel can be resolved by the patch clamp technique. Thus, the kinetics, the sequence of the open and closed conformational states, of a single protein molecule can be studied.

Hodgkin and Huxley proposed a model of ion channels kinetics in the 1950s. This model was consistent with a considerable amount of electrophysiological data and the ideas of chemical kinetics at that time. The mathematical form of their model was based on a Markov process that assumes the channel protein has 1) *a few, discrete, stable conformational states*, 2) these states are *separated by significant, sharply defined energy barriers*, 3) the stable states and the energy barriers that separate them are *constant in time*, and 4) the parameters that characterize these states and thus determine their properties are *independent*.

We showed how *fractal concepts can be used to develop a new method to analyze channel data* in which the probability of switching between the open and closed states is evaluated at different time scales. This is done by determining the effective kinetic rate constant, k_{eff}, which is the conditional probability the channel changes states given that it has remained in a state for a sufficient time, t_{eff}, to be detected in that state. A plot of $\log k_{eff}$ versus $\log t_{eff}$ is very useful because different processes have different forms on such a plot. These plots demonstrate that many channels have simple or complex scalings characteristic of fractal behavior.

The implications of the existence of simple or complex fractal scalings are that the channel protein has 1) *a broad continuum of many conformational states*, 2) these states are separated by *continuous distributions of energy barriers*, 3) the states and the energy barriers that separate them may *vary in time*, and 4) the parameters that characterize these states are *related by a scaling*, so that *different processes are cooperatively linked together*.

The analysis based on fractal concepts reveals that different channels have different types of kinetics, and that one channel may have different types of kinetics at different time scales. These types cover a continuous range. At one end of this range the open (or closed) time histograms are nearly *exponential*. At the other end of this range the open (or closed) time histograms are nearly a *power law*. In the *dynamic interpretation* this range corresponds to different rates of change in time of the energy barrier between open and closed states. When the energy barrier changes very little in time, the histograms are single exponential. When the energy barrier rises rapidly with the time the channel has remained in a state, the histograms are power laws. In the *structural interpretation* this range corresponds to the breadth of the distribution of energy barriers between the open and closed conformational substates. When this distribution is very narrow the histograms are single exponential; when this distribution is very broad, the histograms are power laws.

The experiments have not yet been done to determine if ion channel proteins have the physical properties implied by the Markov or fractal models because of the experimental difficulties in studying membrane-bound proteins, such as ion channels. However, it is likely that membrane proteins share many properties in common with globular proteins, and many experiments and simulations demonstrate that globular proteins have physical properties that are *consistent* with the *fractal model* and *inconsistent* with the *Markov model*.

Future Directions

Most of the present analysis of patch clamp data has centered on the statistical analysis of the distributions of the closed and open durations. New information from such analysis seems limited. Thus, new types of statistical analysis are needed. For example, the higher-order correlation properties of the data need to be studied. Other nonstatistical approaches also need to be tried, such as methods from nonlinear dynamics. How the kinetics properties of each channel at a microscopic scale collectively contribute to the macroscopically measured currents through whole cells and tissues also needs to be developed.

Many channels have different fractal dimensions in different time regimes. These different dimensions correspond to different processes, which operate at either narrow or broad ranges of time scales. Many other systems have a similar form. An analysis tool is needed that can determine from the data the number of processes and their ranges of time scales.

The statistical properties of a protein switching from one state to another are very poorly understood. In fact, the statistical properties of any complex system switching between different states are very poorly understood. We must understand the properties of such behavior if we want to be able to interpret and understand the protein data. Thus, studies of model systems with such switching behavior are particularly urgent. These include chaotic systems, self-organizing critical systems, percolation systems (Stanley and Ostrowsky, 1986), and neural networks such as Boltzmann machines (Aarts and Korst, 1989).

At this time there is no complete description of the action potential of a neuron in terms of the fractal model. What is needed are descriptions of the time- and voltage-dependence of the conductances of at least the sodium and potassium channels over the full range of potentials that can be explored with membrane patch clamp and neuron voltage clamp techniques. This ought to be the most demanding test of the fractal concept. It may fail, for of all the channels that have been well characterized, these and the cardiac "slow inward" calcium channel are the ones that appear to have clearly defined few-state processes. A likely end conclusion is that some channels will be dominant in a few conformational states with a multitude of minor secondary states, while others will lack dominant states and appear more fractal.

Most channel data have been limited to electrophysiological experiments. Other types of biophysical measurements, such as nuclear magnetic resonance (NMR), would be more sensitive to motions within the channel, the number of conformational states, and the distribution of energy barriers. Such information is needed to resolve the controversy about the physical nature of the channel. The three-dimensional structure of channels is not well understood, because membrane proteins are difficult to crystallize, which is required for high-resolution X-ray diffraction studies. We await future success in crystallizing these channel proteins for X-ray diffraction studies, including perhaps time-resolved X-ray studies using high-intensity synchrotron sources.

Background References

Ion channels and the patch clamp technique are reviewed in books by Sakmann and Neher (1983) and Hille (1984). The fractal description of ion channel kinetics is detailed by Liebovitch et al. (1987a) and Liebovitch and Sullivan (1987), and the models proposed to explain the scalings seen in the data are reviewed by Liebovitch and Tóth (1990a, 1990b). The debate between the few-state Markov and fractal interpretation of channel data may be found in Liebovitch et al. (1987a), Liebovitch (1989a), Liebovitch and Tóth (1990a, 1990b, 1990c), French and Stockbridge (1988), Stockbridge and French (1989), McManus et al. (1988), Korn and Horn (1988), Horn and Korn (1989), McManus et al. (1989), and Sansom et al. (1989).

9

Fractals in Nerve and Muscle

... a circuit of nervous fluid, like electric fire.

Luigi Galvani (translated by Montraville, 1953)

9.1   Spread of Excitation

Mechanical vibrations, chemical concentrations, and action potentials are all forms of information that can be transmitted from one point in space to another by means of various mechanisms. We are familiar with the sound waves that enable us to hear the distant roll of thunder and light waves that enable us to see the lightning flash, but we are less familiar with the waves found in excitable media that enable us to transform these sensory inputs into messages usable by the brain. In practice, these latter waves result from the rhythmic timing of spatially distributed pacemakers such as found in the heart, intestine, kidney, uterus, and stomach. These pacemaker cells are nonlinear biological oscillators that are capable of spontaneous excitation, which can also be entrained by external excitation.

The generation and propagation of the excitation of nerve and muscle cells involves electrochemical processes localized in the membranes of those cells. The movement of the nerve pulse coincides with the movement of small ions, such as sodium, potassium, and chloride, into and out of, the nerve cell known as the neuron. The neuron is quite similar to other cells in that it contains a nucleus and cytoplasm. It is distinctive in that long, threadlike tendrils emerge from the cell body, and those numerous projections branch out into still-finer extensions. These are the dendrites that form a branching tree of ever more slender threads like a fractal tree. One such thread does not branch and often extends for several meters even though it is still part of a single cell. This is the axon, which is the nerve fiber in the typical nerve. Excitations (depolarization waves) in the dendrites usually travel toward the cell body, whereas in the axon they travel away from the cell body.

9.2   The Fractal Heart

Physiological fractals include the vascular tree, the tracheobronchial system, and the multiply-enfolded mammalian brain. But perhaps the most compelling evidence for

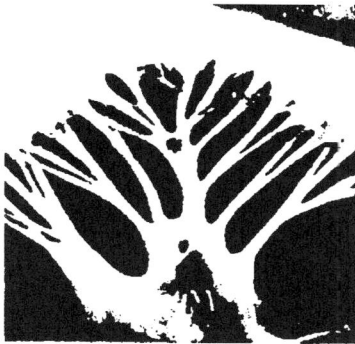

Figure 9.1. Chordae tendineae of heart valve. (Figure courtesy of Edwards, 1987, with the permission of the Mayo Foundation.)

the ubiquitous nature of fractal geometry is provided by cardiac anatomy, where fractals appear at all levels from the branching coronary vascular tree beginning on the epicardium to the trabeculated, irregular endocardial "coastline" of the ventricles (Goldberger et al., 1985a; West and Goldberger, 1987). The chordae tendineae appear to be fractal canopies (Fig. 9.1) tethering the AV value leaflets to the papillary muscles. Embedded within the myocardium is the His-Purkinje conduction system (Fig. 9.2), an irregular but self-similar dichotomous branching system of conduction tissue that gives rise to multiple generations of daughter branches on progressively smaller scales.

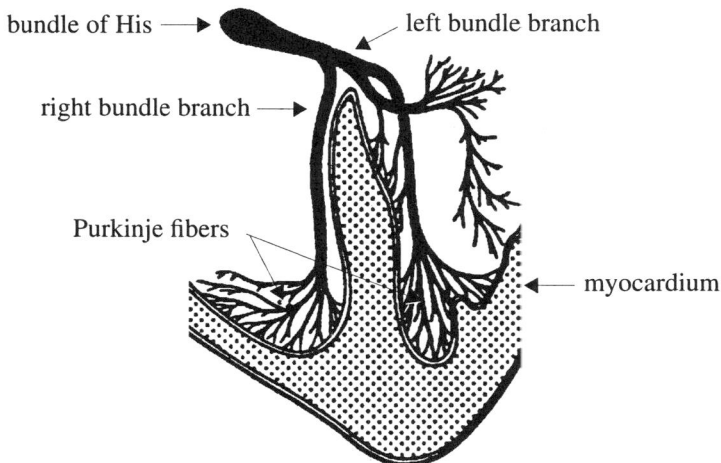

Figure 9.2. Irregular but self-similar pattern of branchings seen in the His-Purkinje network is characteristic of fractal structures. (From Goldberger et al., 1985.)

Cardiac Conduction and Fractal Dimension

What is the functional consequence of depolarizing the ventricles via a fractal conduction network? Normally, each heartbeat is initiated by a stimulus from pacemaker cells in the sinus node in the right atrium. The activation wave then spreads through the atria to the AV junction. Following activation of the AV junction, the cardiac pulse spreads to the ventricular myocardium through a ramifying network. The His-Purkinje conduction system is strongly reminiscent of the bronchial fractal we discussed in Chapter 3. In both structures, one sees a self-similar tree with finely scaled details on a "microscopic" level. The spread of this depolarization wave is represented on the body surface by the QRS-complex of the electrocardiogram. Spectral analyses of the QRS time trace reveals a broadband frequency spectrum with a long tail corresponding to an inverse power law in frequency. To explain the inverse power law spectrum Goldberger et al. (1985) have conjectured that the repetitive branchings of the His-Purkinje system represent a fractal set in which each generation of the self-similar segmenting tree imposes greater detail onto the system. At each fork in the network, the cardiac impulse activates a new pulse along each conduction branch, thus yielding two pulses for one. In this manner, a single pulse entering the proximal point of the His-Purkinje network within distal branches generates many pulses at the interface of the conduction network and myocardium. In a fractal network, the arrival times of these pulses at the myocardium are not uniform, because of the different lengths of the conduction network at different scales and because of the variations in the lengths at each scale. The effect of the richness of structure in the fractal network is to subtly decorrelate the individual pulses that superpose to form the QRS complex.

As we have discussed, a fractal network is one that cannot be expressed in terms of a single scale, e.g., a single average decorrelation rate in the present case. Instead, there will be an infinite series of terms needed to describe this decorrelation process. Each term in this series gives the probability of a higher decorrelation rate contributing to the overall process. The distribution of decorrelation rates in the time trace is in direct correspondence to the distribution of branch lengths in the conduction network. For such a fractal process, the distribution of decorrelation rates will take the form of an inverse power law. Furthermore, since the frequency spectrum of the resultant QRS complex depends on the statistics of the arrival times of pulses at the myocardium, the frequency spectrum of the QRS complex should also take the form of an inverse power law.

Goldberger et al. (1985a) demonstrated, using the standard Fourier analysis techniques, that the normal QRS power spectrum is consistent with the fractal depolarization hypothesis. Fourier analysis decomposes a waveform into its constituent frequencies and reveals how much strength each frequency component contributes to the time series. For example, if the QRS were a perfect sine wave its spectrum would consist of only a single frequency component, i.e., the frequency of the sine wave itself. If, on the other hand, the QRS were a narrow spike (pulse) (like the pacemaker pulse) the spectrum would reveal a white noise pattern, i.e., a broadband frequency pattern with approximately equal contributions from low and high frequencies. The normal QRS is of course neither a spike (pulse) nor a sine wave and its power spectrum shows a distinctive broadband pattern with a predominance of power in the low-frequency components, as well as a long, low amplitude tail of higher frequencies (> 100 Hz). Furthermore, if one replots these data in a log-log format on which an inverse power law distribution appears as a straight line having negative slope, an excellent fit is obtained (cf. Fig. 9.3).

Not surprisingly, inverse power law spectra may be characteristic of a variety of other apparently unrelated physiological processes (van den Berg et al., 1975; Musha et al., 1982; Kobayashi and Musha, 1982), for example, the beat-to-beat interval variations of the heartbeats in normal subjects with a superimposed spike at the breathing frequency (Kobayashi and Musha, 1982). This observation suggests that fractal mechanisms may be involved in the regulation of heartbeat variability, giving rise to self-similar fluctuations over multiple time scales (Goldberger and West, 1987a; West and Goldberger, 1987). The details of such a fractal regulatory system are as yet purely speculative but seem likely to involve a hierarchical ("treelike") neuro-humoral control network.

From a pathophysiological viewpoint, perturbation of a fractal control system should narrow the frequency response of the system, just as disruption of the fractal His-Purkinje system leads to narrowing of the QRS spectrum. This spectral shift, reflecting alterations in fractal scaling, may be of diagnostic and prognostic value.

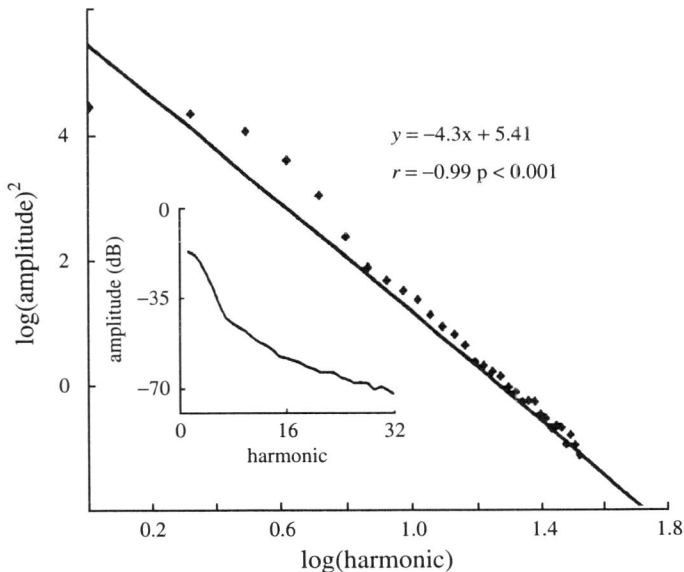

Figure 9.3. The power spectrum (inset) of normal ventricular depolarization (QRS) waveform (mean data of 21 healthy men) shows a broadband distribution with a long, low-amplitude high-frequency tail. Inverse power-law distribution is demonstrated by replotting the same data in log-log format. Fundamental frequency = 7.81 Hz. (Adapted from Goldberger et al., 1985.)

Loss of heart rate variability has already been described in numerous settings, including multiple sclerosis and diabetes mellitus (Neubauer and Gundersen, 1978), fetal distress (Kariniemi and Ämmälä, 1981), bed-rest deconditioning (Goldberger et al., 1986), aging (Waddington et al., 1979), and in certain patients at risk for sudden cardiac death (Majid et al., 1985). Spectral analysis may be useful in quantifying this loss of physiological variability. Presumably, the more severe pathologies are associated with the greater loss of spectral power, analogous to that seen with the most severe arrhythmias, which begin to resemble "sine wave" patterns. Some authors (Goldberger and West, 1987b) have referred to such narrowing of the frequency spectrum as a *loss of spectral reserve*. As noted, fractal scaling pathologies may be associated with a relative decrease in the high-frequency part of the spectrum, indicating the loss of the more responsive characteristics of this system. When plotting log power versus log frequency this loss would shift the regression line in a more negative direction.

9.3   Fractal Neurons

It has been known for some time that the activity of a nerve is based on electrical phenomena. Whether it is the external excitation of a nerve or the transmission of a message from the brain, electrical impulses are observed in the corresponding axon. As pointed out by Kandel (1979), because of the difficulty in examining patterns of interconnections in the human brain, there has been a major effort on the part of neuroscientists to develop animal models for studying how interacting systems of neurons give rise to behavior. There appear, for example, to be no fundamental differences in structure, chemistry, or function between the neurons and their interconnections in man and those, for example, in a squid, a snail, or a leech. However, neurons can vary in size, position, shape, pigmentation, firing pattern, and the chemical substances by which they transmit information to other cells. Here we are most interested in the differences in firing patterns, exemplified in Fig. 9.4. Certain cells are normally "silent," while others are spontaneously active. Some of the active ones generate regular action potentials, or nerve impulses, and others fire in recurrent brief bursts or pulse trains. These different patterns result from differences in the types of ionic currents generated by the membrane of the cell body of the neurons.

The rich dynamic structure of the neuronal firing patterns has led to their being modeled as nonlinear dynamical systems. In Fig. 9.4 the normally silent neuron can be viewed as a fixed point of a dynamical system, the periodic pulse train is suggestive of a limit cycle, and the erratic bursting of random wave trains is not unlike the time series generated by certain chaotic phenomena. This spontaneous behavior of the individual neurons may be modified by driving the neurons with external excitations.

Neurons have two different types of fractal properties. First, the geometric *shape of nerve cells may be fractal in space,* just as we observed for the cardiac conduction

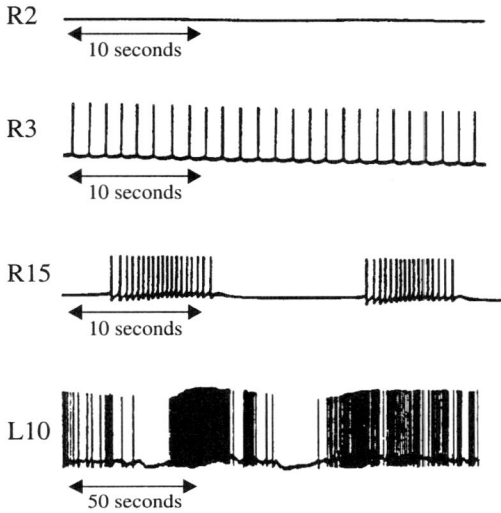

R2

10 seconds

R3

10 seconds

R15

10 seconds

L10

50 seconds

Figure 9.4. Firing patterns of identified neurons in *Aplysia's* ganglion are portrayed. R2 is normally silent, R3 has a regular beating rhythm, R15 a regular bursting rhythm and L10 an irregular bursting rhythm. L10 is a command cell that controls other cells in the system. (From Kandel, 1979, his Fig. 4., with the permission of *Scientific American.*)

system. The fractal dimension has been used to classify the different shapes of neurons and to suggest mechanisms of growth responsible for these shapes. Second, the *time intervals between the action potentials recorded from nerves may be fractal in time*, again as we observed for the interbeat interval distribution in cardiac time series. The statistical properties of these intervals seemed to have very strange properties. For example, collecting more data did not improve the accuracy of the statistical distribution of the intervals measured for some neurons as characterized by the width of the distribution, for example. The realization that these intervals are fractal helps us to understand these surprising properties. This understanding also aids us in designing experiments and in analyzing the data from them. The fractal dimension of these intervals can also be used to classify neurons into different functional types. In this section we discuss how fractals have been used to analyze the shapes of neurons and the timing of their action potentials.

Neural Branchings and Fractal Dimensions

Smith, Jr., et al. (1989) and Caserta et al. (1990b) analyzed the *shapes of the dendritic branching of neurons* growing in vitro *in culture* dishes or in the retina of the eye of a cat. They studied these types of neurons because they are growing as planar structures, so that the images represent the true shape of the neuron. Analysis of the two dimensional image photographed from a neuron that extends in three dimensions would be more difficult. Smith, Jr., et al. measured the perimeter of the neurons at different scales by: (1) using calipers of different sizes to determine the perimeter, (2) widening the border by a given scale size and then determining the perimeter as the total area covered divided by the scale size, and (3) using grids of different sizes and determining the perimeter as the number of grid boxes needed to cover the cell borders. Caserta et al. (1990b) measured: 1) the number of pixels in the neuron images within different radii and 2) the spatial correlations between pairs of pixels. Both groups found that *the shape of these neurons is self-similar.* That is, as shown in Fig. 9.5, they found that the logarithm of the properties measured were proportional to the logarithm of the scale size used for the measurement. The slope of this relationship determines the fractal dimension.

As shown in Fig. 9.6 *the fractal dimension can serve as a useful tool to characterize the shape of the neuron.* Smith, Jr., et al. (1989) found that the fractal dimension of cultured vertebrate central system neurons ranged from 1.14 to 1.60. Caserta et al. (1990b) found that the average dimension for 11 cultured chick retinal neurons was 1.41, and for 22 neurons in the cat retina was 1.68. It is important to

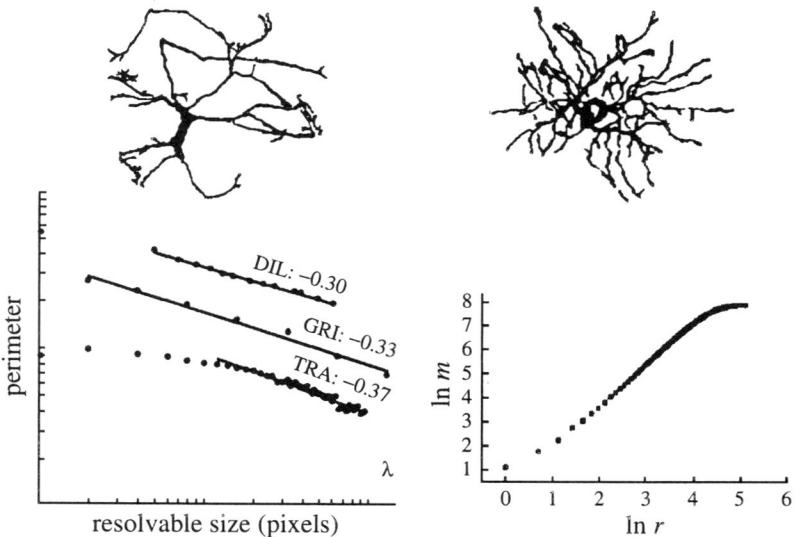

Figure 9.5. *Left:* Cultured neuron analyzed by Smith, Jr., et al. (1989). *Right:* Ganglion cell in cat retina analyzed by Caserta et al. (1990). At the top are shown the images of the neurons. At the bottom are shown how the measured perimeter (*left*) or number of pixels within a given radius (*right*) varies with the scale of the measurement. The straight line on these log-log plots indicates that the shape of the neurons is fractal. The fractal dimension is equal to one minus the slope on the plot at the left and equal to the slope of the plot on the right. (*Left:* Smith, Jr., et al., 1989, Fig. 4. *Right:* Caserta et al., 1990, Figs. 1a and 1c.)

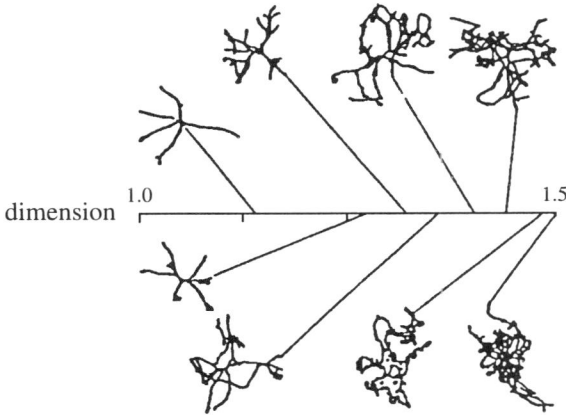

Figure 9.6. The fractal dimension is useful in characterizing the shape or form of neurons. When the neuron is less branched, the fractal dimension is closer to one, and when it is more branched the fractal dimension is closer to two. (Smith, Jr., et al., 1989, Fig. 5.)

remember that the *fractal dimension alone does not uniquely specify neuronal shape,* as Fig. 9.7 shows. H. Eugene Stanley suggested (personal communication) that one way to remove this ambiguity and provide more information about the shape is to use the spectral dimension. This dimension depends on the statistical properties of a random walk on the fractal object. Fractals with the same dimension

Figure 9.7. Although useful in characterizing the shape of neurons, the fractal dimension does not uniquely specify the shape. Both these neurons have the same fractal dimension. The dimension of the neuron at the left primarily reflects its rugged borders, while the dimension of the neuron on the right primarily reflects its many branches. (Smith, Jr., et al., 1989, Fig. 7.)

but different properties, such as are shown in Fig. 9.7, have different spectral dimensions. Lowen and Teich (1991) define the spectral dimension, D_s, as dependent on the fractal dimension of the object, D_f, and the exponent D_d describing the power law variation of the diffusion constant with distance:

$$D_s = 2D_f / (2 + D_d) . \tag{9.11}$$

D_d depends on the Euclidean dimension, being $^1/_2$ for $E = 1$, 1 for $E = 2$, and $^3/_2$ for $E = 3$. For a two-dimensional retinal neuron with $D_f = 1.4$, then $D_s = 2.8/(2 + 1)$ $= 0.93$, ostensibly the same for both neurons in Fig. 9.7. Contrarily the routes available to a diffusing particle are quite different on these two neurons, so while Stanley's idea sounds sensible, it is not yet worked out.

Diffusion-limited aggregation (DLA) is a physical model developed to describe the morphogenesis of complex irregular structures using simple physical and mathematical arguments. Such structures are formed using two basic ingredients: (1) a monomer of a given type undergoes a random walk in a space described by a lattice; (2) when this monomer comes in contact with another of its species its sticks to it and ceases moving. To create a DLA one places a "seed," i.e., a stationary monomer in the vicinity of the origin of a given lattice, on a simple cubic lattice, for example. A monomer is then released from a random point in space far from the origin and allowed to take steps of equal probability to all adjacent sites at each point in time. This distant walker will eventually "diffuse" (random walk) to the origin and there it will stick to the seed-monomer. Once this dimer is formed, another monomer is released far from the origin and diffuses toward the origin, where it will eventually form a trimer with a large number of possible spatial configurations. This process is repeated again and again until the DLA structures, such as depicted in Fig. 9.8, are obtained. See also Chapter 11 on growth processes.

Caserta et al. noted that the fractal dimension measured for the neurons in the cat retina is the same as that of the diffusion-limited aggregation model. Thus they suggested that *the shape of the neuron may be determined by diffusion-limited phenomena, such as electrical fields, chemical gradients, or viscosity differences between the neuron and its environment, which can influence the growth of the neuron.* It is not known if the fractal dimension of a neuron remains constant as it grows. However, preliminary data indicate that the fractal dimension does remain constant for neostriata neurons grown in culture from neonatal rats.

Schierwagen (1990) suggests using the diffusion-limited growth model as a satisfactory description of dendritic pattern formation in nerve cells. He argues that if a given neuron corresponds to a fractal, then the relation between the total length N of all branches inside a sphere of radius R should follow the power law

$$N(R) \propto R^D , \tag{9.1}$$

as discussed in Chapter 2 and diagrammed in Fig. 9.9. If the thickness of the individual branches is neglected, then the number of branches $n(R)$ at a given distance R from the source is given by

Figure 9.8. A typical DLA cluster generated using Fractint (Tyler et al., 1991).

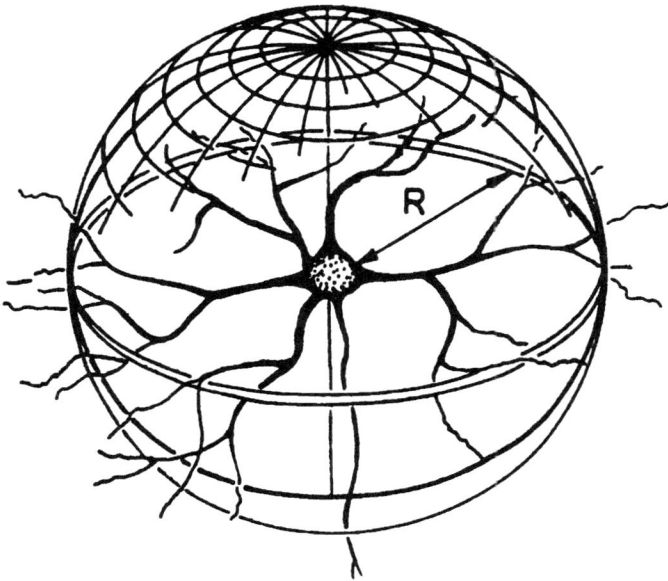

Figure 9.9. Determination of fractal neuronal dimension. With the soma in the contour of a sphere of radius R, the number of branches for different values of R allows calculation of the exponent D (Eq. 9.1). (From Schierwagen, 1990, Fig. 8.)

$$n(R) \propto \frac{dN(R)}{dR} \propto R^{D-1}. \tag{9.2}$$

Therefore, the validity of Eq. 9.1 becomes an experimental question in which the number of branches for different values of R provides information about the exponent D. In Fig. 9.10 the fractal dimension of various neuron types are shown using the above measuring scheme.

The diffusion-limited growth models demonstrate that one can start from initially undifferentiated "seeds" and generate very specialized branched structures due to the action of global, diffusion-based growth rules. Schierwagen (1990) draws the theoretical conclusion that differentiation of cellular morphology is the decisive process underlying organismic morphogenesis resulting from such general laws. The existence of these laws implies that neuronal form generation can be explained by physical properties programmed by the genes rather than direct genomic governance of the pattern itself. This could be regarded as a contemporary version of Roux's "causal scheme" of morphogenesis.

Fractal Dimension of Interspike Intervals

Statistical approaches: Several different groups have found that the durations of the times between action potentials for some neurons have fractal properties. Each

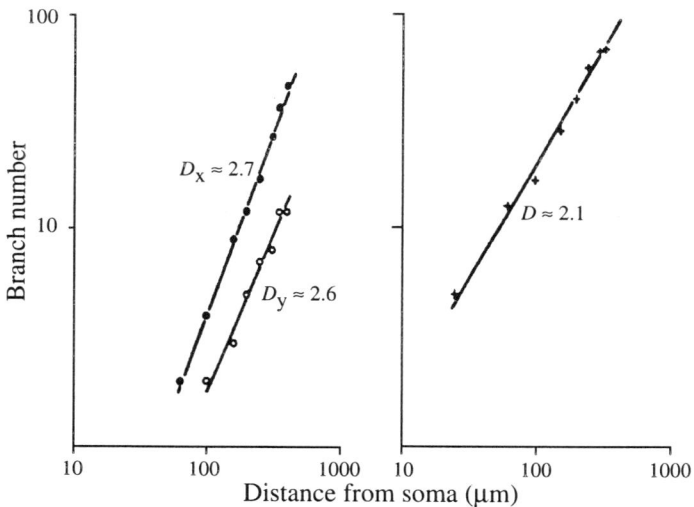

Figure 9.10. Logarithmic plot of branch number $n(R)$ versus distance, R. *Left:* For allo- and indiodendritic neurons direction-related exponents can be derived: D_x refers to the dominant orientation of dendrites, D_y to a perpendicular direction. *Right:* Isodendritic neurons are characterized by a single fractal exponent D. (From Schierwagen, 1990, Fig. 9.)

group has used different techniques to analyze the data and proposed different models to interpret their results. Each of these approaches, in its own way, helps us to better understand the patterns found in the action potentials recorded from neurons. Here we describe some of these different approaches.

Gerstein and Mandelbrot (1964) used two techniques to analyze the statistical properties of the intervals between action potentials recorded from different neurons. First, they analyzed how often an interval of a given duration t_1 is followed by an interval of another duration t_2. Second, they plotted "scaled interval histograms." That is, they determined the distribution of times between consecutive action potentials, between every other action potential, between every fourth action potential, every eighth action potential, etc. Such analyses can be used to reveal if the distribution of action potentials looks self-similar. More technically, if the pattern of action potentials is fractal, then the sum of successive intervals has the same probability density, to within a scale factor, as the intervals between every n^{th} action potential.

The action potentials of the neurons they studied displayed three different types of statistical properties. 1) Some neurons are well described by a *Poisson process*. For these neurons, the probability that there is an action potential in a time interval Δt is proportional to Δt and the durations of subsequent intervals are statistically independent. 2) Some neurons fired almost, but not exactly, in a constant rhythm. These are well described by a *Gaussian model*, where the intervals have a small dispersion around a mean value. 3) *Some neurons showed self-similar fractal patterns on the scaled interval histograms*, as shown in Fig. 9.11.

They studied the statistical properties of the neurons with fractal patterns of action potentials. These neurons occasionally had very long intervals between action potentials. These long intervals occurred frequently enough that they were the most important component in the determination of the average interval length. As more and more data were collected, longer and longer intervals were found. Thus, the average interval increased with the duration of the data record analyzed, in such a way that if an infinitely long record could be analyzed, then the average interval would be infinite. The variance, the dispersion around the average interval, had similar proprieties. Such infinite moments are characteristic of a type of stable statistical fractal process called a Lévy process.

As described in Chapter 2, Lévy processes have infinite variances and thus the *central limit theorem in the form usually quoted is not applicable* to them. *This technical mathematical point has two very important practical consequences for the interpretation of the experimental data.* 1) As more and more data are analyzed the values found for the average interval and the variance in the average increase. That is, in the limit of an infinitely long data record, both the average and its variance can become infinite. Since the variance is infinite, the *averages measured for different segments of the same data record may be markedly different,* as noted by Mandelbrot (1983). It is commonly thought that if the moments, such as the average, measured from data are constant in time, then the parameters of the process that generated the data are constant, and that if these moments vary, then the parameters that generate the data also vary. This commonly held notion is invalid. *Processes, especially fractal processes, where the generating mechanism remains unchanged can yield time series whose moments, such as the average, vary with*

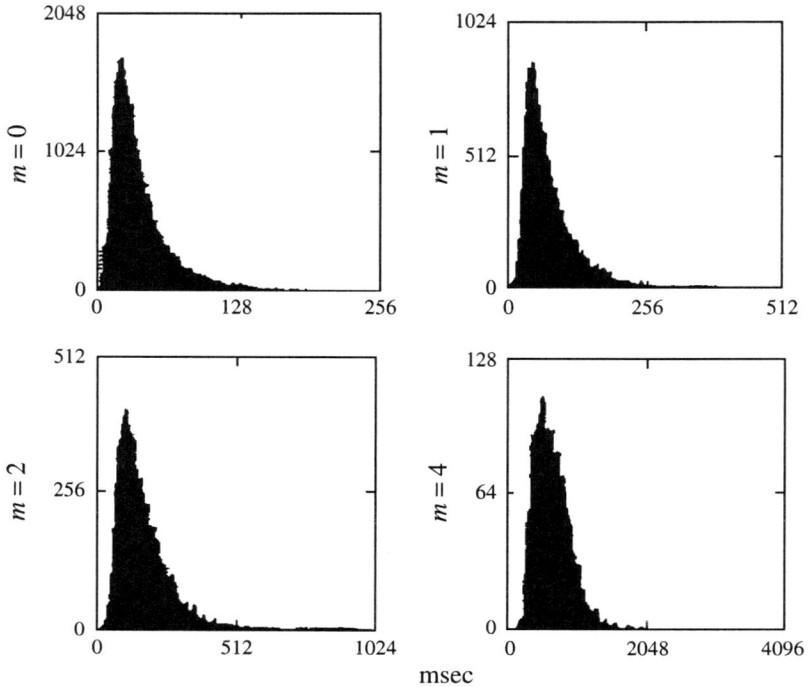

Figure 9.11. Scaled interval histograms recorded from a single neuron. Histograms of the intervals between every ($m = 0$), every other ($m = 1$), every second ($m = 2$), and every fourth action potential ($m = 4$) are shown. The similarity of these distributions indicates that this is a self-similar fractal process. (Gerstein and Mandelbrot, 1964, their Fig. 2c.)

time. 2) Because the usual form of the central limit theorem is not applicable to these processes, they have a surprising property that is counterintuitive but not illogical. As additional data are analyzed, increasingly long intervals are found. Hence the inclusion of these intervals *increases* rather than decreases the variance in the measured distribution. That is, the *statistical irregularities in the measured distributions become larger as more data are collected*. As stated by Gerstein and Mandelbrot (1964),

> Thus, in contradiction to our intuitive feelings, increasing the length of available data for such processes does not reduce the irregularity and does not make the sample mean or sample variance converge.

Gerstein and Mandelbrot used a process that is known to have the above fractal properties to model the timing of the action potentials. They assumed that random excitatory events in the dendritic arborization incrementally increase the electrical potential of the neuron by a small amount and inhibitory events decrease the potential by a small amount. Thus, the potential continues to fluctuate until it crosses the threshold to fire another action potential. Then the potential resets to its initial resting value. This is equivalent to a *one-dimensional random walk*, illustrated in Fig. 9.12. The potential starts at a certain value and takes equal-size steps toward or away from the threshold. When it reaches the threshold it fires an action potential and resets. Technically, the probability density of the first passage times in a one-dimensional walk that begins some fixed distance from an absorbing barrier is fractal. This model describes the qualitative but not the quantitative properties of data. They found a better quantitative fit to the data when they modified the model so that the rates of excitatory and inhibitory events are not equal, called a random walk with drift. They also explored random walks in dimensions greater than one. In all their models, memory of the previous potential is lost when the neuron resets after the action potential. Thus, there is no explicit correlation between successive intervals of action potentials.

Wise (1981) analyzed the distribution of intervals between action potentials recorded from single neurons in the somatosensory cortex and the respiratory system. He found that it was *very helpful to plot the logarithm of the number of intervals versus the logarithm of the size of the interval.* These log-log plots revealed two major types of distributions. 1) Some neurons had relatively few long intervals and could be well fit by *log normal distributions.* 2) Some neurons had a relatively large number of very long intervals. This is the behavior reported by Gerstein and Mandelbrot (1964) for some neurons. As shown in Fig. 9.13, these distributions are straight lines at long intervals on these log-log plots, namely, they are *inverse power laws* that have fractal properties. That is, the probability density

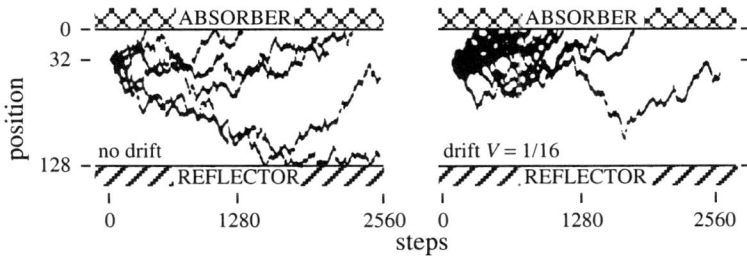

Figure 9.12. One-dimensional random walk models predict intervals between action potentials with statistical properties similar to those found in the neuron in Fig. 9.5. The electrical potential of the neuron increases incrementally with excitatory inputs and decreases incrementally with inhibitory inputs. When the potential reaches the threshold the action potential fires and the potential is reset. The interval between action potentials is proportional to the number of steps in the random walk. In the model at left there are equal rates of excitatory and inhibitory events, while in the model at right the rate of excitatory events is greater than that of the inhibitory events. (Gerstein and Mandelbrot, 1964, their Fig. 3.)

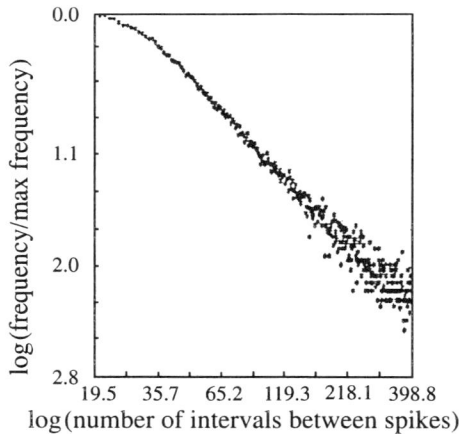

Figure 9.13. Distribution of action potential intervals from a neuron in the somatosensory cortex of a rabbit. The logarithm of the number of intervals is plotted versus the logarithm of the interval duration. (Intervals were 0.312 ms long.) The straight line indicates that the distribution can be represented by a power law that has fractal properties. (From Wise, 1981, Fig. 4.)

$f(t)$ of the occurrence of intervals having durations between t and $t + dt$ is proportional to $t^{-\alpha}$. Current interpretations and models would not have anticipated such inverse power law behavior. Thus, Wise (1981) remarks that,

> A disturbing but inevitable conclusion seems to be that very many theoretical papers on modeling spike interval distributions would and should have not been written if the authors had known about these power laws.

The random walk model proposed by Gerstein and Mandelbrot (1964) predicts that the distribution of intervals between action potentials is a power law with exponent $\alpha = 1.5$. In fact, a random walk interpretation is possible if α is in the range $1 < \alpha < 1.5$. However, Wise (1981) found that the value of α measured from different neurons varied from 1 to 7. He showed that α could be greater than 1.5 if the *threshold at which the neuron fired fluctuated in a random way* with a dispersion around a mean value. Nonetheless, Wise emphasizes that the experimental distributions are complex and cannot be fit by any simple model with a given set of parameters. Rather, the pattern of action potentials seems to consist of *"mixtures" of distributions of the same form but with different parameters* for each distribution. That is to say, the parameters themselves are random variables corresponding to higher-order random processes.

For some neurons the durations of successive intervals between action potentials are not correlated. This can happen if the intervals arise from a *stochastic* process such as a Poisson process. In that case, in each small time period, there is the same probability that there will be an action potential. *It has always been assumed that the observed lack of order in such data is due to stochastic processes.* However, as explained in Chapters 6 and 7, which describe chaos, deterministic processes can also mimic such "random" behavior. Glass and Mackey (1988) have developed an elegant *deterministic model of intervals between action potentials.* In their model, the duration of each interval t_n is entirely determined by that of the previous interval t_{n-1}, namely,

$$t_n = -(1/R) \ln\left|1 - 2 \, e^{-Rt_{n-1}}\right|, \tag{9.3}$$

where R is a constant, for example, $R = 3 \, s^{-1}$. Thus, their model is completely deterministic. Yet the durations of the predicted sequence of intervals mimics random behavior. The model predicts a distribution of intervals $f(t)$ given by

$$f(t) = R \, e^{-Rt}, \tag{9.4}$$

which is identical to that predicted by the Poisson model, which is based on stochastic processes. Thus, the statistical distributions alone cannot differentiate between stochastic and deterministic processes, as we already observed in Chapter 6. As they emphasize: *"Therefore, observation of exponential probability densities is not sufficient to identify a process as a Poisson process."* Moreover, the Glass and Mackey (1988) model also has the same mean, variance, power spectrum, and autocorrelation function as the Poisson model. That is, such linear properties cannot be used to differentiate these two very different types of models.

Action potentials can be recorded from the *primary auditory neurons that transmit information about sound* and from the *pulse vestibular neurons that transmit information about head position* to the brain (Teich and Khanna, 1985; Teich, 1989). Teich and Khanna (1985) analyzed their data by determining the *pulse number distribution,* that is, *the entire data record is divided into time window lengths, and the number of window lengths that have 0, 1, 2, 3, . . . etc. action potentials is measured.* As shown in Fig. 9.14, as the length of that time window was increased in recordings from vestibular neurons, these distributions became smoother with smaller variance. However, as the length of that time window was increased in recordings from the primary auditory neurons, these distributions became more irregular with larger variance.

As described above, such behavior is the signature of the fractal nature of Lévy stable processes. Thus, as more data were collected the variance increased. This happened because there are very small correlations between action potentials, but these correlations become increasingly manifest over longer time window lengths. These longer time window lengths contain more details about the correlations in the data, and thus have larger variance. Because the variance increased with the time window length, the mean number of action potentials per unit time did not converge to a finite value. Rather, it sometimes followed "a trend in which it consistently

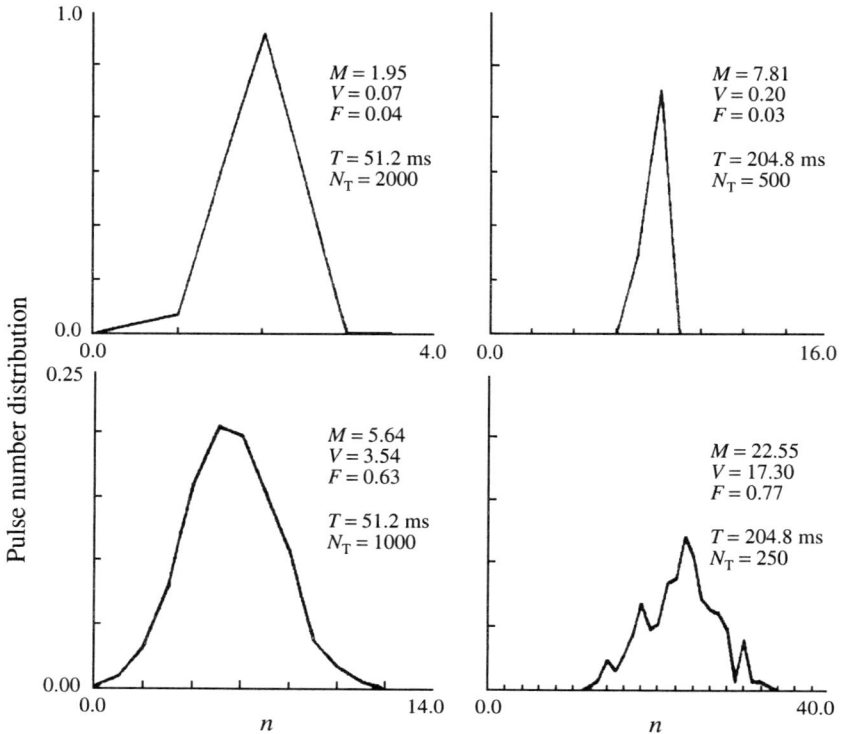

Figure 9.14. Recordings from auditory neurons in the cat were used to determine the distributions of the number of action potentials within a given time. *Left:* Distributions from 51.2 ms window lengths of data. *Right:* Distributions from 204.8 ms of data, that is, a recording time window length that is four times as long as that on the left. *Top:* Data from a vestibular neuron. As the time window length is increased, the peak is smoother and more narrow, that is, the variance of the distribution decreases. *Bottom:* Data from a primary auditory neuron that transmits sound information from the inner ear to the brain. As more data are recorded the peak is broader and more irregular, that is, the variance of the distribution increases. The data from this neuron show a hierarchy of clusters of clusters of action potentials. Thus, as more data are analyzed, larger groups of correlated action potentials are found. This causes the local peaks at those numbers of firings and broadens the distribution, increasing its variance. The degree of hierarchical clustering of action potentials is measured by the fractal dimension, which varies from 0.5 to 1.0. (From Teich, 1989, Figs. 5a, 5b, 3a, 3b.)

decreases, for example, only to reverse direction and consistently increase for a period of time" (Teich, 1989). Again, as emphasized above, this does not mean the process that generated the data is changing in time. *Processes, especially fractal processes where the generating mechanism remains unchanged, can yield time series whose moments, such as the average, vary with time.* Teich (1989) concluded:

... that the irregular nature of the long count pulse number distributions does not arise from statistical inaccuracies associated with insufficient data, but rather from event clustering inherent in the auditory neural spike train.

He added,

> The presence of spike pairs and clusters in the nervous system is greeted with skepticism by some and with enthusiasm by others.

Teich (1989) found that to analyze the data it is very helpful to determine the *Fano factor*, *F(T)*, which is the *variance divided by the mean*, measured in time window lengths of duration T:

$$F(T, n) = \frac{(\text{Variance of the average } x(t) \text{ over intervals of length } T)}{(\text{Mean of the averages over } T)} . \quad (9.5)$$

The fractal D is $d \log F(T, n)/d \log T$ for a chosen number of intervals, n, of the differing durations T:

$$F(T, n) \, \alpha T^{D} . \quad (9.6)$$

In practice, the slope may be calculated from points at two large values of T, T_1 and T_2:

$$D = \frac{\log [F(T_2)/F(T_1)]}{\log (T_2/T_1)} . \quad (9.7)$$

An example of the firing rate of an auditory neuron (from Teich, 1992) is shown in Fig. 9.15, where each rate is the number of spikes occurring within the window of duration T. The values of $F(T)$ are shown for each of three values of T in Fig. 9.15. The power law relationship shows only at long T; at short T's the neuron's refractory period prevents firing.

If the action potentials were *periodic*, then the variance would decrease and the Fano factor would approach zero as the time window T is increased. If the action potentials were a *Poisson* process, that is, if they occurred with equal probability per unit time, then the Fano factor would approach unity as the time window T is increased; this is the same thing as saying that the standard deviation of a single exponential density function (the square root of its second moment) is equal to its mean (its first moment). For a Lévy stable *fractal process*, the *Fano factor increases as a power law of the window T*, i.e., $F(T) \propto T^{D}$, where D is the fractal dimension. As shown in Fig. 9.15, at the shortest windows, the Fano factor is slightly less than unity as a result of the refractory period that limits the interval between successive action potentials. At intermediate windows, the Fano factor is approximately unity because the action potentials constitute a Poisson process. At longer windows, the Fano factor increases as a power law of the window duration $\propto T^{D}$, which indicates

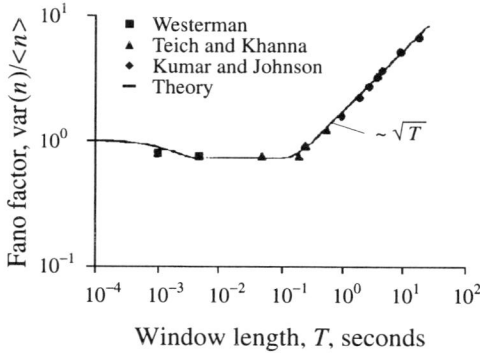

Figure 9.15. The Fano factor, the variance divided by the mean measured in a time window of duration T, is plotted versus the time (or window length) over which the number of action potentials were recorded from a primary auditory neuron. For the shortest window lengths, the Fano factor is slightly less than unity because of the refractory period that limits the interval between successive action potentials. For intermediate window lengths, the Fano factor is equal to unity, indicating the action potentials are not correlated. For the longest window lengths, the Fano factor increases as a power of the duration of the window, indicating the distribution of action potentials is a fractal, with the slope of this line, which is the fractal dimension, found to be between 0.5 and 1.0, depending on the neuron and the stimulus used to excite it. This increase in variance occurs because these longer window lengths reveal the rarer clusters of action potentials that increase the variance and thus the Fano factor. (From Teich, 1989, Fig. 4a.)

that the timing of the action potentials is fractal with dimension D. For the data in the figure, $D = 1/2$. The fractal dimension varied between 0.5 to 1.0, depending on the neuron and the stimulus used to excite it.

Many natural sounds have low-frequency fractal behavior (Voss, 1988). Teich (1989) speculates that a similar fractal pattern of action potentials may be able to match and thus efficiently sample such natural fractal sounds. He also suggests some mechanisms that could produce this behavior. These include: (1) the fractal opening and closing of cell membrane ion channels, (2) the fractal diffusion of neurotransmitter across the synaptic junction, (3) the fractal distributions of intervals between action potentials produced by random walks of excitatory and inhibitory steps, and 4) chaotic oscillations of the intracellular potential.

Chaotic dynamical approaches: We now examine a completely different model leading to the observed distribution of interspike intervals. Recall the association we made earlier between the types of dynamic attractors and the firing patterns of different neurons. Rapp et al. (1985) speculate that transitions among fixed-point, periodic, and chaotic attractors can be accomplished by varying system control parameters and may be observed clinically in failures of physiological regulation. There seems to be evidence accumulating in a number of physiological contexts to support the view that the observed rich dynamic structure in normal behavior is a consequence of chaotic attractors, and the apparent rhythmic dynamics are the phase coherence in the attractors (West, 1990). Rapp et al. (1985) present experimental evidence that spontaneous chaotic behavior does in fact occur in neurons.

In their study Rapp et al. (1985) recorded the time between action potentials (interspike intervals) of spontaneously active neurons in the precentral and postcentral gyri (the areas immediately anterior and posterior to the central fission) of the squirrel monkey brain. The set of measured interspike intervals $\{t_j\}$, $j = 1, 2, \ldots, M$ was used to define a set of vectors $X_j = (t_j, t_{j+1}, \ldots \ldots, t_{j+m-1})$ in an m-dimensional embedding space. These vectors were used to calculate the correlation integral of Grassberger and Procaccia, discussed in Chapter 7.

To determine the correlational dimension from the interspike data one must determine a scaling region in the correlation function $C_m(r)$ between the noise at small separation distances r and the constant value of unity for larger r. The plateau in the slope, which determines the dimension of the phase space set versus the embedding dimension E_m in Fig. 9.16, defines the scaling region. In Fig. 9.16, upper left, we observe a plateau region for the dimension in the interval $(-4.5, -2.5)$ for $m = 15$ to $m = 20$. In Fig. 9.16, lower right, we see that no such plateau region is reached up to an embedding dimension of $m = 40$, indicating that this time series cannot be distinguished from uncorrelated random noise. Of ten neurons measured, three could clearly be described by low-dimensional chaotic time series, two were ambiguous, and five could be modeled by uncorrelated random noise.

Rapp et al. (1985) drew the following two conclusions from the study:

> 1 . . . the spontaneous activity of some simian cortical neurons, at least on occasion, may be chaotic; 2 . . . irrespective of any question of chaos the dimension of the attractor governing the behavior can, at least for some neurons for some time, be very low.

For these last neurons we have the remarkable result that as few as three or four variables may be sufficient to model the neuronal dynamics if in fact the source of their firing pattern is a low-dimensional attractor. It would have been reckless to predict this result, but we now have evidence that in spite of the profound complexity of the mammalian central nervous system the dynamics of some of its components *may* be describable by low-dimensional dynamical systems. Thus, even though we do not know what the dynamical relations for these neuron systems might be, the fact that they do manifest such relatively simple dynamical behavior, bodes well for the eventual discovery of the underlying dynamical laws.

9.4   Spatiotemporal Organization

The spread of excitation through the excitable cells of the heart, or the excitable neurons of the nervous system, is also analogous to the spread of excitation through

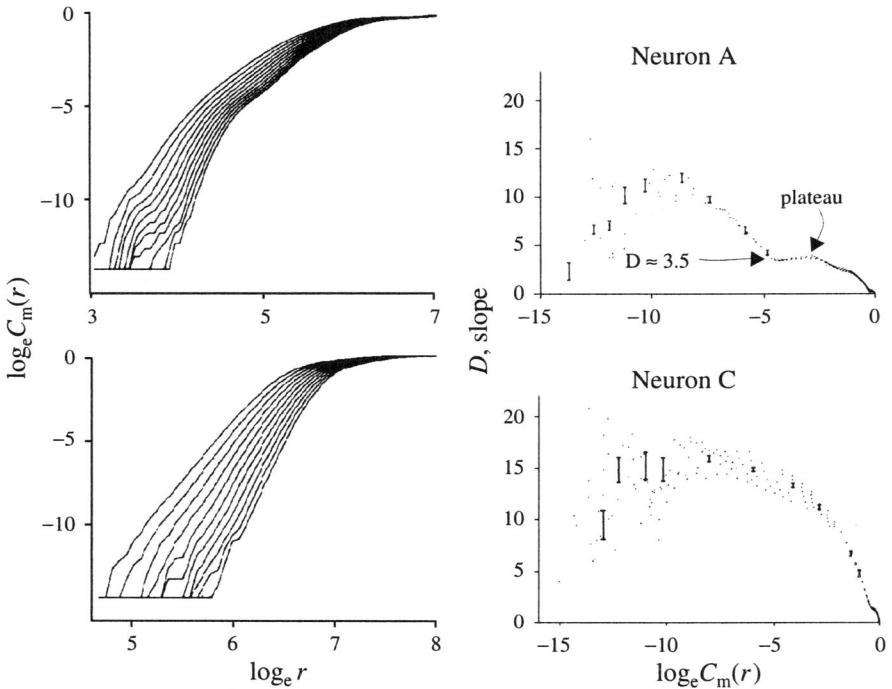

Figure 9.16. Estimation of attractor dimension. *Upper left:* Plot of $\log C_m(r)$ versus $\log r$, $m = 10$ to $m = 20$, for neuron A. *Upper right:* The corresponding plot of $d \log C_m(r)/d \log r$ as a function of $\log C_m(r)$, $m = 15$ to $m = 20$, for neuron A. The plateau exists between -4.5 and 2.5. The dimension of the attractor is estimated to be 3.5. *Lower left:* Plot of $\log C_m(r)$ versus $\log r$ for neuron C, $m = 20$ to $m = 40$ in steps of two. *Lower right:* The plot of $d\log C_m(r)/d\log r$ as a function of $\log C_m(r)$ for neuron C, $m = 30$ to $m = 40$ in steps of two. No plateau is found up to embedding dimension $m = 40$. (From Rapp et al., 1985.)

other systems, such as the spread of chemical reactions. Our modern understanding of wave propagation in such reactions may be traced back to the chemical oscillator of Boris Pavlovich Belousov in 1951. His oscillator consisted of cerium and inorganic bromate in a solution of sulfuric acid. The intent in combining these chemicals was to construct an inorganic analogue of the Krebs cycle for which Hans Adolf Krebs was awarded the Nobel Prize in Chemistry in 1953. Belousov's oscillator cycles rhythmically in time: the yellow of oxidized cerium ions was visible for approximately one minute, then the yellow faded as the cerium ions oxidized the citric acid; the color returned in the next minute, as the bromate ions oxidized the reduced cerium (see, e.g., Winfree, 1987, for a more complete discussion of the history and chemical mechanisms involved). The history itself is of interest inasmuch as Belousov's articles on this topic were consistently rejected by journals on the basis that such chemical oscillations violated thermodynamical laws. They rejected his experimental results as due to obvious fraud or scientific incompetence, neither of which was true. It was not until the 1960s that a graduate student named Zhabotinskii investigated the chemical recipe of Belousov and made some refinements on the original recipe. He was able to duplicate the oscillating behavior of the chemical reaction, and Belousov's original work was vindicated.

Spatial-temporal organization in biological, chemical, and physical systems are manifest through the propagation of traveling waves in excitable media. Such wave propagation depends on the trade-off between the local dynamical properties of the medium—i.e., the local bio-oscillator behavior—and the diffusive coupling of neighboring spatial regions—i.e., the coupling of these bio-oscillators over space. The typical behavior of the local dynamical behavior of a piece of excitable tissue consists of a rest state that is stable against small perturbations; however, when the perturbation exceeds some critical level, the system responds by initiating a cycle of excitation and recovery to the rest state. The level of excitation is independent of the magnitude of perturbation i.e., independent of how far the perturbation exceeds the critical level. The local excitation can now spread through the medium via the diffusive coupling mechanism, i.e., the excitation travels by means of some version of the random walk process. This excitation process is described by means of a continuous partial differential equation in space and time.

Chemical Waves

Chemical reactions have manifest time behavior that can be constant, periodic, and chaotic. In the spatial domain it was observed in the 1970s that the chemical media of Belousov and Zhabotinskii (BZ) could support waves of chemical activity. As pointed out by Winfree (1972, 1974), the waves emerge periodically from tiny regions of chemical rotation. Here we point out that the interactions between adjoining bits of membrane through which information is passed, and the interactions between adjacent parcels of liquid in the chemical solution are analogous. In the BZ reaction the information transfer is accomplished by molecular diffusion. In the case of cell membranes it is the electrical potential that diffuses from adjacent regions of higher to lower voltage. Thus, electrical potential is carried by currents of charged ions. The coupling is mathematically the same, and so too is the resulting excitation propagation.

As stated so effectively by Winfree (1987):

> Propagation implies direction. The direction is determined by history: one region fired first, turning to unexcitable ashes temporarily, and its excitable neighbor fired later, infected from the first. This asymmetry (excitable quiescence ahead, refractory ashes behind) gives direction to the waves so it cannot backfire. Thus when two waves collide, they should snuff one another: neither can propagate into the ashes left behind by the other.

Zaikin and Zhabotinskii (1970) used ferroin dye to enhance the visibility of regional variations of chemical activity in the BZ reaction. In Fig. 9.17 are shown

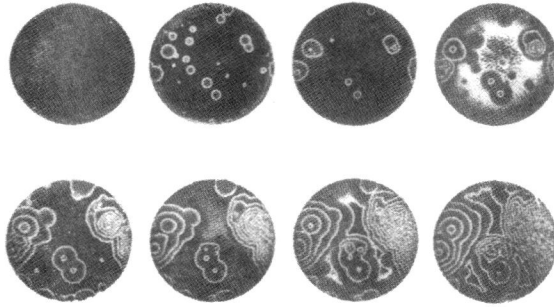

Figure 9.17. Eight snapshots taken one minute apart showing blue (light gray as reproduced here) waves of oxidation in a 66 mm dish of orange (dark gray) malonic acid reagent. Waves erupt periodically from about ten pacemaker nuclei scattered about the dish. In the end only those of shortest period survive. The tiny circles are growing bubbles of carbon dioxide. (From Zaikin and Zhabotinskii, 1970.)

typical wave fronts. Waves are intermittently emitted by points within the solution that Winfree (1987) suggests are analogous to ectopic foci on the heart. At those points the chemical clock ticks faster, turning blue (light gray in Fig. 9.17) before their surroundings, and the red-to-blue (dark-to-light) transition propagates in concentric rings. At first sight these rings appear to be like ripples in a pond, but unlike water wave ripples these waves of chemical activity annihilate each other on contact rather than passing through one another like water waves. The wave front extends only a fraction of a millimeter, and on the time scale of milliseconds concentrations can change by factors of a thousand. One difference between these chemical waves and those of sound and light, mentioned earlier, is that the excitable media supporting the propagation continuously supplies energy to the wave. Therefore, unlike light and sound waves, which attenuate as they traverse a medium, the chemical wave like other excitable waves maintains its initial strength as it propagates.

Winfree (1972) demonstrated that an apparently minor variation of the Zhabotinskii recipe would suppress the spontaneous oscillation but would not inhibit the excitability of the medium. That is to say, the capacity to propagate chemical signals was not diminished. This property of BZ reactions is quite similar to that of the neurons discussed earlier. The analogy is more than a superficial one, however, extending to the level of the partial differential equations that describe the physical mechanisms in the BZ reaction and the electrical membranes in living organisms; see Winfree (1974) and Troy (1978). As Winfree (1987) states:

> . . . They are so close in form and in qualitative behavior that the malonic acid reagent might be fairly called "aqueous solution of the equations of electrophysiology" (pun intended). As such it provides a convenient tool for teaching and even for research into the implications of excitability in two-dimensional and three-dimensional contexts.

Fig. 9.17 shows one such implication: propagating circular waves.

Nonlinear Wave Equations

The description of wave propagation in nonlinear media has been one of the most successful applications of the concepts of nonlinear dynamics systems theory. Periodic waves traveling in excitable biological and chemical media provide a dramatic illustration of spontaneous organization in space and time. Let us assume that the system is represented by an excitation variable u and a recovery variable v, and that the former changes on a much faster time scale than the latter and the two interact locally according to the ordinary differential equations,

$$du/dt = f(u, v)$$
$$dv/dt = g(u, v) . \tag{9.8}$$

Nullclines are defined by the loci $f(u,v) = 0$ and $g(u,v) = 0$ and are plotted in Fig. 9.18 in the (u,v) phase space. The interaction of the two nullclines defines a unique rest state for the excitable medium. Large perturbations trigger long excursions, whereas small perturbations are damped out. As described by Gerhardt et al. (1990): "(i) the excitation variable increases rapidly, causing (ii) a slower and temporary increase in v, followed by (iii) rapid extinction of u and (iv) slow decrease in v back to the rest state."

In the BZ reaction the above cycle gives rise to the oscillating change in color, where the excitation variable is bromous acid and the recovery variable is ferroin. In the same way for cardiac oscillations we can identify u with the membrane potential and v with the slow transmembrane ionic currents. What distinguishes the theory of wave propagation in excitable media from linear wave propagation are the effects of *dispersion* and *curvature*. The wave velocity normal to the phase front is given by

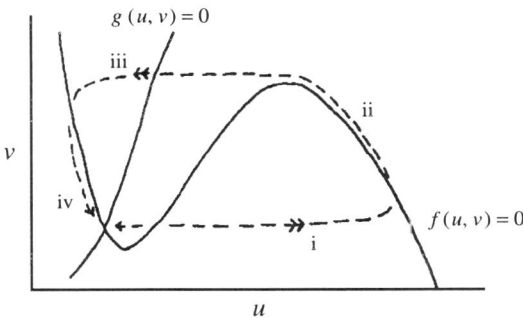

Figure 9.18. Phase plane diagram for an excitable medium. The excitation variable u and the recovery variable v interact dynamically. (From Gerhardt et al., 1990, their Fig. 1, with permission from *Science*.)

$$V_{\mathrm{N}} = C + DK, \tag{9.9}$$

where C is the speed of the planar wavefront, D is the diffusion coefficient of the excitation variable, and K is the curvature of the wavefront. The dependence of the wave speed C on the period of the signal is called the dispersion relation. Eq. 9.9 has been verified experimentally by Foerster et al. (1988, 1989) for a chemical excitable medium. A transition from a stable cycling mode to a chaotic mode in a computational representation of a chemical reaction system is depicted in Fig. 9.19.

9.5   Summary

Nerves and muscles provide for spatial spread of depolarization by undergoing regenerative depolarization: the wave front initiates sufficient depolarization to trigger an increase in ionic conductance, allowing a large current flow and causing rapid and complete depolarization, which in turn initiates the same sequence of events in the neighboring still-polarized region, and has no influence on the depolarized region behind the wave front. The rate of recovery of excitability, along

Figure 9.19. Chaotic spiral waves in a computer-generated "chemical" reaction. After a stable beginning (upper panels), the wave forms become chaotic (lower panels). (From Gerhardt et al., 1990, their Fig. 5, with permission from *Science*.)

with the rate of spatial propagation of the wave front, govern the form of events to follow: fast recovery allows earlier re-excitation. Chaotic waves of excitation can occur in two- or three-dimensional spaces. Cardiac arrhythmias are sometimes chaotic, as in fibrillation, and serve as a rich source of data in developing the theories of the nonlinear spreading of excitation, as exemplified in the book by Jalife (1990) and papers of Winfree (1990, 1991), Courtemanche and Winfree (1991), and Jalife et al. (1991).

10

Intraorgan Flow Heterogeneities

> Some parts are homogeneous . . . others dissimilar, heterogeneous . . .
>
> William Harvey (1616)

10.1   Introduction

The variation in a physiological property, characteristic, or measure is often found to be important to organ function. Variations in histologic staining, in receptor densities, in NADH fluorescence, etc., are noted in many current articles. Most such variations that have been revealed by morphometric or chemical methods are spatial variations. Advances in technical accuracy allow us to distinguish between true spatial variation and experimental errors in the values measured at different spatial positions. These observations raise questions as to the cause of variation. In this chapter we shall use measures of regional blood flows to illustrate the approaches, while remembering that the same methods may also be used to evaluate spatial heterogeneity for other attributes, such as the density of binding sites for β-blockers or the concentrations of ATP or lactate dehydrogenase.

The standard textbooks and handbooks listing the flows of blood to organs tend to fall into two classes. Medical texts give the organ blood flows, usually for a 60-kg man, in units of ml/min absolute flow. This is good enough to give an indication of where the cardiac output goes, but it is not very useful for understanding tissue metabolism. More refined tables may list tissue blood flows in various species in units of flow per unit mass of tissue (for example, $ml\ g^{-1}\ min^{-1}$). No indication of intraorgan variation will be found, and the ranges of values given implicitly reflect either interanimal or temporal fluctuation.

The heart provides a good example, because so many data are now available. The prejudice that flow per unit mass of cardiac tissue should be uniform is based on the essential monofunctionality of the heart. It contracts almost simultaneously throughout the organ with each beat, so one may easily succumb to the idea that myocardial blood flows might be uniform in space and constant over periods longer than several heartbeats. Thus, only when a physician views images of thallium deposition showing a "defect," a region of less-than-average flow, does he see the nonuniformity, but then he attributes it to a pathological condition.

Normal flows are broadly heterogeneous, as is illustrated by autoradiographic mappings and by the range of local flows in the probability density functions shown in Fig. 10.1. *Such nonuniformity has been observed in all organs examined.*

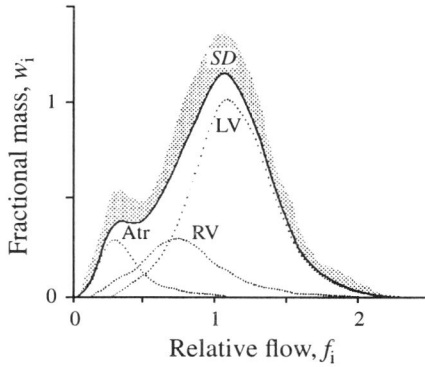

Figure 10.1. Probability density function of regional myocardial blood flows in thirteen baboon hearts. The fractional mass w_i, having a relative flow between f_i and $f_i + \Delta f_i$ is plotted versus f_i. Flows were estimated from the local deposition densities of 15 μm microspheres in 208 pieces per heart of about 70 mg size after injection of the spheres into the left atrium or left ventricular apex. Atr = atrium; RV = right ventricle; LV = left ventricle; SD = the standard deviation for the values measured in the whole heart. (From King et al., 1985, Fig. 4, with permission.)

Commonly, when the microsphere deposition technique was used to measure tissue blood flows only a few large samples were taken, with the result that even when variation was recognized, its magnitude was underestimated. This emphasizes what was discussed in Chapter 4, the generality that observations on large tissue samples give a false sense of uniformity, but smaller sizes reveal more of the heterogeneity. In this chapter we explore and extend the utility of the fractal relationship between the degree of heterogeneity, described by the coefficient of variation of the regional flows, which we call the relative dispersion, RD, and the mass of the observed tissue elements, m:

$$RD\,(m) \;=\; RD\,(m_0) \cdot (m/m_0)^{\,1-D}, \tag{10.1}$$

where m_0 is a chosen reference mass, usually 1 g. (See discussion in Chapter 4.)

10.2   Methods of Measuring Regional Flows

For years there was a generally held opinion that the variation in apparent regional blood flows was due to unavoidable experimental errors in the methods. The prime method was the local deposition of *tracer-labeled microspheres,* which are injected into the artery and become trapped in organs. Studies on methodological variation by Archie et al. (1973), Buckberg et al. (1971), Yipintsoi et al. (1973), Utley et al. (1974), King et al. (1985), and Nose et al. (1985) showed that the errors were much less than the observed variation. The "molecular microsphere," a nearly 100% extracted tracer-labeled compound ^{131}I-desmethylimipramine, was assuredly delivered and deposited in proportion to the local flow (Little and Bassingthwaighte, 1983) and therefore served as a reliable regional flow marker with minimal methodological variation (Bassingthwaighte et al., 1987a, 1990a). Using the molecular microsphere, Bassingthwaighte et al. (1990a) affirmed the results obtained with the standard microspheres, that there was broad variation in local flows. They also showed that the methodological error in the microsphere technique was systematically greater in low flow regions, where fewer microspheres were deposited. The tracer microsphere method is well described in detail by them and by Heymann et al. (1977). The essence of the method is to inject tracer-labeled 15-μm-diameter microspheres into the left atrium, where they mix with the blood and distribute throughout the body into each region or organ in proportion to the fraction of cardiac output going to the region. (For the lung, right atrial injection is used.) A succession of different tracer labels may be used in one animal to estimate flows at different times prior to sacrifice and to measure changes in local flows.

For obtaining measures of the local concentrations of the deposited microspheres one may in theory use imaging techniques such as positron emission tomography or, for X-ray absorbing markers, reconstructive X-ray computed tomography (Shu et al., 1989). The simplest and most accurate method is to remove the organ, section it finely, and measure the microsphere radioactivity in each piece.

10.3   Estimating the Fractal D for Flow Heterogeneity

Analysis using the standard approach to estimating the fractal dimension for myocardial flow distribution is outlined in the following steps. Details follow later.

1. Inject labeled microspheres into the left atrium. They mix with atrial and ventricular blood and are distributed through the body to each organ and tissue fed in direct proportion to the flow. (This does not apply to organs supplied by secondary portal systems.) Microspheres of diameters 15 μ are nearly 100% trapped, and therefore their local intratissue concentration is proportional to flow.

2. After sacrifice remove the heart and section it into uniformly sized small pieces. For each piece record the position, weight, and microsphere radioactivity for each of the tracers used. Ideally, the pieces should be of equal weight, m, which is also the total heart weight divided by the total number of pieces.

3. Calculate the relative deposition density, that is, the radioactivity per unit mass in each piece divided by the mean radioactivity per unit mass in the

whole organ. The relative deposition density of the microspheres is taken to be the relative regional flow per gram, that is, the local flow per gram divided by the mean flow per gram for the heart.

4. Using the relative deposition density of the individual pieces, obtain the histogram of the mass fraction of the heart with different relative deposition densities. The abscissa, as in Fig. 10.1, is the relative flow, f_i, which is the local flow in ml/min per gram of tissue divided by the mean flow for the heart, which is the total coronary blood flow, ml min^{-1}, divided by the mass of the heart in grams. The ordinate is the fraction of the heart mass w_i having a flow f_i within a selected class of width Δf_i, per unit mean flow. Thus the mean of the distribution is always unity,

$$\sum w_i f_i \, \Delta f_i = 1, \tag{10.2}$$

and the area of the probability density function, $\Sigma w_i \Delta f_i$, is also 1.0. A polygonal representation of the histogram is constructed by joining the midpoints of each element to provide a smooth look, as in Fig. 10.1. (See Appendix 1, "Probability Density Functions.")

5. At this piece size, m, determine the RD, the relative dispersion for the regional flows, which is the *standard deviation divided by the mean* and is the coefficient of variation for the regional flows.

6. Next, reduce the level of resolution by increasing the size of the elements. Lump each two adjacent pieces together and repeat steps 3, 4, and 5. In a cubic lattice each piece has six immediate neighbors to choose among (up, down, right, left, and front, behind). Since opposite sides give similar pairing, we used only the three unique arrangements of the double-size pieces. For each arrangement repeat steps 3, 4, and 5 and calculate the mean RD and its SD. The average aggregate size, m_2, must be $2m$.

7. Obtain larger aggregates of nearest neighbors and repeat steps 3, 4 and 5. Since there is more than one way of forming larger aggregates for larger values of m, each arrangement gives an estimate of the RD. Use the mean of these. (One could plot all of the estimates at a given resolution, as in "pox" plots for the Hurst rescaled range analysis described in Chapter 4, and then use all the individual estimates in the regression analysis explicated later in this chapter.)

8. Plot the estimates of RD versus the mean mass m at each level of resolution on a log-log plot. Calculate the slope, $1 - D$, and the intercept, $\log RD(m_0)$, where m_0 is a handy reference unit size such as 1 g:

$$\log RD\,(m) = \log RD\,(m_0) + (1 - D) \cdot \log\left(\frac{m}{m_0}\right). \tag{10.3}$$

D is the fractal dimension. It must be between one and two for this one-dimensional approach to the analysis:

$$D = 1 - \frac{\log\left[RD(m)/RD(m_0)\right]}{\log\left[m/m_0\right]}. \tag{10.4}$$

This one-dimensional approach assumes that the tissue is isotropic, that is, there is no directionality in the variation, and that the near-neighbor correlation is the same in all directions. The analysis of myocardial flows shows a fractal D of about 1.2. Actually, since the organ is a three-dimensional object, not one-dimensional, the fractal dimension of the property is actually something greater than three, not simply greater than one. The Hurst coefficient, $H = E + 1 - D$, or $4 - D$ in the three-dimensional case, is unaffected by choosing a one-dimensional versus a three-dimensional analysis.

[Aside]

Details of the calculation of the observed heterogeneity of regional blood flow. We now describe some of the foregoing steps in greater detail.

Step 3: relative deposition density. The relative deposition density (d_j) is calculated from the amount of radioactivity in a piece or an aggregate of pieces:

$$d_j = \frac{a_j/m_j}{A/M}, \tag{10.5}$$

where a_j is the activity measured in the piece, m_j is the mass of the piece, A is the total activity in all the pieces of that heart, and M is the total mass of the heart. Then $f_j = d_j$ is the flow in piece j relative to the mean flow for the whole heart.

Step 4: Histogram of the masses with different relative deposition densities. A probability density function is constructed from the d_j's. This function determines the fractional mass of the heart that has pieces with a relative deposition density in a given range. Since in our experiments (King et al., 1985; Bassingthwaighte et al., 1990a) all the d_j's were in the range 0.0 to 3.0, this range was divided into thirty intervals of width $\Delta d = 0.1$. The pieces were then sorted into groups according to the interval into which d_j for the piece fell. The thirty groups are denoted by the subscript i with $i = 1$ to 30. (The subscript i denotes groups or intervals making up the histogram, while subscript j refers to individual pieces.) The probability density for the ith group is the sum of the masses of the pieces in the ith group divided by the total mass of the heart and the interval width. Mathematically this is expressed as

$$'w_i = \frac{\sum 'm_j \text{ for all pieces for which } ('d_i - \Delta'd/2) < 'd_j < ('d_i + \Delta'd/2)}{\Delta'd \cdot M}. \tag{10.6}$$

This gives the probability density function in the form of a finite interval histogram. The area of the histogram is unity and its mean, $\bar{d}$, is also unity.

(Because the mean of the d_j's in each group is not necessarily equal to the midpoint of the interval, the d_i's, these relations are imperfect, but in practice the errors are less than 1%.) The RD of the density function is $SD/\bar{d}$, and, since $\bar{d} = 1$, the RD is equal to the standard deviation, SD. The RD may be calculated directly from the individual pieces,

$$RD = \left(\frac{1}{M} \sum_{j=1}^{J} m_j \, (f_j - 1)^2 \right)^{1/2} ; \tag{10.7}$$

or from the groups forming the histogram,

$$RD = \left(\sum_{i=1}^{I} w_i \cdot (f_i - 1)^2 \right)^{1/2}. \tag{10.8}$$

The former method is slightly more accurate than the latter. It is this relative dispersion of the probability density function that is used as a measure of the heterogeneity of regional myocardial blood flow. When more than one histogram is plotted on the same axis the histograms are difficult to distinguish, so instead we use polygons joining values at the midpoints of the intervals.

Calculation of the Observed Heterogeneity at Different Sample Sizes

The average activity of each aggregate larger piece is the mass-weighted average of the activities of its component pieces. An example of the results of this procedure is shown in Fig. 10.2. For each grouping, a single value of RD is obtained, but adjacent pieces can be opposed in several different ways, so that for the same sizes of groups or masses of aggregate pieces, the RDs can be calculated in an increasing number of different ways as more and more pieces are put together. This dependence of the estimated dispersion on the element size is a truism: finer division always reveals greater heterogeneity unless the elements are internally uniform.

Using this approach gives both an estimate of the dispersion at each effective piece size and an estimate of the variance of the dispersion. These data are plotted as open circles in Fig. 10.3. A complicating factor not written into our fractal expressions is that there is also variation in the piece size at each level, since the aggregates were put together in a pattern fashion rather than in a fashion designed to achieve a particular mass. (The analysis has been applied only to ventricular myocardium, and not the atria, where this problem would occur.)

Linear least-squares regression lines were obtained for the logarithm of the relative dispersion versus the logarithm of the average mass of the aggregate pieces, at each level of division, using Eq. 10.3. The logarithmic slope is $1 - D$; since the regression analysis also can provide the estimate of the error in the slope, the confidence limits for D can be obtained directly. Excluded from the regression were aggregates weighing more than 2 g. (Because the correlations were high, there is no

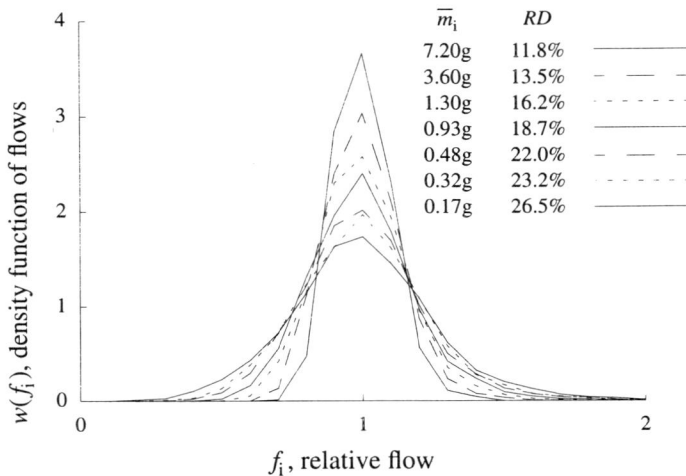

Figure 10.2. Composite probability density functions for regional flows at seven different element sizes in the left ventricles of 11 awake baboons. The order of the average element sizes, m_i, is the same, from top to bottom, as the heights of the peaks of the density functions. Bin width $= 0.1$ times mean flow. The spread of the probability density function of regional flows is dependent on the size of the elements used to make the observations. (Figure reproduced from Bassingthwaighte et al., 1989, their Fig. 4, left panel.)

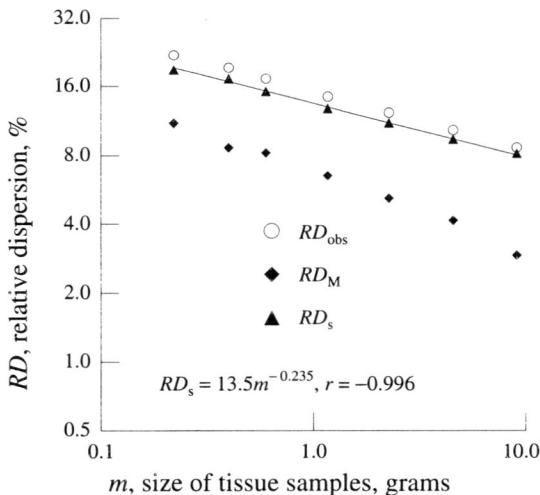

Figure 10.3. Fractal regression for spatial flow variation in left ventricular myocardium of a baboon. Plotted are the relative dispersion of the observed density function (RD_{obs}), the methodological dispersion (RD_M), and the spatial dispersion (RD_s) at each piece mass calculated using Eq. 10.10. (Figure reproduced from Bassingthwaighte et al., 1989; Fig. 5, left panel.)

important difference between the log-log regressions and the optimized best fits of Eq. 10.3 against the data using linear least-squares, and the expedient process of using the linear least-squares fit of the logarithms was considered acceptable in this circumstance, as discussed by Berkson (1950).)

Correction for measurement error. When the deposition of a radioactive tracer is used to measure the relative dispersion of flows, the observed dispersion, RD_{obs}, is the composite of at least two dispersive processes, which we distinguish as spatial dispersion, RD_s, and methodological dispersion, RD_M. The *spatial dispersion* is the true variation due to the heterogeneity of blood flow. The *methodological dispersion* is the variation of the mean in each sample around the true mean. It is the spatial dispersion for which we need to uncover the basis. If variation due to the method and due to the spatial heterogeneity are independent processes, then the total variance is the sum of the variances of the components. Since all the distributions are normalized to have a mean of unity, this relationship can be summarized in terms of the RDs:

$$RD^2_{obs} = RD^2_s + RD^2_M.$$ (10.9)

The methodological error can be estimated from the differences in deposition of simultaneously injected tracers (Bassingthwaighte et al., 1990a) or, taking the error to include temporal fluctuations, from the variation observed in multiple depositions separated in time (King et al., 1985). From this we calculate the true spatial dispersion RD_s to be the observed dispersion minus the methodological dispersion, assuming the latter is independent of the true dispersion:

$$RD^2_s = RD^2_{obs} - RD^2_M.$$ (10.10)

RD_M is composed of radioactivity counting error, which is inversely proportional to the square root of the number of disintegrations per minute, and statistical error reciprocal to the number of microspheres deposited, and weighing error. The methodological error is not likely to be an inherently fractal phenomenon. [An aside: If RD_M is nonfractal, that is, does not follow Eq. 10.3, then the combination of Eq. 10.3 and Eq. 10.9 can still hold true for both RD_s and RD_{obs}. Eq. 10.9 must always be true, but Eq. 10.3 cannot be applicable to more than two out of RD_s, RD_M, and RD_{obs}, as can be seen by substituting. However, since RD_{obs} contains the methodological error, the logic used in judging RD_M to be not necessarily fractal should also apply to RD_{obs}.] Consequently, the spatial heterogeneity as characterized by RD_s has the best possibility of being fundamentally fractal, and observed dispersions were corrected for methodological dispersion using Eq. 10.10 and values of RD_m such as shown in Fig. 10.3 to give RD_s before fractal analysis was applied.

Relative dispersion in lung flows. The regional flows in the normal lung have been examined in a similar way by Glenny and Robertson (1990) using both a slice-by-slice gamma camera imaging method (about 5 mm resolution) and the "slice and dice" technique, as described above for the heart, giving about 2-cm cubic resolution. The *RD* versus element size relationships in Fig. 10.4 show that the fractal slopes are lower for the lung than for the heart, that is, fractal *D*s for the lung are closer to 1.1 than to 1.2. There is a suggestion of a superimposed cycling, reminiscent of that in tracheal diameters shown in Fig. 3.9.

Spatial correlation is characterized by the fractal dimension. A useful and important feature of the fractal relationship is that it spells out the degree of correlation between neighbors in a spatial domain. In this section we consider the regional flows as a one-dimensional intensity, analogous to a voltage as a function of time. Really we should be considering the regional flows as the intensity of a property in three-dimensional space, just as we would consider the density of a cloud of chlorine to give a measure of its toxicity at different positions in a three-dimensional domain. In reconstructing the heart or lung from the smallest pieces aggregated into larger and larger pieces, we were careful always to group together the nearest neighbors to form the aggregates of a larger size. When pieces of a given size are cut in half, then the increase in apparent variation will be greater when the two halves of each piece are uncorrelated than when they are correlated. Van Beek et al. (1989) worked out the general relationship, derived in Chapter 4, for nearest-neighbor correlation, r_1:

$$\text{Correlation coefficient, } r_1 = 2^{3-2D} - 1, \quad \text{or} \quad r_1 = 2^{2H-1} - 1. \quad (10.11)$$

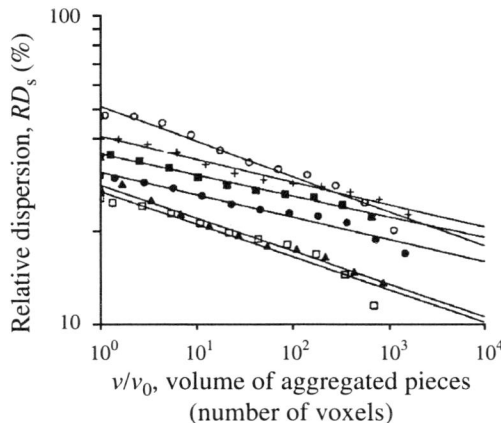

Figure 10.4. Heterogeneity of lung blood flows. Relative dispersion (*RD*) of regional pulmonary blood flows plotted as a function of volume of aggregated lung pieces. Smallest regions (v_0) are "voxels" (volume elements) from a planar gamma camera in which voxels are $1.5 \times 1.5 \times 11.5$ mm or 24 (this is millimeters cubed). (Figure reproduced from Glenny and Robertson, 1990.)

When $D = 1.5$, then $r = 0$, and when $D = 1.0$, $r_1 = 1$, fulfilling expectations. With $D = 1.2$, the average for our sheep and baboon hearts, then $r_1 = 2^{0.6} - 1$ $= 1.52 - 1 = 0.52$. The effect of doubling the resolving power, cutting pieces in half for example, increases the RD predictably. If the fractal dimension is 1.2, halving the voxel sizes increases the relative dispersion by only a relatively small factor, $2^{1/5} - 1$ or 15%. (This is obtained by $RD(m_2)/RD(m_1) = 1/2^{1-D} = 1/2^{-0.2} = 2^{1/5}$.) The correlation between pieces of a given size is expected to decrease as the distance between the pieces increases; this is analogous to the decay of electrical potential away from a point source of current. The description of the diminution with distance or the number of intervening units is given by an expression derived (Chapter 4) for infinite numbers of units in an organ:

$$r_n = \frac{1}{2} [|n + 1|^{2H} - 2|n|^{2H} + |n - 1|^{2H}] , \qquad (10.12)$$

where n is the number of units (of any chosen uniform size) between points of observation. The absolute value signs allow noninteger values of n, negative values, or values less than one. This can be seen to fit heart data at two different piece sizes in Fig. 10.5. This is the self-similarity or "double-bind" test described in Chapter 4. Using this method the value for D is a little higher than that obtained by the RD method, presumably because methodological noise is included in the estimate. (The method should be improved to correct for measurement error.)

Eq. 10.12 is a general rule for fractal, correlated characteristics in a domain of infinite size. The slow diminution in correlation over long distances is its hallmark, fractal extended-range correlation. In examining the correlation falloff within the heart and lung over more than a few centimeters, however, a contradictory phenomenon is observed: the correlations fall, through zero, to become *negative* at larger distances. This is to be expected in a finite domain: the corollary to short-range correlation in local flows is that there must be inverse or negative correlation

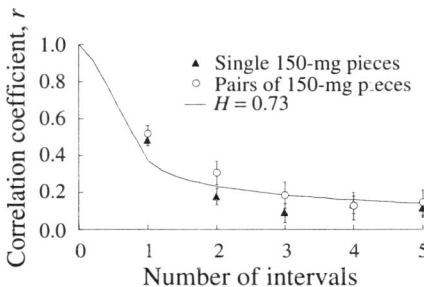

Figure 10.5. Correlation in flows in spatially separated regions. The correlation falls off similarly for 150-mg units (▲) and 300-mg units (○), even though the latter are actually twice the distance between centers. A value of $H = 0.73$ or $D = 1.27$ describes the falloff. (From Bassingthwaighte and Beyer, 1991, Fig. 7, right panel.)

at long distances. If the flow in one part of the organ is lower than average, and is similar in its neighboring parts, then in distant parts it must be higher than the average. At this time we do not have a theoretical expression to state the principle more precisely.

Correlation in organs of finite size. The fractal extended-range correlation expressed by Eq. 10.12 really applies only to domains of infinite size. When applying the idea to flows within an organ of finite size the concept *must* fail for the simple reason that if nearby regions have similar flows, being for example much higher than the mean flow, then there must be regions more distant from each other that are negatively correlated, one having flow above the mean and the other below the mean. Likewise, in an organ with constant total flow, if the flow in one region increases, the flow in some other region must decrease. The consequence is that Eq. 10.12 fails: it fits the data well over several intervals when those intervals are small compared to the total domain size, but the correlation falls to zero at distances around half the length through the organ. At distances greater than half the organ thickness, values of r tend to show more scatter but are mainly negatively correlated. Even if the system is truly fractal when observed at high magnification, the statistical correlation cannot show exact self-similarity tested for as in Fig. 10.5. In the dispersional analysis the corollary of the negative correlation at long distances is the smaller values of RD at large element sizes, m, so that there is downward curvature of RD versus m at large m.

10.4   Fractal Vascular Anatomy

The arterial network is composed of a series of segments of cylindrical tubes joined at branch points. The downstream, daughter branches are smaller than the parent. Many vascular beds show anastomoses or arcades, but the importance of these varies from tissue to tissue and from animal to animal. In the hearts of pigs and humans, the coronary artery system is basically one of dichotomous branching, two daughters from a parent, repeated recursively down to the terminal arterioles. Ninety-eight percent of branchings are dichotomous (Kassab et al., 1993). At the terminal arteriolar endings in the heart the story changes. Instead of dichotomous branchings, there tends to be "arteriolar bursts," where the arteriole goes through two to four sets of branchings within a short distance, 50–100 μ. Within these "bursts," the branchings are not necessarily dichotomous, but the parent may give rise to three or four branches. The "burst" or "flower" arrangement on the end of the arteriolar stalk then feeds a multitude of capillaries. Capillaries are not a part of the fractal network but may be regarded as a swamp. In the heart they lie in long parallel arrays, one capillary per cell in cross-section, and a single capillary may be traced for many millimeters, even centimeters. This hugely dense network composed of 2000–4000 capillaries/mm^2 in cross-section is fed by many arteriolar bursts scattered in three-dimensional space and drained by roughly twice as many venular confluences. Venules, larger and commonly oval in cross-section, travel with the arterioles, two venules per arteriole, through much of the network. This is not so on the epicardium, where there are many venules unaccompanied by arterioles, which tend to lie below the surface.

What Is the Evidence That the Arteriolar Network Is Fractal?

Van Bavel (1989) observed approximately logarithmic relationships between the diameters of parent and daughter vessels in dog myocardium. Strikingly detailed and persuasive studies were accomplished by Suwa et al. (1963) and Suwa and Takahashi (1971) in the mesenteric and renal arterial beds. They observed log-log relationships that were apparently linear over 2 to 3 orders of magnitude in dimensions for vessel radii, branch lengths, and wall thicknesses. The logarithmic relationships illustrate the approximate constancy of the ratio between parent and daughter dimensions. The relationship (which is typical of many of those illustrated in their papers) is shown in Fig. 10.6. The fact that there is a single log-log slope without an obvious break suggests that either there is a single recursion rule, or if there is more than one recursion rule then all of the rules are repeated at each generation. In the figure shown, the rule might be "at each branch point, make two daughter branches 70% as long as the parent branch." This of course provokes the question "How in the world would the daughter branch know the length of the parent branch?" So obedience to such a rule is no trivial matter, and the mechanisms for adhering to such simple descriptive rules require much research.

In a branching arterial system, or a tree, it is easier to understand that there might be constancy in the diameter–length relationship. Van Bavel (1989) observed approximately logarithmic relationships. In a very detailed analysis of the pig

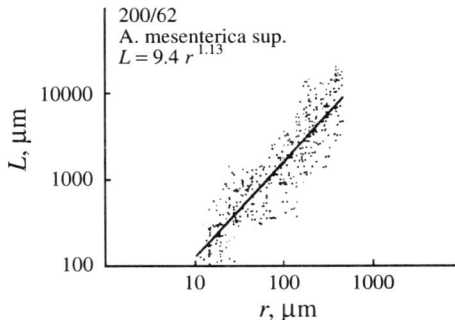

Figure 10.6. Relationship between the length, L, of the branches of the superior mesenteric artery plotted against radius, r, of the branch on a log-log scale. The approximate constancy of the ratio over the large range of lengths is compatible with a fractal recursion rule. (Figure reproduced from Suwa et al., 1963, their Fig. 10, with permission.)

coronary arterial system, Kassab et al. (1993) provide data which we find to fit simple log-log scaling relationships such as in Fig. 10.7. The slope is consistent with the renal and mesenteric artery data from Suwa et al. (1963).

The vascular beds of organs are not very treelike. As a generality, the vascular beds of plants tend toward filling a sunlight-gathering two-dimensional surface rather than a three-dimensional space. Even so, there are similarities and in some ways the circulation of the mesentery can be equated to that in veins of a leaf: there are dichotomous or trichotomous branching systems with anastomosing arcades. In organs, as in leaves, the vascular system grows as the organ grows. The primary branches enlarge, but their topology is maintained, even as they undergo remodeling with respect to the details of their form, length, wall thickness, and so on. It is likely that the major arteries to an organ develop in some nonfractal fashion just as the aorta and pulmonary artery develop from the primitive branchial arches. At the other extreme, at the solute delivery end of the network, the form of the capillary bed is determined by the nature of the tissue, the arrangement of the cells and the requirement for close proximity of capillaries to the functioning cells, so the capillary geometry is secondary to the arrangement of the cells. Only the intermediate network of arteriole-feeding networks and venular-draining networks can really follow simple recursion rules.

10.5   Dichotomous Branching Fractal Network Models for Flow Heterogeneity

Vascular trees seem to display roughly the same patterns at different levels of scale (Suwa et al., 1963; Suwa and Takahashi, 1971), a property found in fractal structures (Peitgen and Richter, 1986; Peitgen and Saupe, 1988). Mandelbrot (1983)

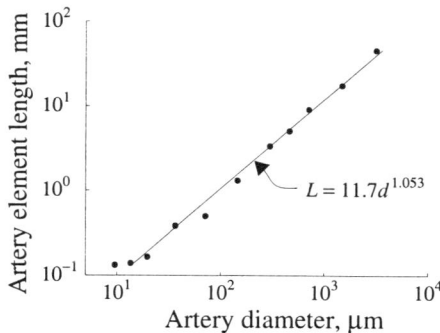

Figure 10.7. Analysis of casts of the coronary left anterior descending arterial tree of a pig heart: length, L, of average element (i.e., segments in series) of a given generation versus the average diameter, d, of the lumen. The logarithmic slope is 1.053. (The data points are from Kassab et al., 1993a.)

suggested that fractal bifurcating networks mimic the vascular tree. Lefèvre (1983) approximated physiologic and morphometric data on the pulmonary vasculature in describing it as a fractally branching structure; the model minimized the cost function of energy and materials in delivering blood to the peripheral lung units. West and Shlesinger (1990) used fractal concepts to describe length and diameter relationships in the bronchial tree. Dichotomous branching tree structures have been considered as algorithmic descriptors of the coronary arterial network by Pelosi et al. (1987) and Zamir and Chee (1987); those of Pelosi et al. allowed for asymmetry such as was observed in the pial arteriolar microvessels by Hudetz et al. (1987).

The dichotomously branching and randomly asymmetric variation models of simple bifurcating networks of blood vessels that are described below prove adequate to describe the dependence of the relative dispersion of the flow distribution on the size of the supplied region, even though they give overly simple descriptions of the vascular network. In each case the form of the microvascular network is linked to an experimentally measurable variable, local blood flow, and its heterogeneity in the normally functioning organ. While the approach has been tried only on the heart and the lung, it seems safe to predict that such fractal approaches will be useful in describing other systems with heterogeneous flow distributions.

Dichotomously Branching (Bifurcating) Models

Constant asymmetry model. The basic element is a dichotomously branching structure in which the distribution of blood flow between daughter branches is characterized by an asymmetry parameter, γ. This model is *model I* of van Beek et al. (1989). Flow, volume/time in the parent branch is F_0, the flow to one daughter branch is $\gamma \cdot F_0$, and flow to the other daughter branch is $(1 - \gamma) \cdot F_0$ (Fig. 10.8, left).

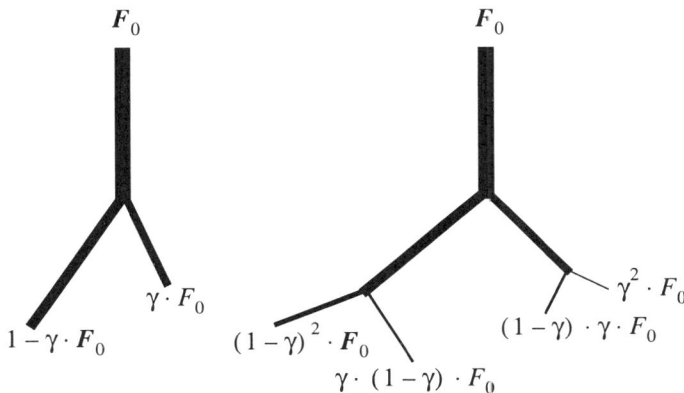

Figure 10.8. Dichotomously branching fractal model. *Left panel*: Basic element in which fractions γ and $1 - \gamma$ of total flow F_0 are distributed to daughter branches. *Right panel*: Flow at terminal branches in network of two generations. (Figure after van Beek et al., 1989.)

A γ of 0.5 will produce equal flows in both daughter branches and a uniform flow distribution in a network. As γ diverges from 0.5, flow becomes more asymmetric.

A second generation of branches is produced by appending this basic element to the end of each terminal branch (Fig. 10.8, right). The fraction of flow, γ and $1 - \gamma$, distributed to the daughter branches remains constant at each bifurcation. The asymmetry of flow in a parent branch affects the flows in all subsequent generations down the vascular tree.

A model of the pulmonary vascular tree can therefore be constructed with n bifurcations or generations and thus 2^n terminal branches or perfused regions of lung. These terminal branches have discrete values of flow given by the equation

$$ F = \gamma^k \cdot (1 - \gamma)^{n-k} \cdot F_0 \tag{10.13} $$

after n generations. F_0 is the total flow into the vascular tree, the generation number k has integer values from 0 to n, and the frequency of occurrence of each value for flow is $n! / [k!(n-k)!]$ (van Beek et al., 1989). Fig. 10.9 shows two examples of the frequency distribution of flows for a γ of 0.47 and 0.45 with n equal to 15. Fifteen generations produce a model with 32,768 terminal branches, which approximates the number of pieces in each of the experimental data sets on the lung (Glenny and Robertson, 1990). The flows are presented as mean normalized flows where the mean flow per branch after n generations is $F_0/2^n$. The flows generated by the network are similar to those obtained from experimental data in that they have a distribution which is skewed to the right.

The flow heterogeneity can be characterized by the RD. An analytic solution for RD as a function of γ and n has been obtained by van Beek et al. (1989) and is given by

$$ RD = \sqrt{2^n \cdot [\gamma^2 + (1 - \gamma)^n]^n - 1} . \tag{10.14} $$

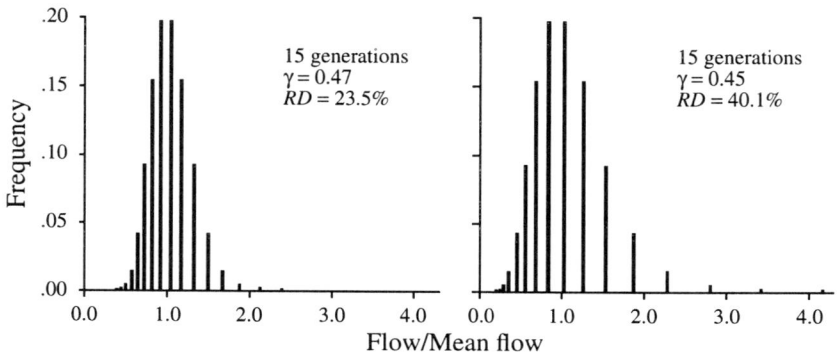

Figure 10.9. Frequency distributions of flow in network with fifteen generations. Flows are normalized to mean flow. (From Glenny and Robertson, 1991.)

Eq. 10.14 can now be related to the fundamental fractal Eq. 10.3, giving the relative dispersion for any n generation vascular tree. The RD after one branching is equivalent to partitioning the tree into two pieces. If the RD is determined after two branchings, the vascular tree is partitioned into four pieces. The volume of these pieces can be determined by arbitrarily selecting m_C equal to a single terminal branch. Because there are 2^n terminal branches in an n generation tree, the size of the pieces (m) after one branching will be $2^n m_0/2$ and the size of the pieces after two branchings will be $2^n m_0/4$. Generalizing this for all of the branchings in the vascular model with n bifurcations, $m = 2^n m_0/2^i$ for the ith generation, where i assumes integer values from 1 to n and

$$RD\,(m) \;=\; \sqrt{2^i \cdot [\gamma^2 + (1-\gamma)^2]^{\,i} - 1}\,. \tag{10.15}$$

Although Eq. 10.14 cannot be reduced to a power law equation of exactly the form of Eq. 10.3, the relationship between the heterogeneity of flow from the model (RD_{model}) and the perfused piece size (m) appears to be close to a power law relationship, which should be a straight line on the log-log plot. Fig. 10.10 shows the RD_{model} as a function of piece size for a fifteen-generation vascular tree with a γ of 0.45. The individual points represent the relative dispersion calculated from Eq. 10.15 as a function of m. When plotted on a log-log scale after deleting the four largest pieces, the relationship appears almost linear, being only slightly concave

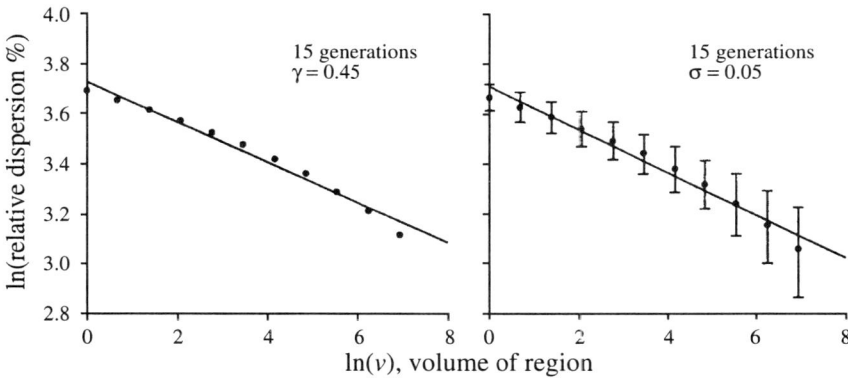

Figure 10.10. Relationship of relative dispersion (RD) to size of theoretical perfused region or generation level of the tree; the four largest pieces were excluded from fractal fit. *Left:* Log-log relationship of RD versus m for network constructed with constant asymmetry. Line is least-squares linear fit to data and represents theoretical fractal relationship of Eq. 10.3. *Right:* Log-log relationship of RD versus m for network constructed with random variation asymmetry with $\gamma = 0.5$ and $\sigma = 0.025$. Line again is least-squares linear fit to data and represents theoretical fractal relationship. *SD* bars represent variability in RD for each of 10 simulations. (Figure reproduced from Glenny and Robertson, 1991.)

downward, suggesting that it is approximately fractal over the range of observations (Glenny and Robertson, 1991).

The largest pieces are not used in the analysis because they do not fit the theoretical fractal relationship of Eq. 10.3 as well as the smaller pieces. Eq. 10.14 produces a slightly curved relationship between $\log m$ and $\log RD(m)$, while Eq. 10.3 gives a straight log-log plot (van Beek et al., 1989). The fractal dimension D, defined by the slope of $\log RD(m)$ versus $\log m/m_0$, is theoretically constant over the range of observations. However, D will gradually decrease as m decreases for the relationship of Eq. 10.15, and the fractal dimension is therefore best defined by $RD(m)$ for the smallest volumes.

Randomly Asymmetric Variation

This model is identical to *model III* of van Beek et al. (1989) and allows variability in the anatomical predicted asymmetry of flow as advocated by Horsfield and Woldenberg (1989). There is a different γ at each branch point; its value is selected from a Gaussian distribution with a mean $\gamma = 0.5$ and a standard deviation σ. With $\sigma = 0.0$ there is a uniform distribution. This model does not have an analytic solution (van Beek et al. 1989). A simulated vascular network is constructed by randomly assigning γ values of $0.5 \pm \sigma$ to each bifurcation. The RD can be calculated at each generation. Because each γ is randomly chosen, the flow distributions are different each time the simulation is run. An average distribution of flow can be estimated by running the model several times. Fig. 10.10 right shows the average RD (after ten runs) as a function of piece size for a fifteen-generation vascular tree with a $\gamma = 0.5$ and $\sigma = 0.05$. When plotted on a log-log scale, the relationship is curved, not linear, but a straight line fit is close enough that the result could be considered as statistically fractal as well as algorithmically fractal.

The distributions of flow are no longer discrete in this model. A pseudo-continuous distribution can be constructed by plotting frequency histograms of flow with narrow intervals. Fig. 10.11 shows frequency distributions of flow for two networks of fifteen generations with σ values of 0.03 and 0.05. The flow distributions are right-skewed and have RD values of 23.0 and 39.7%, respectively.

The curvature in the log-log plots of RD versus m using the branching models allowed just as good fits to data from heart and lung as did the statistical model of Eq. 10.3. In Fig. 10.12 (left) are shown the two branching model solutions for the heart, and in Fig. 10.12 (right) for the lung. The asymmetry of the heart is slightly greater than that for the lung, but at the higher degrees of resolution, the heart and lung have similar fractal dimensions, about 1.15.

10.6   Scaling Relationships within an Organ

What do fractal relationships do for us in understanding the nature of heterogeneities of properties or functions of the cells or tissues of an organ? Can we

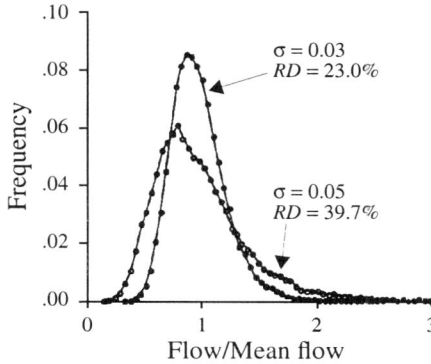

Figure 10.11. Probability density functions of local flow. Each distribution is a single realization of randomly asymmetric model for fifteen generations. (From Glenny and Robertson, 1991, their Fig. 4, with permission.)

compare the relative dispersion between completely and partially sampled organs, between different organs, and between the same organs in different animals and different species?

Incomplete sampling of a single organ is the simplest case. The question would be, for example, if only half the organ is sampled, what estimate of heterogeneity could one make for the whole organ? The approach is based on the presumption that

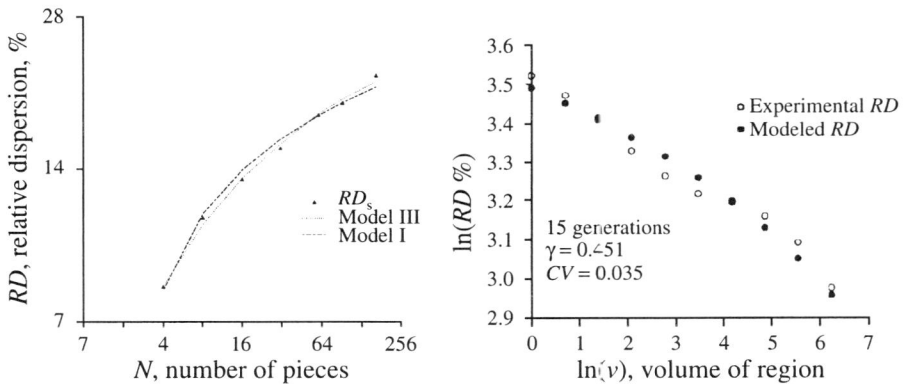

Figure 10.12. Fitting the fractal branching models to data on regional flow heterogeneity in the heart and lung. *Left panel:* Composite data from the hearts of ten baboons adapted from the analysis of van Beek et al. (1989). For constant asymmetry γ was 0.46; σ for random asymmetry was 0.043. *Right panel:* Data from one dog lung from Glenny and Robertson (1991). For constant asymmetry γ = 0.451; for random variation around γ = 0.5, a value of σ = 0.042 gave the result (modeled *RD*).

the same fractal D holds for the whole organ as might be estimated from the sampled half. Here's one approach:

Cut the available half organ into 64 pieces (or a handy number) and measure the property in each weighed piece. Attempt to keep the pieces all the same size, since the influence of variable sample size has not been worked out. Calculate the RD, as described in Chapter 4, for this smallest piece size, designating this as $RD(m_1)$.

Aggregate pairs of nearest neighbors together to obtain the mean value of the property per gram of tissue in each of these double-size pieces. Calculate the $RD(2m_1)$. Repeat for quadruple-size pieces, etc. Each aggregate may be formed in a few different ways, always with nearest neighbors.

Estimate the fractal D from the slope of the log RD, log piece size regression.

Assuming that the same degree of correlation applies to the unsampled half of the organ, then for any piece size, the RD for the whole organ is simply

$$\frac{RD \,(\text{whole})}{RD \,(\text{half})} = 2^{D-1}. \tag{10.16}$$

The more general relationship is

$$RD_2 \,(m) = RD_1 \,(m) \cdot (W_2/W_1)^{D-1}, \tag{10.17}$$

where RD_2 is the RD in mass W_2 using piece size m and similarly for RD_1. Thus the fractal relationship for the whole organ for RD versus piece size parallels that for the part, but is higher by a predictable ratio, as diagrammed in Fig. 10.13.

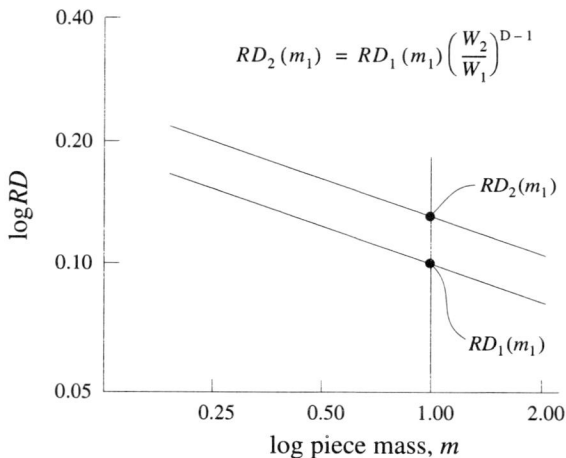

Figure 10.13. Fractal relationship for RD versus piece size in a large tissue mass, W_2, compared with the fractal RD for a subset of the same tissue. For this construction $D = 1.25$, $W_2 = 50$ g, $W_1 = 10$ g, $RD_1(m_1) = 10\%$, and $RD_2(m_1) = 0.1 \times 5^{0.25} = 0.15$.

10.7   Scaling Relationships from Animal to Animal

The range of heart masses in mammals is huge, from the mouse with a 0.1-g heart to a whale with a 60-kg heart (Crile and Quiring, 1940). This is a range of 600,000 times. The overall relationship for the hearts of individual animals of different size might all be assumed to have the same fractal D, 1.2, or $H = 0.8$. This means a nearest-neighbor correlation coefficient, $r = 2^{2H-1} - 1 = 2^{0.6} - 1 = 0.51$. A further assumption might be that the heterogeneity *within* a 1-g volume of tissue might be the same for all animals. Taking a standard resolution level such as $m_0 = 1$ g, then there is a general relationship for which one equation might describe both the heterogeneity at various degrees of spatial resolution within one animal and the relationships for the whole hearts of animals of different size.

Writing the $RD(W, m)$ as the expected RD for a heart of total mass W at uniform sample size m, we restate Eq. 10.1 for a heart of mass W_0:

$$RD(W_0, m) = RD(W_0, m_0)(m/m_0)^{1-D}, \qquad (10.18)$$

and for comparing this with a heart of mass W, we restate Eq. 10.17:

$$RD(W, m) = RD(W_0, m_0)\left(\frac{W}{W_0} \cdot \frac{m_0}{m}\right)^{D-1}. \qquad (10.19)$$

The corollary of this is that for such a set of hearts obeying this relationship, the observed $RD(W, m)$ is the same for all hearts that are cut in N equal-size pieces regardless of size, such that when $N = W/m = W_1/m_1$,

$$RD(N) = RD(N_0) \cdot (N/N_0)^{D-1}, \qquad (10.20)$$

regardless of whether an $N \neq N_0$ is obtained by piecemealing organs of the same size to a different level of spatial resolution or piecemealing organs of different sizes. A set of such scaling relationships is shown in Fig. 10.14. The hypothesis expressed by Eq. 10.20 and the figure postulates exceedingly large variance in regional flows for large animals.

Can the heterogeneities in regional myocardial flows be similar to those in Fig. 10.14? A more precise version of this question can be based on the expectation that the microvascular unit sizes in the mouse and the elephant probably do not differ by more than a factor of four. We think of the microvascular unit as being the set of capillaries fed by one terminal arteriole and the tissue associated with these capillaries. The cardiac muscle cells in the elephant are perhaps twice as large, the intercapillary distances accordingly 40% or 50% greater, and the oxygen consumption per cell perhaps half of those in the mouse, so it is likely that the size of the region supplied by a single terminal arteriole is within four times that in the mouse. Then the question is, "Can the fractal relationship in the elephant for regional flow heterogeneity be a nice linear one over the whole range of piece sizes

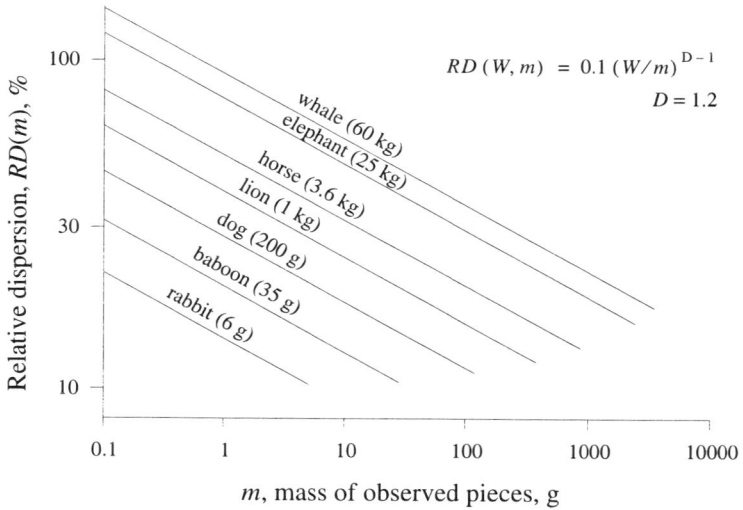

Figure 10.14. Scaling relationships hypothesized for mammals assuming that all have the same fractal D, 1.2, and the same RD *within* a 1-g volume of tissue, $RD(1\ g) = 10\%$. The graph raises questions about the hypothesis, since the RD would be extraordinarily large in large mammals.

and have a fractal D similar to that in the mouse?" The answer is surely no; the hypothesized relationship is seemingly out of the question. The data on hamster hearts (Stapleton et al., 1988) would indicate a relative dispersion about 55% at the microvascular unit size of 50 to 100 μg. The fractal D is not known exactly, but this observation would fit a fractal D around 1.18 for this size heart (0.5 g). If a 10-kg elephant heart had the same dispersion relationship within 0.5-g pieces of myocardium, the overall RD would be 590%. (The calculation is $(10000/0.5)^{0.18}$.) Consequently, one can say that either the distribution is not remotely symmetric and is highly right skewed, or that the log-log fractal relationship is not valid. Since only fairly symmetric distributions of flows have been found in animals up to the size of dogs (200-g hearts), the suspicion is that this fractal relationship doesn't hold. The curvilinear relationships suggested by the dichotomous branching algorithms certainly curve in the appropriate direction, but are really only convenient descriptions rather than fundamental anatomically-based structural analogs.

10.8   Do Fractal Rules Extend to Microvascular Units?

The slopes of the power law relationships for RD versus m differed from animal to animal, but in each set of observations in a given species, baboons, sheep, and rabbits, there was a consistent phenomenon. Those animals that exhibited large relative dispersion at the 1-g reference size had lower fractal dimensions or slopes than did those that had small relative dispersions. Contrary to the implications of Fig. 10.14 and Eq. 10.20, there was no statistical influence of heart size on $RD(1\ g)$, but the range of heart sizes was less than twofold. From this we surmised that the data for a given species might be represented by a family of lines intersecting at a common point. This idea was applied by Roger et al. (1993) as illustrated for the sheep data in Fig. 10.15.

The point of intersection at $RD = 102\%$ at a volume element size of 75 µg is shown in Fig. 10.15. The lines through the sets of data have the common equation

$$RD = 102\% \cdot (m/(75\mu g))^{1-D}. \tag{10.21}$$

The fractal dimension gives the individual lines; each animal has a separate and distinct fractal D, which can be calculated from the RD observed with 1-g pieces:

$$D = 1 - \frac{[\log RD(1)/1.02]}{\log[75 \times 10^{-6}]}. \tag{10.22}$$

This relationship then gives a common representation of all the data for the ten baboons and eleven sheep shown in Fig. 10.16.

This intersection at a 75-µg piece size might suggest the common microvascular unit is about 75 µg or 75 nanoliters in volume. This estimated mass for a functional unit, supplied by one arteriole, is about 1/13 mg (or 1/13 mm^3). This is compatible

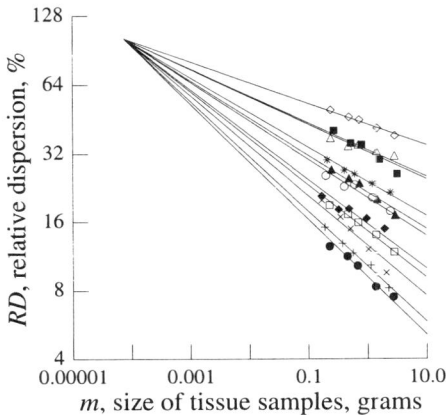

Figure 10.15. Projection of the fractal relationships for the relative dispersions of regional flows for eleven sheep hearts through a common point. The best fit was obtained with the intersection point at $m = 75$ µg and $RD = 102\%$ ($r = 0.975$). (From King et al., 1990, their Fig. 12, with permission.)

with the estimates of Eng et al. (1984), Kaneko (1987), and Bassingthwaighte et al. (1987a), but larger than those which can be calculated from Batra and Rakusan (1990). However, at an *RD* of 102%, the apparent heterogeneity at this projected microvascular unit size is large compared with the direct observations of Stapleton et al. (1988), who obtained estimates of *RD* under 60% from high-resolution measurements in hamster hearts. They used autoradiography to assess flow heterogeneity in element sizes of greater than 100 microns cubic, using the "molecular microsphere," iododesmethylimipramine developed by Bassingthwaighte et al. (1987). While the fractal relationships were not well determined in Stapleton's observations, due to limitation in the statistical assessment, this provoked a reassessment. After all, the projection of the data on the sheep and baboons was an extrapolation from the smallest observed pieces of about 100 mg down to pieces of less than 100 μg, a three order of magnitude extrapolation, leaving a lot of room for doubt.

Van Beek et al. (1989) sought a combination of an explanation for the flow heterogeneity and one that gave a more reasonable, that is, smaller, estimate of the *RD* at the unit size. They found that small asymmetries in a dichotomously branching system could give rise to the appropriate relative dispersions as described above. The plots of relative dispersion versus number of branchings or piece size from such networks are curvilinear, not straight, even though their basis is perfectly fractal. These curvilinear relationships gave slightly better fits to the data on the individual sheep and baboons than did the power law relationships.

The branching-algorithm, curvilinear relationships could also be extrapolated, in the same fashion as were the power law, straight-line relationships to an intersection

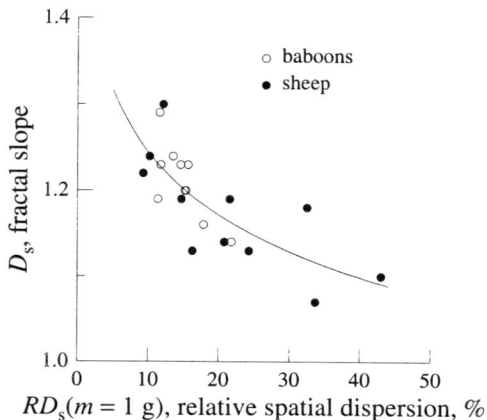

Figure 10.16. Scatter plot of the fractal dimension versus the relative dispersion of regional flows (at $m = 1$ g). The line is the best-fitting parameterization of Eq. 10.19 for the data from ten baboons and eleven sheep (the coefficient of variation is 0.032), using the equation for a common point of intersection of the fractal relationship. (Figure reprinted from Bassingthwaighte et al., 1990b, their Fig. 4, with permission.)

at a common point surmised to be the functional microvascular unit. Fig. 10.17 shows data on ten baboons, where the branching algorithms extrapolate to an estimated microvascular unit size of 50 µg. This extrapolation gives an estimate of the projected overall heterogeneity of a distribution of flows at a standard deviation of 55% of the mean. This is entirely compatible with the direct observations on the hamster hearts using the molecular flow marker. Data on larger hearts have not been obtained at this level of observation. The observation and projection also raise the issue of whether the heterogeneity in larger hearts is, for any given unit size, larger than in small hearts. It would seem reasonable that it might be, since there must be a larger number of generations of branching to supply a given unit size in a large heart than in a small heart. An argument against this is that the metabolic requirements of a given unit size might in fact be the same in all hearts, at least when normalized against total body metabolic rate, since small animals normally have higher metabolic rates than large animals (Schmidt-Nielsen, 1984). The ratio of heart mass to body mass varies greatly among species: for example, dog hearts are about 1% of the body mass while sheep hearts are only 0.3%. Such differences in ratios of heart mass to body mass presumably reflect the metabolic activity of the heart and the maximum ability of the animal to exercise (McMahon and Bonner, 1983).

10.9 Fractal Flows and Fractal Washout

The foregoing sections provide good evidence that both regional flow distributions and vascular networks have fractal characteristics. The observed fractal flow

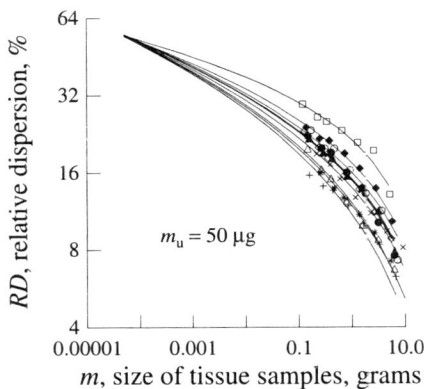

Figure 10.17. Projection of a dichotomously branching fractal arterial network model to fit the data from the hearts of ten baboons. The common point of intersection shows a relative dispersion of 50% at a common unit size of 50 µg (coefficient of variation = 0.045). (Figure reproduced with permission from Bassingthwaighte et al., 1990b, their Fig. 6, with permission.)

relationships extend over only a 200-fold range, and there are no data that are continuous of the range from large regions down to the capillary level. Clearly the vascular branching fractals cannot extend to generations beyond the arteriolar level; however, it is possible that the capillary network behaves as a percolation field, which would also distribute flows in a different, but yet fractal, manner.

The prediction from such data is that if regional flows and network structures are fractal then the time course of tracer washout after pulse injection into the arterial inflow should also be fractal. From isolated blood-perfused rabbit hearts after injection of ^{15}O-water into the inflow Deussen and Bassingthwaighte (1994) obtained both the residual tracer content of the organ, $R(t)$, and the outflow concentration-time curves, $C(t)$, as in Fig. 10.18. Clearly the washout is *not* monoexponential, $R(t)$ and $C(t)$ being curved on semilog plots (left). The tails of both $R(t)$ and $C(t)$ are straight power law functions (right), and therefore fit the prediction of being fractal. The relation is $R(t) \sim t^{\alpha}$, where α is the exponent characterizing the self-similarity over the time scale. For $R(t)$, α is -2.12 in this case, a less steep slope than for $C(t)$, -3.57, as must be the case when $R(t)$ is concave up on the semilog plot (left). The exponent α is not at this point directly translatable in a descriptor of the network fractal either in terms of flow distributions or vascular patterns; as pointed out by Havlin (1990), the interpretation of α is dependent on a combination of the network structure and the nature of the flow and diffusion phenomena. In any case, the vascular structures, the regional flow, and tracer kinetics are all apparently fractal.

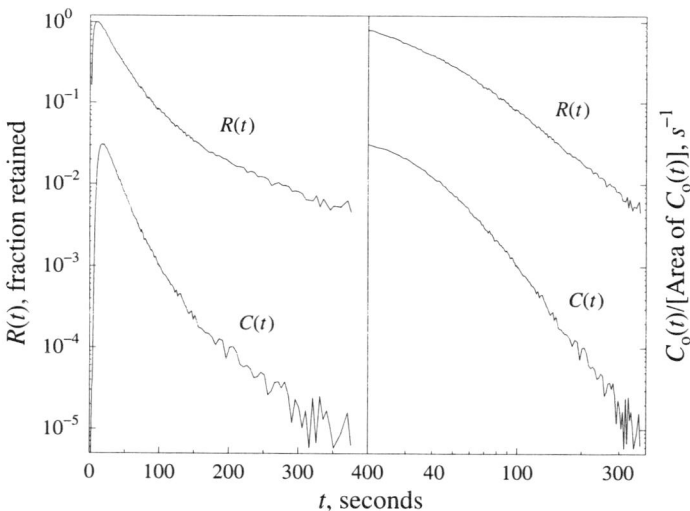

Figure 10.18. Residue curves, $R(t)$, and outflow concentration-time curves, $C(t)$, from rabbit after injection of ^{15}O-water into the coronary inflow. Both $R(t)$ and $C(t)$ have tails that are powerless functions of time, that is, they are straight on log-log plots (*right panel*) and clearly not monoexponential, being concave up on semilog plots (*left panel*).

10.10   Summary

What the Fractal Analysis Tells about Fractal Intraorgan Heterogeneity

Fractal analysis is useful for discussing flow heterogeneity in the myocardium at three levels. We do not know if these concepts also apply to tissue properties other than local flow. The first level is simply *descriptive*, giving a common relationship for flow heterogeneity over a large range of sizes of the observed tissue samples. This description is important because it allows data of one laboratory to be interpreted by another, independently of the sizes of the samples. The second level is one of *explanation*, namely, that fractally branching arterial networks *can* give rise to the observed heterogeneity. Whether the dichotomous networks giving the fits shown here are close approximations to the real arterial networks is as yet unknown, since sufficient data have not yet been acquired on coronary arterial branching patterns. A third use of the fractals is in projecting the heterogeneity to the *functional microvascular unit size*. The extrapolation is over three orders of magnitude, an uncomfortably large range, but nevertheless the estimated dispersions are compatible with those obtained by making observations in another species in pieces within the range of the expected microvascular unit sizes. Verification is not secure when it is from another species, and so the question remains, and needs to be resolved by observations in larger hearts made with pieces ranging from many grams down to pieces of 20 to 100-µg sizes.

One tends naturally to presume that high flow within a tissue element means that the local metabolic rate is high. To date no data bear directly on this point, although some point in this direction, and of course whole organ data support the idea—increasing metabolic demand increases flow. Some sparse data indicate that the amount of machinery for transporting and metabolizing substrates is proportional to the local flow capacity. Upsher and Weiss (1989) noted that adrenergic receptor densities correlated moderately well ($r = 0.6$) with regional myocardial flows when the metabolic demand was high. Caldwell et al. (1991) noted that the deposition of tracer-labeled fatty acid, the main substrate for energy production in the heart, was almost proportional to flow in the hearts of dogs exercising strenuously. For this to occur, the transport capacity of the membranes is also required to be almost proportional to flow. The suggestion is that flow, transport capacities, and metabolism match locally. If so, since local flows are spatial fractals, then so also would be local metabolic rates.

Future Directions

While fractal relationships have not been proven to be the basis of flow heterogeneity in the myocardium, the use of fractal approaches gives rise to some ways of making observations and asking questions that seem useful in the field. Further studies need to be accomplished before the assessment is complete.

Consideration of the fractal nature provokes examination of the spatial heterogeneity, its anatomic and functional basis and the relation to temporal fluctuations. If metabolic drive dominates the long-term regulation of regional flows, what are the mechanisms that lead to the stable differences between high- and low-flow regions? Is capillarity higher in high flow regions? Are arteriolar diameters greater? If coronary flow reserve, defined as the ability of a region to decrease its resistance in response to a vasodilator infusion, is only present in nonischemic regions, as Coggins et al. (1990) interpret their data, then one is driven to accept metabolic drive as the dominating one at low flow. But what is the relative importance of myogenic regulators, of endothelial-derived relaxing factor (EDRF), of arteriole-to-arteriole coordination, and of shear-dependent vasodilation in the normal operational state? Such considerations go beyond anatomic branching patterns and into consideration of the dynamics.

Background References

This chapter has summarized recent research, principally on the heart and lung. The roots lie in three sources, the wealth of data on arterial structures (Suwa and Takahashi, 1971), the early observations of intraorgan flow heterogeneity (Yipintsoi et al., 1973), and the structural/functional relationships of Lefèvre (1983). Recent work is described in the articles by Bassingthwaighte (1991a, 1991b), Glenny et al. (1991a), and Wang et al. (1992).

11

Fractal Growth

> Organic form itself is found, mathematically speaking, to be a function of time . . . We
> might call the form of an organism an event in *space-time*, not merely a *configuration in*
> *space*.
>
> D'Arcy Thompson (1942)

11.1 Introduction

Humans have but 100,000 genes made up from about 10^9 pairs of bases (purines or
pyrimidines) in DNA. There are about 250 different cell types in the body and each
has a multitude of enzymes and structural proteins. The numbers of cells in the body
is beyond counting. The numbers of structural elements in a small organ exceeds the
numbers of genes; the heart has about 10 million capillary tissue units, each
composed of endothelial cells, myocytes, fibroblasts, and neurons. The lung has
even more. Consequently, the genes, the instruction set, must command the growth
of cells and structures most parsimoniously, and end up with functioning structures
that last for decades.

The background for the fractal descriptions of growth processes comes from the
mathematicians and physicists as much as from the biologists. D'Arcy Thompson
was one of the path finders, setting forth the principles of scaling of animal form
with respect to form and function. His book *On Growth and Form* (1961, the
abbreviated version of the 1925 original) covered the self-similar forms of spiral
growth of snails and conches. The form is the logarithmic spiral (see Fig. 11.1),
labeled the "Spira Mirabilis" by Bernoulli; the form is $r = A^\theta$, where r is the radius,
A a constant, and θ the angle of rotation, so that the straight line from the center
intersects the arc at a constant angle independent of θ. The curve has the same shape
at all magnifications. Thompson (1961) traces back the heritage of these ideas to the
seventeenth century, observing, "Organic form itself is found, mathematically
speaking, to be a function of time . . . We might call the form of the function an
event in *space-time*, and not merely a *configuration in space*."

Lindenmayer (1968a, 1968b) developed a formalism for describing
developmental processes. These ideas evolved, like Thompson's and Darwin's,
from observation. "In many growth processes of living organisms, especially of
plants, regularly repeated appearances of certain multicellular structures are readily
noticeable . . . In the case of a compound leaf, for instance, some of the lobes (or
leaflets), which are parts of a leaf at an advanced stage, have the same shape as the

Figure 11.1. Shell of the chambered Nautilus, halved, showing the log spiral form. (Figure reproduced from Ghyka, 1977, with permission.)

whole leaf has at an earlier stage." His L-systems were developed first for simple multicellular organisms, but developed a theoretical life of their own, serving as the basis for computational reconstructions of beautiful portrayals of plants and flowers (Prusinkiewicz et al., 1990). The idea of a leaf being composed of its parts is the basis of Barnsley's Collage theorem, which he used to construct, for example, a wonderfully realistic fern (Barnsley, 1988b).

Vascular branching patterns were being explored without putting them in terms of fractals, and even now it is not proven they must be fractal. Horsfield and colleagues (Singhal et al., 1973) embarked on a magnificent series of studies related to the lung vasculature and airways, following some of the strategies taken from Strahler's geomorphology (Strahler, 1957). The work reflected approaches initiated by the renowned anatomist-morphometrist Ewald Weibel with Domingo Gomez, who had an innovative flair for mathematical applications (Weibel and Gomez, 1962).

The growth patterns in mammals are neither simple nor primitive. The huge number of evolutionary steps has led to vascular growth processes that are highly controlled, well-behaved patterns precisely matched to the needs of the organ. A glance at a casting of the microvasculature reveals to the eye a pattern that uniquely identifies the organ, even though none of the cells remain to show the fundamental structure of the organ. In general, there is a capillary beside every cell of the organ, to bring nutrient and remove waste with minimal diffusion distance. The principles of optimality applied to growth raise questions. Which functions are to be optimized or minimized? Vascular volume? Delivery times for oxygen? The mass of material in the vascular wall? The energy required to deliver the substrates to the cell? As

Lefèvre (1983) expressed it, one must find the cost function for the optimization. The cost function is the weighted sum of the various features that are to be minimized; we need to figure out the definition of the cost function in order to gain deeper understanding of the total processes involved in growth, deformation, remodeling, dissolution, and repair.

What is clear is that growth of a particular cell type requires the collaboration of its neighbors. A heart does not develop well from the mesodermal ridge unless the nearby neural crest is also present (Kirby, 1988). The endoderm seems also to be necessary (Icardo, 1988); this is not so surprising, for we now recognize the importance of endothelial-derived growth factors.

The majority of explorations of branching fractals have used dichotomous branching, variants on the binary tree. There are plants that have triple-, quadruple-, and higher-branching ratios, particularly near the flowers or terminal leaf groups, but the trunks of deciduous trees tend toward binary branching. From the algorithmic point of view it probably does not matter much, because a ternary branch can be regarded as approximating two binary branches with a short link between them.

No matter how the system forms, branching is the hallmark of fractal systems, from watershed geomorphology to neural dendrite formation. In this chapter we explore some of the growth mechanisms and patterns. In concentrating on spatial structuring we ignore the dynamical fluctuations in flows within the vasculature (King and Bassingthwaighte, 1989; Bassingthwaighte, 1991b).

11.2   Primitive Growth Patterns

Diffusion-Limited Aggregation

When particles floating in a fluid medium are oppositely charged, or have sites that are mutually attractive, then there is a tendency for the particles to aggregate into doublets, triplets, etc., and multiparticle clumps. The processes involved are the diffusion (or possibly convection, as in stirring chemical reactants), and binding. The binding process may be specific—as, for example, requiring rotation and docking of a molecule to fit into the binding site on the recipient molecule, as for a substrate into a stereospecific site on an enzyme, or an agonist onto a receptor—but these local maneuvers tend to occur so rapidly that many aggregation events can be considered to be purely diffusion-limited, that is, the kinetics of the aggregation process are dominated solely by the rate of diffusion of the particles.

It is in this context that we think of diffusion-limited aggregation, DLA, as having features analogous to simple growth processes. If an epithelial cell in a glandular organ is growing, perhaps dividing, it secretes a growth factor that is a stimulant to vascular growth. The secreted molecule diffuses through the interstitial matrix randomly, and if it is not lost or degraded, randomly hits a target cell, e.g., an endothelial cell of a capillary or small venule. If the stimulus is sufficient, or repeated frequently enough, the endothelial cell responds, leading to endothelial bulging into the interstitium, protrusion to form a capillary bud, and extension to form a new connection allowing blood to flow. In the simplest of the DLA analogs of this process, a single "particle" touching the surface of the previously developed network results in a growth event.

Witten and Sander (1983) introduced an algorithm that mimics a large number of such processes. The original incentive was to consider the deposition of a metallic cation from a solution of low concentration onto a negative electrode. Diffusing randomly in the solution, the ions stick to the electrode when they hit it, and so coat the electrode. The surface coat is not smooth, but has a treelike structure. Their model was this: in a lattice plane, place a seed particle to which others will stick, then release another particle into the field at a distant point and let it diffuse randomly. If it should happen to hit the seed particle, let it stay in the adjacent position in the lattice. Repetition of this process leads to the formation of complex clusters of intriguingly general form.

The algorithm is diagrammed in Fig. 11.2, from Meakin (1983). The seed or origin is placed in a fixed location in the center of the field and a particle is released from a random position on a circle centered on the seed. The radius of the circle can be quite small because any particle released randomly at a greater distance would encounter this circle at a random point, so all that is required is that the circle be larger than the cluster growing from the seed. The diagram shows the movement of two particles on a square lattice: each particle moves randomly up, down, right, or left until an event occurs. In the illustration one particle wanders off into the distance without striking the cluster. The algorithm stops calculating a particle's movements after it vanishes across some prescribed distance from the center. The other particle diffuses randomly on the matrix until it lands on a position adjacent to one of the particles in the fixed cluster, at which time it also becomes fixed. In this algorithm the cluster does not move. Simple rules for the radius of the initiating

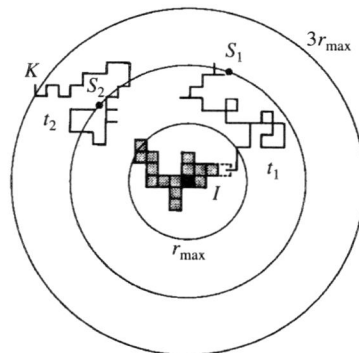

Figure 11.2. A square lattice model for diffusion-limited aggregation. This figure shows two typical particle trajectories (t_1 and t_2) at an early stage in a simulation. This model is discussed in the text. (Figure reproduced from Meakin, 1983, his Figure 1, with permission.)

circle might be the cluster radius plus five lattice units and for the outer radius beyond which the particle is ignored, three times the cluster radius.

A typical small cluster is shown in Fig. 11.3, taken from Witten and Sander (1983). The clusters have multiple branchings, and at this size show no obvious influence of the form of the lattice. Are such structures like parts of the interstitial matrix? What is evident is that the recent depositions are virtually all at the periphery of the cluster; none of the recently released particles reach points close to the center of the cluster but by their random movements first touch surfaces on the arms of the cluster. The profiles of likely next contact are portrayed nicely by Meakin (1986b) and by Mandelbrot and Evertsz (1990) in the beautiful Plate 5 of Guyon and Stanley (1991). The number of sites on the perimeter available for growth, N_p, is proportional to the total number of sites. Both scale with the radius of gyration of the cluster, which is nearly the diameter, L, of the cluster:

$$N_p \sim L^D,$$ (11.1)

where D is the fractal dimension, about 1.7 for two-dimensional clusters. (For DLA D is 1.7 in two dimensions, 2.5 in three, and 3.3 in four (Meakin, 1983).) Because protruding points are more likely to be contacted than points on a stem, the cluster tends to extend outward from its center. The radius of gyration, R_g, and the diameter increase in proportion to the number of particles, N, raised to a noninteger power, β:

$$R_g \sim N^\beta,$$ (11.2)

from which it follows that the fractal dimension is

$$D = 1/\beta.$$ (11.3)

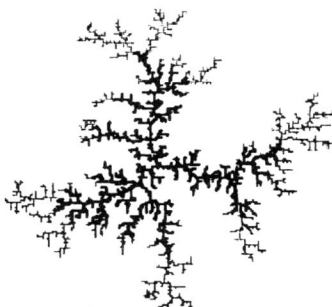

Figure 11.3. A small cluster of 3000 particles formed by diffusion-limited aggregation (DLA). The first 1500 points are larger, and the paucity of small dots attached to the earlier deposits (larger dots) indicates that very few particles penetrate into the depths of the cluster, but are caught nearer the growing terminae. (Figure reproduced from Witten and Sander, 1983, with permission.)

To determine the fractal dimension D from the observations at a fixed point in time one can use this expression in a slightly different way, applying the "mass–radius method" for estimating D. Plot the number of particles (or the surface area of the cluster) for a sequence of radii up to the radius of the cluster; the mass enclosed by each r, $M(r)$, is proportional to the radius of the fractal dimension:

$$M(r) \sim r^D, \tag{11.4}$$

thereby defining D as the slope of the log-log plot of $M(r)$ versus r.

The concentration of particles is therefore a function of the distance from the center of the cluster. Fig. 11.4 from Meakin (1983) shows how the surface density within a given distance from the center depends on the value of that distance. The irregularity of this relationship, especially for small clusters, inhibits its use in estimating the fractal D.

The autocorrelation function of densities can also be used to estimate the fractal dimension. The density $\rho(r)$ is the number of particles per unit surface area in the local area circumscribed by a circle of radius r from the center of the cluster. The density–density correlation function, $C(r)$, is the product of the density at radius r' with that at radius $r' + r$:

$$C(r) = N^{-1} \sum_{r'} \rho(r') \rho(r + r'), \tag{11.5}$$

where N is the total number of particles.

A result from Meakin (1983) is shown in Fig. 11.5. The correlation function yields a satisfactory power law relationship,

$$C(r) \sim r^{-\alpha}, \tag{11.6}$$

where $\alpha = 2 - D$, $C(r) = r^{D-2}$. The fact that $C(r) = r^{D-2}$ follows from the self-similar character of the cluster. The power law scaling of Eq. 11.6 holds well for r greater than a few lattice spacings, but the correlation falls off when r approaches the radius of the cluster itself, because there are fewer measures, and the local densities are more nearly a function of the randomness of the deposition process.

11.3   Influences of Matrix Structure on the Form

The DLA in Fig. 11.6 takes on a diamond shape, a square at 45 degrees to the lattice lines, reflecting the underlying rectangular structure of the matrix grid used to grow it. But as they enlarge they tend to lose their diamond shape and the sides become more concave, as in Fig. 11.7.

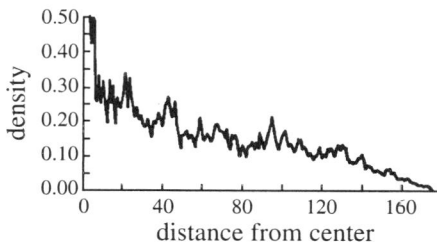

Figure 11.4. Dependence of the surface density on distance from the origin for a two-dimensional DLA cluster of 10,200 particles. (Reproduced from Meakin, 1983, his Figure 6, with permission.)

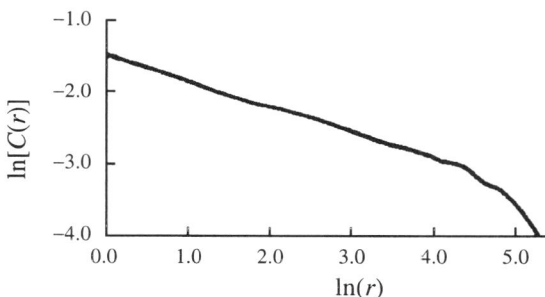

Figure 11.5. Density–density correlation function for a DLA. $D = 2 -$ slope $= 1.62$. (Figure reproduced from Meakin, 1983, his Figure 3, with permission.)

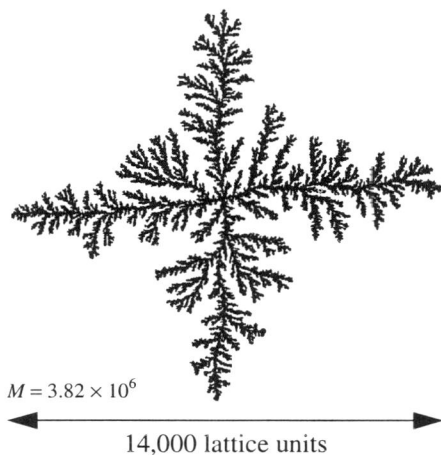

$M = 3.82 \times 10^6$

14,000 lattice units

Figure 11.6. A cluster of 3.8 million particles grown on a square lattice. (Figure reproduced from Meakin, 1989, with permission.)

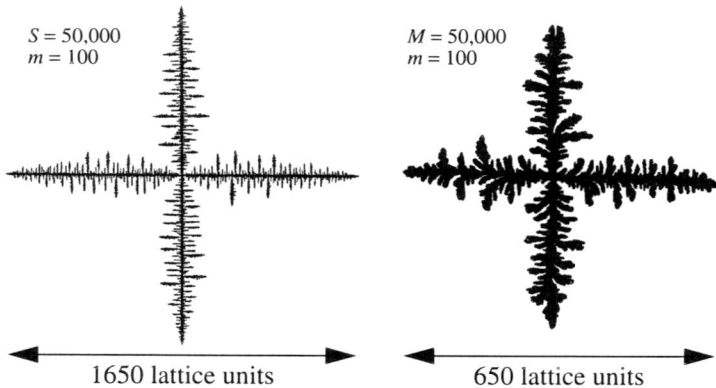

Figure 11.7. Square lattice DLA clusters with noise-reduction parameter $m = 100$ for two growth algorithms. *Left panel:* Same algorithm as Fig. 11.2. The effect of requiring multiple contacts reduces randomness, revealing the square lattice structure. *Right panel:* Requiring the particle to take a final step *onto* a site already occupied on the cluster and then occupying the adjacent occupied site makes it much denser and much less governed by the matrix, even though it is still clearly square. (Figure reproduced from Meakin, 1989, with permission.)

This influence of the matrix structure becomes much more evident if more than one contact is required to add a particle to the cluster. Requiring several contacts of the particle to the cluster has the effect of reducing the randomness of the diffusion, and the number of contacts, m, can be considered a noise-reduction parameter. Fig. 11.7, left panel, shows that using $m = 100$ transforms the diamond into a cross with practically all of the branchings occurring at right angles along the lattice lines. The right panel of Fig. 11.7 illustrates that an apparently minor change in the rules has a large influence on the form of the cluster. The rule change was that instead of the particle hitting a given eligible site 100 times, the particle was assigned to the last perimeter site it occupied before moving onto the cluster to be "absorbed." This form is much more like those obtained by percolation of a medium by a viscous fluid, so called viscous fingering. (See van Damme, 1989).

11.4   More General Types of Aggregation Processes

The DLA beginning from a fixed point is an excellent vehicle for gaining an understanding of the results of the random diffusion and binding because everything, in theory, can be known about the events. Even so, not all the theory is worked out, and thus it isn't surprising that more complex cases are less well understood. The next most studied is the aggregation at a line from two-dimensional diffusion of particles in a plane (a two-dimensional problem) and aggregation at a surface (a three-dimensional problem). Treelike structures evolve from surface aggregation, since the diffusion of particles is from a biased source; in the simplest algorithms it is merely a line source at some distance from the aggregate, just like the circle for the single-seed DLA. The difference is that having a line for attachment allows for growth from many sites along the line. The process is still quite similar when the initial conditions are that the solution has a uniform concentration of particles that are then all allowed to diffuse, as shown in Fig. 11.8. As the particles near the surface are used up the fluid concentration diminishes so that there is a net concentration gradient toward the surface. The individual trees look much like the arms of the DLA in Fig. 11.3 but have shorter side branches since they all grow in parallel, like trees in a forest. In these cases there was no aggregation of particle to particle prior to attachment to the fixed particles. One biological analog is the attachment of platelets to the damaged surface of a vessel, following which other platelets bind to those already deposited.

Deutch and Meakin (1983) showed that the cluster radius for multiparticle DLA growth occurred at a predictable rate, which depends not only on the diffusion coefficient but also on the fractal dimension (which is in turn dependent on the Euclidean dimension, since D is 1.7 in two dimensions, 2.5 in three dimensions, etc.). The cluster radius increases with time:

$$R(t) \sim t^{[1/(D-d+2)]},\tag{11.7}$$

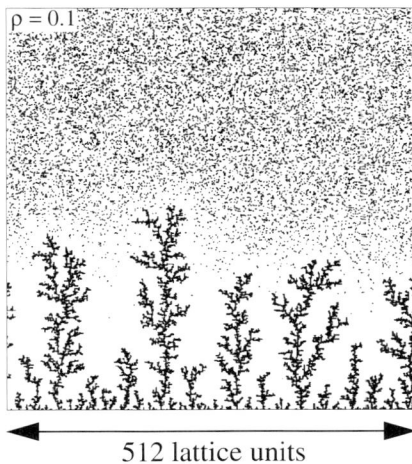

512 lattice units

Figure 11.8. Structure generated using a two-dimensional model for diffusion-limited aggregation of particles from a solution onto a lattice linear array of 512 elements across the bottom. The original solution concentration was 1 particle every 10 lattice units on a square array. Half of the original particles in solution had been aggregated when the simulation was stopped. (Figure reproduced from Meakin, 1989, with permission.)

where d is the Euclidean space. For $d = 2$, the result is that $R(t) \sim t^{1/D}$ or the mass of the aggregate, $M(t) \sim t$, approximately, which is consistent with simulations. Again the theory is imperfect. This probably would not apply well to platelet aggregation, since it is observed in test situations that the platelet aggregates do lose chunks from their surfaces, which are thus forming small emboli.

Another related phenomenon is cluster–cluster aggregation. When one starts with a suspension of single unaggregated particles and allows particle-to-particle contact to result in attachment, aggregates form, coalesce with each other to form ever larger cluster–cluster aggregates. In a closed volume without surface adherence, eventually only one large cluster remains. Such clusters are considerably more irregular than are the simple DLAs, as shown in Fig. 11.9 (from Weitz and Huang, 1984) by colloidal gold clusters at different magnifications. In spite of the irregularity, one gets the impression of self-similarity over the range of scales. By light and small angle neutron scattering Weitz et al. (1985) found a cluster fractal dimension of $D \cong 1.79$, a result similar to that obtained from the electron

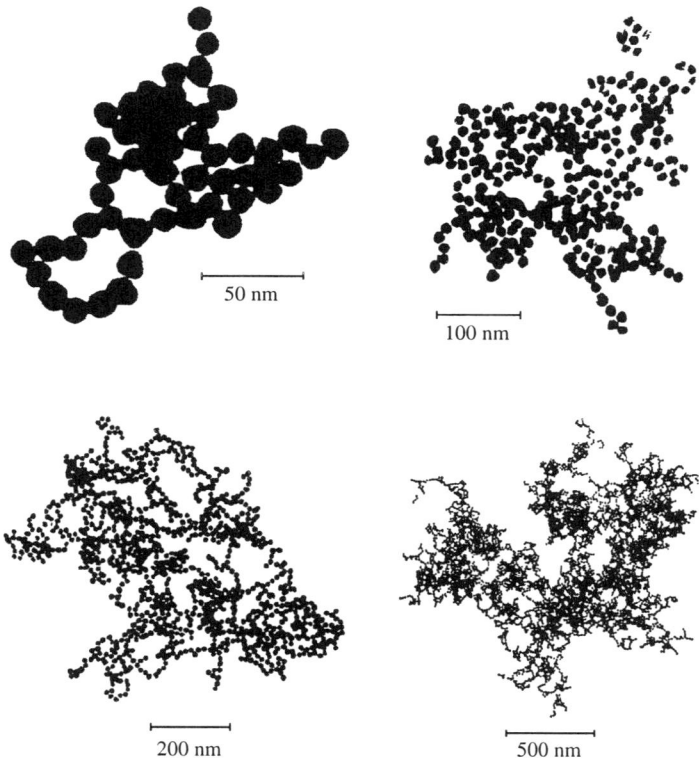

Figure 11.9. Transmission electron micrographs of gold clusters of different sizes. (Figure reproduced from Weitz and Huang, 1984, with permission.)

micrographs in the figure, which was 1.7. The clusters are loosely packed; while we would expect a fractal dimension of 2.5 for clusters in three dimensions, the estimates for these averaged only 1.75, presumably because they use two different techniques, scattering and mass–radius methods, both in a two-dimensional context, augmenting the effect of the looseness or lack of compactness of the clusters.

A better approach was used by Feder et al. (Section 3.1 in Feder, 1988) to estimate the fractal dimension of immunoglobulin proteins of the IgG type using light scattering. They estimated the radii of clusters as they aggregated in response to raising the temperature, again using light scattering. They obtained a cluster fractal dimension of 2.56 ± 0.3, which appears indistinguishable from the theoretical expectation for a DLA in three dimensions. What was particularly convincing was that the curves of apparent light scattering radius versus time obtained at different temperatures could all be superimposed by normalizing the time scale with respect to the rate constant; this is a mark of similarity rather than self-similarity, so the self-similarity is in the clustering itself.

11.5   Neuronal Growth Patterns

The diffusion-limited aggregation (DLA) pattern is remarkably universal, and a good many variants may be seen in Vicsek's book (1989) and in an attractive small picture book edited by Guyon and Stanley (1991), ranging from electrical discharge patterns to percolation clusters, mountain ranges, and galaxies. Neurons are fairly similar. Neuronal growth has been examined with increasing clarification by Smith et al. (e.g., 1989) and Stanley and colleagues (e.g., Caserta et al., 1990). An example is shown in Fig. 9.5, a retinal neuron, which also lies in a plane. The fractal dimension is about 1.7. Ongoing explorations have revealed that the fractal dimension of neurons does not give a unique description of their degree of dendritic branching and deviation from it; Fig. 9.7 shows two very different neurons with the same fractal D. But perhaps even this one-parameter measure of neuronal complexity may be useful. One of the hallmarks of the premature aging of Alzheimer's disease is the reduction in the richness of branching of the neurons. This is not a situation in which repeated biopsies will be used to assay the effects of a therapeutic agent, but nevertheless a good method for describing the densities of dendritic processes will be useful in examining pathological specimens.

11.6   Algorithms for Vascular Growth

It is not obvious that DLA algorithms are close analogs to vascular growth, but they are nevertheless strikingly similar in certain features. Certainly the methods that are useful for estimation of fractal dimensions from DLAs can be readily applied to vascular patterns, so at a minimum they provide a useful measurement technique.

Retinal vessels delineated by fluorescein provide a nice test since they lie on a surface. These were analyzed by Family et al. (1989) and Mainster (1990). An angiogram analyzed by Family et al. (1989) is shown in Fig. 11.10. The fractal dimension was determined by the mass–radius measure (Eq. 11.4) and the density–density correlation technique (Eq. 11.5), being about 1.7. This is a good example of exploring the characterization by two different techniques. Family et al. went on to suggest that the results, $D = 1.7$, indicated that the underlying process leading to the configuration should therefore be a DLA-type process. This could be true, but since probably quite a number of stimuli could result in similar forms, the fractal D can scarcely be counted on to prove the mechanism involved.

Neuronal growth does not appear to be fractal through all stages. Neale et al. (1993) made observations on neurons growing in culture during the first week. They photographed the different types of neurons periodically and calculated the fractal dimension. Neurons with two main branches are type 2, with three or four branches, type 3/4, and with five or more branches type 5. Type 5A has multiple branches tightly clumped and 5B has long extended branches. Neale et al. (1993) interpreted their data as exhibiting an exponential rise in estimated fractal D until a plateau was reached. Their estimated Ds are plotted in Fig. 11.11 over a week's growth. All of these neurons had small fractal dimensions less than a two-dimensional DLA. Since for all types of the neurons the estimated Ds actually continue to rise through time 48 to 168 hours, the exponential model remains unproven, and another model leading slowly to a higher plateau might be better.

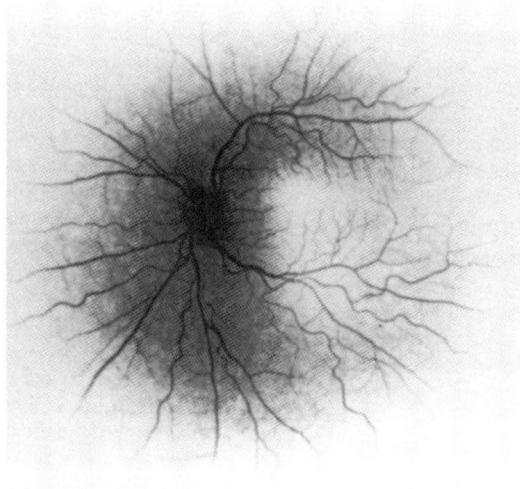

Figure 11.10. Fluorescence angiogram of a fractal pattern in the human retinal vasculature. (Figure reproduced from Family et al., 1989, with permission.)

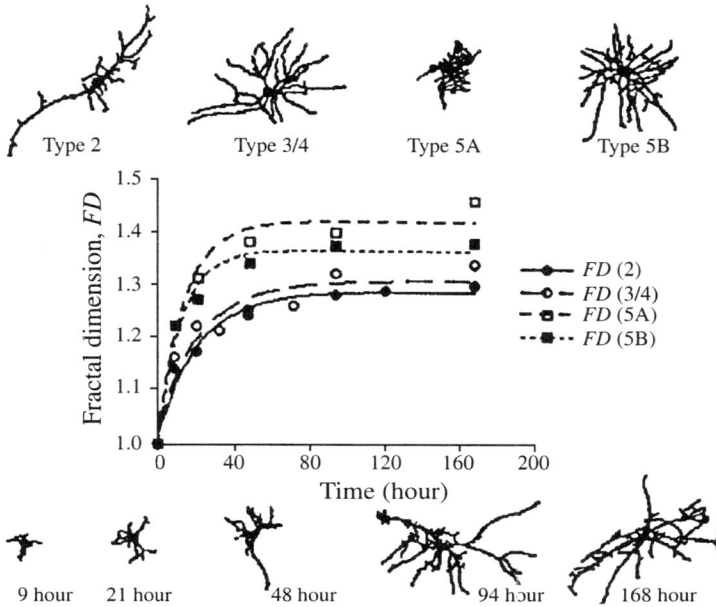

Figure 11.11. Change in fractal dimension and morphology of spinal cord neurons with time in culture. Spinal cord neurons were grouped (see text) into four types, all illustrated in the *upper* panels. Fractal dimensions were analyzed at various times during the first week in culture and plotted for each morphological type in the *middle* panel. The development of morphology over this same interval is shown for type 3/4 neurons in the *lower* panels. (Reprinted courtesy of Smith and Neale, 1994, with the permission of Birkhäuser Verlag.)

Their idea of using an exponential function appeared well based in fractal theory. Following Mandelbrot (p. 51 ff., 1983), they showed that line-replacement algorithms applied to simple initiators such as a square or triangle were not fractal in their early iterations; they are *pre*fractals. At each iteration in the application of the algorithms they estimated the fractal D just as they had for the neurons, and observed an exponential increase toward the expected fractal dimension for these classical fractal objects, shown in Fig. 11.12. Perhaps neurons are a little too complex to obey the behavior of a simple prefractal, but the idea is a good one.

The multiplication of the number of growth algorithms seems to be faster than the growth of the vasculature in an embryo. The challenge is to find algorithms that have simple rules and yet reproduce the general form of vascular structures. The task is nontrivial, and although there are some real successes, no truly satisfactory algorithms are yet in hand. This is in part because each tissue is different from all others: a casting of the vasculature of a particular organ can be identified as to its origin without any tissue being present; the form of the vasculature alone is enough of a clue. Stated conversely, the structure and arrangement of the cells in a tissue so

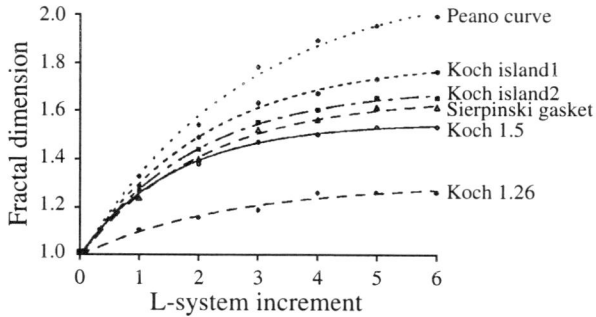

Figure 11.12. A plot of the fractal dimension as a function of iteration number for a number of known fractal objects listed in the upper left portion of the figure. The curves of the points fit an expression of the form $D(i) = A + B\{\exp(i/n)\}$ (Reprinted courtesy of Smith and Neale, 1994, their Fig. 6, with permission from Birkhäuser Verlag.)

dominate the form of the microvasculature that the casting identifies the cell arrangement. What this means is that vascular growth algorithms cannot successfully mimic vascular form without mimicking the cell arrangements in the tissue. Thus the successes so far are limited, but on the other hand it is to be admired that they can go as far as they do, handicapped as they are by lacking a representation of the tissue.

Dichotomous branching is a natural starting point (Mandelbrot, 1983). Branching of trees and blood vessels is very nearly purely dichotomous, and trichotomous branching can even be regarded as mathematically equivalent to having two dichotomous branchings close together. Simple symmetric dichotomous branching gives rise to a binary tree; it is not interesting, because it gives rises to a uniform distribution of flows at any given level or generation of branching. However, simple asymmetric branching can give rise to heterogeneous flow distributions; an example from van Beek et al. (1989) is shown in Fig. 10.8. The fraction of flow entering one branch is γ, and that entering the other is $1 - \gamma$. Van Beek et al. (1989) found that a γ of around 0.47 created enough asymmetry to match the observed variances in flows in the heart, as did Glenny et al. (1991a) and Glenny and Robertson (1991) in the lung. Several variations on the theme worked almost equally well, and small degrees of random asymmetry served as well as did fixed degrees of asymmetry. See, for example, Chapter 10 on flow heterogeneities.

The simplest of dichotomous algorithms are unrealistic in the sense that they disregard the constraints imposed on vascular patterns, specifically to be confined within a three-dimensional domain, and to fill all of that domain. This is illustrated for a two-dimensional setting in Fig. 11.13. The algorithm is described by the following two steps: (1) From an arbitrary starting point on the perimeter project a line in a direction that divides the area into halves; grow the branch from the starting point along the line to a point at a fraction α of the total line length. (2) From this

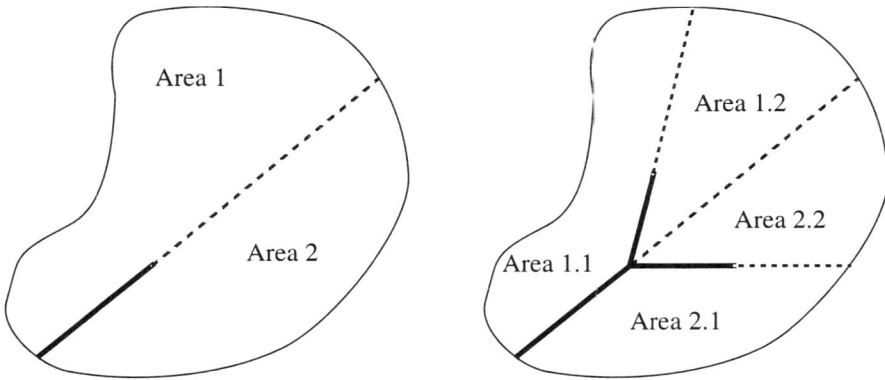

Figure 11.13. Area-splitting algorithm for two-dimensional dichotomous branching system filling the space available. The rule is to draw a line to split the area available, and sprout along the line to a point at some specific fraction of its length (stage 1). Repeat this starting from the specified fractional point to split each half into quarters. Repeat either for N generations or until some density or distance criterion is met. (From Bassingthwaighte, 1992, Fig. 5, with permission.)

point extend two lines to halve each region, and again grow branches to a fraction α of each line length. Repeat these steps for a chosen number of generations. This is the algorithm used by Wang et al. (1992) as an example; it needs to be done in three dimensions rather than two, but did have the virtue of looking fairly realistic, as well as being space-filling. Fig. 11.14 shows the results of such an algorithm plotted for different fractional lengths of growth: values of fractional length of 0.2 and 0.25 look unnatural compared to those using 0.3 and 0.35. The short fractions give rise to excessively long terminal branches; fractional lengths of 0.4 or higher give rise to initial branches that are disproportionately long and terminal branches that are too short. The algorithm could be improved by adding more rules, for example, the ratios of lengths should lie within a specified range over several generations, as has been observed in real systems (e.g., Suwa and Takahashi, 1971).

Improvements in this algorithm can be made in many ways. A major problem with the algorithm is that the domain shape and size were preordained, which is not the way growth occurs. Gottlieb (1991) proposed an algorithm (Fig. 11.15) that accounts for growth by symmetric enlargement, and for vascular budding by stimulus from tissue (as if by a growth factor) when there is no capillary sufficiently close to a tissue element. The algorithm provides for an n-fold expansion of the size and shape of the $(i-1)^{th}$ configuration of the vascular tree, VT_{i-1} (shown in the left lower panel), where g_i is the growth rule for this generation. In the figure the rule applied to the square in the left lower corner is to double all lengths to form VT_i (the right panel). The next phase, indicated by the $U(\sigma)$, is to augment the vascular tree so that it might fulfill the needs of the tissue. In the example, the algorithm determines which growing points (old and new) are farther than a specified distance

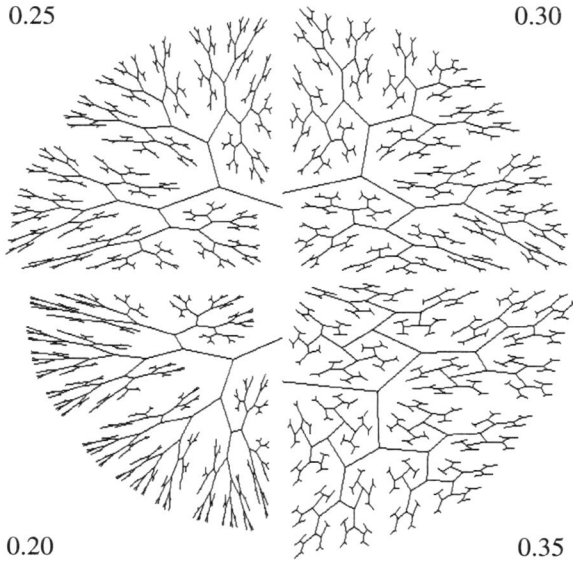

Figure 11.14. Area-dividing branching algorithm applied to quarter circles. The patterns are strikingly dependent on this value of the fractional distance to the far boundary. (Figure provided courtesy of C. Y. Wang and L. B. Weissman, reprinted from BMES Newsletter, with permission).

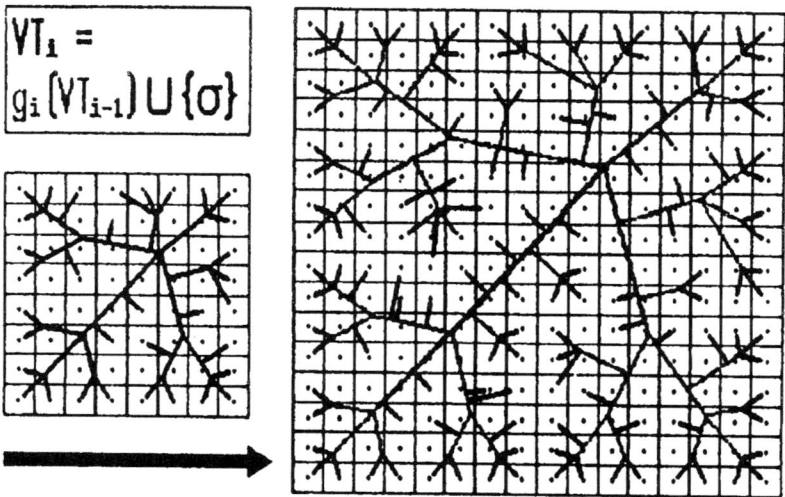

Figure 11.15. Tissue expansion followed by vascular sprouting. In a given iteration the cells (the squares) divide and grow (1 cell → 4 cells here); the matching growth function g resizes the existing vessels, then new vessels are located in regions of insufficiency and the sprouts added to the structure. (Figure reproduced from Gottlieb, 1991, with the permission of IEEE.)

from a vessel, finds a nearest point on a vessel, and then grows a sprout from this point toward the growing point. In a version where the fraction of the distance sprouted can be random up to 1.0 and where sprouts can occur from vessels on opposite sides to the growth center, then vascular anastomoses can develop. Whether these algorithms account for the pressure and flow distributions in the microvasculature is not known, but Gottlieb has produced remarkable good likeness to a variety of vascular beds.

Fenton and Zweifach (1981) mapped the vasculature of the conjunctivae of humans and the omentum of rabbits in order to develop a statistical base for describing the network geometry and to calculate pressure and flows. They found branching ratios greater than 2.0, namely 2.3, more than dichotomous; from the length and branching ratios they created networks that gave appropriate pressure distributions as a function of position along the length of the vessels. Dawant et al. (1986) augmented these probabilistic network formulations so that they could describe the influences of variability in lengths versus diameters on the distributions of flows in the small terminal branches; Levin et al. (1986) made a further extension, to show the expected variability in hematocrits, and the degree of separation of plasma and RBC's in traversing the capillary bed.

The theories so far developed are incomplete. Gottlieb's growth algorithm is the only one allowing anastomoses at a range of levels. We classify the anastomosing mosaic network of Kiani and Hudetz (1991) as a different type of structure, based as it was on branching angles and resulting in randomly sized intercapillary spaces of polygonal form in a plane square. (Plane-tiling polygons average, mathematically, six sides, as described by Grünbaum and Shephard, 1987). Missing are the constraints of how the tissue cells grow, divide, and arrange themselves to serve the organ's function. The general rule of one capillary beside each cell is a good one for highly vascular tissues, even if not for bone and cartilage, and will serve as a rough measure of how many capillaries are needed by a tissue. The questions of how many capillaries there are per arteriole, and what size the terminal microvascular unit is, remain unanswered for most tissues.

11.7   Patterns of Vascular Branching

While all of the algorithms mentioned above are fractal, by virtue of having constant (more or less) ratios of lengths, diameters, etc., only a few have been carefully compared to the actual anatomy, and of these almost none to the actual distributions of flows. A fractal algorithm recently developed by Krenz et al. (1992) for the lung microvasculature uses branching ratios greater than two and through the incorporation of randomization of the resistances at each bifurcation gives rise to the degrees of flow heterogeneity seen in nature. Thus, the model can be based on average ratios for branching, length, and diameters, and also provide flow heterogeneities equivalent to those modeled by van Beek et al. (1989) and Glenny et al. (1991a) and Glenny and Robertson (1991).

Capillaries follow different rules than do large vessels. The hallmark of capillary growth patterns is that their arrangement follows that of the cells of the tissue. Where there are either long cylindrical muscle cells as in skeletal muscle (Mathieu-Costello, 1987), or cells in syncytia in series, as in the heart (Bassingthwaighte et al., 1974), the capillaries are arrayed in long parallel groups serving a set of parallel muscle fibers within a bundle. Likewise, in a glandular organ the capillaries are arrayed around the secreting cells, forming an acinus, as modeled by Levitt et al. (1979). These are not fractal, and even though there is self-similarity of a sort in the glandular system of the gastrointestinal tract, as illustrated by Goldberger et al. (1990), the capillaries themselves form the terminal units of the vascular system. While both arterial and venous systems may show self-similar branching the capillaries show only branches connecting to equal-size or larger vessels. Nevertheless, we pay much attention to capillary growth since it is from these vessels in embryonic and later growth phases that the arteries and veins develop. As Wiest et al. (1992) put it, "Physiological growth of arteries in the rat heart parallels the growth of capillaries, but not of myocytes." The point is not a subtle one, for even though the capillary budding is surely stimulated by the growth and nutrient requirements of the myocytes, and the capillaries provide for the flow of nutrients and the removal of metabolites, it is the arterioles and venules that serve the capillaries, and in turn the larger arteries and veins that serve the microvasculature.

The answer to the chicken-and-egg question is clear with respect to vascular growth: the tissue's demands are the stimulus. The serial nature of the processes of growth of the vasculature in the heart is discussed by Hudlická (1984) and Hudlická and Tyler (1986). In her comprehensive review, Hudlická (1984) illustrated that growth factors can have a dramatic effect on the form of the branching. A growth factor diffusing from a small orifice in a plate tends to promote a growth pattern that is very close to that produced by DLA. A sequence in the development of a network of arterioles is shown in Fig. 11.16, taken from the work of Folkman and Cotran (1976). Their observations are certainly in line with the general view of vascular growth proposed by Meinhardt (1982). As parenchymal cell growth occurs there is budding of capillary endothelial cells (Rhodin and Fujita, 1989), initially protruding into the interstitial space, then developing a bulge that forms the end of a plasma tube like the bottom of a test tube, and the tube lengthens so that an erythrocyte may be seen to oscillate within it. By some magic, perhaps related to the way that endothelial cells grow to confluence in monolayers, the sprouting capillary reaches out toward another capillary and joins to it, allowing the flow of plasma and then erythrocytes. Remodeling occurs; as vessels grow there are rearrangements of the flow paths, and occasionally, as shown in one of Hudlická's (1984) figures, a capillary may disappear. It is commonly said that smooth muscle cells "migrate" from the larger arterioles into the smaller ones. It seems more likely that cell differentiation occurs because of local stimuli such as pressure oscillations and flow-dependent growth factors (Langille et al., 1989). The intricate mechanisms by which such growth and transformation is controlled are slowly becoming identified, but go far beyond the scope of this book.

Vascular growth usually occurs in parallel with tissue growth. Tumors that outgrow their vascular supply are the exception. The biochemical rules for this may

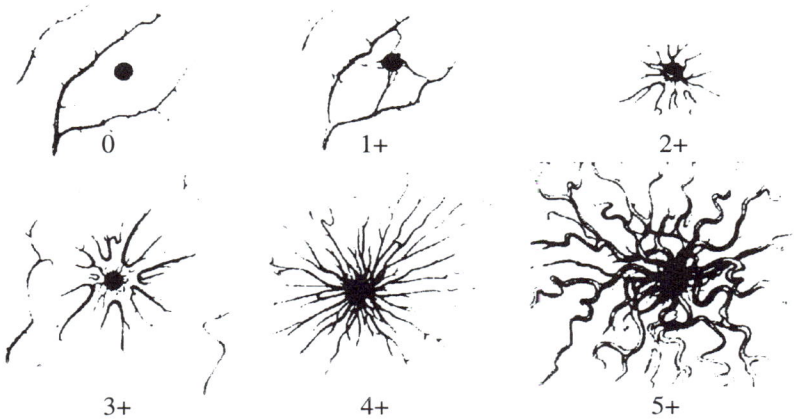

Figure 11.16. New vascular loops consisting of capillaries and venules, after converging on the site of a tiny needle hole in the chick chorioallantoic membrane over which a 1-mm filter disk soaked in tumor angiogenesis factor has been placed, assigned a density grade of 5+. (Figure provided courtesy of Folkman and Cotran, 1976, with the permission of *Int. J. Exp. Pathol.*)

be complex, but the simplest elements of the process are not: the buddings that start as capillaries are simple, dichotomous branchings. It is no wonder, then, that dichotomous branchings are the commonest seen in the adult tissue. Thus, the algorithmic rules might be reduced, albeit too simplistically, to be

  (i) bud (or branch) and extend into the tissue,

  (ii) repeat the operation recursively.

Gottlieb (1991) provides many illustrations of vascular patterns that can be rather realistically matched by his growth and sprout algorithm, which demonstrates that very simple algorithms can go a long way toward description even if they do not go far toward explanation.

It is these types of growth which presumably give rise to the fractal relationships between the degree of apparent heterogeneity of local tissue blood flows within an organ and the resolution of the measurement (the size of the tissue elements over which an average flow is measured). Bassingthwaighte et al., (1989) found the heterogeneity-element size relationship in the hearts of baboons, sheep, and rabbits to be linear on a log-log plot, i.e., a power law relationship the slope of which gives a fractal dimension, D, of 1.2. This fractal D, being greater than one, indicates that the system is self-similar over the observed range and that the heterogeneity is *not* random. The value of D is 1.5 for random processes. Bassingthwaighte and Beyer (1991) show that this translates into a measure of correlation falling off with distance. Van Beek et al. (1989) show that fractal branching processes can explain the heterogeneity.

The 1962 study of Weibel and Gomez pioneered the quantitative analysis of the lung's vascular and airway systems. In his 1963 work, Weibel portrayed the hexagonal nature of the structures. (Incidentally, Weibel and coworkers did the first analysis of mammalian tissue that was labeled "fractal," using the approach for estimating surface areas within mitochondria (Paumgartner et al., 1981)). Horsfield (1978) characterized the branching of the vascular system, developing a classification that differed a little from that of Strahler (1957). Working on the geographical structuring of watersheds and lungs, Woldenberg (1970) devised a system for examining spatial structures in a hierarchical fashion, wherein the hexagonal structures that dominated hills and watershed regions were seen to have repetitive structuring, a statistical self-similarity. The constraint that spaces or surfaces must fit together gave rise to a numerical way of defining the number of fields or regions as a logarithmic function of the fineness of the divisions used. These numerical relationships applied also to the lung vascular and airway systems, and the liver vasculature. Woldenberg and Horsfield (1986) extended analysis of the lung data to consider the cost minimization in structuring the vascular system, that is, minimizing arterial and venous surface areas, volume, power, and drag, and framing these considerations in terms of the physics of the branching, in tune with the approach of Lefèvre (1983).

  These efforts are an approach to the inverse problem, which is to discover in retrospect what nature's growth rules might be. If one starts with a simple view, as above, one soon discovers the shortcomings of that view, and so modifies it to fit better with reality. Stark's (1991) explorations of drainage networks are quite analogous to our explorations of vascular growth. His route to the recognition that there is an asymmetric force influencing stream growth (Fig. 11.17) is exemplary; in fact, the directionality of vascular flows and the self-avoiding path generation are very similar. Continuing in this earthy vein, it is also sensible to keep in mind that it is most unlikely that a single fractal rule governs a system: Burrough (1983a, 1983b) shows how a hierarchical set of fractal processes, each extending over a

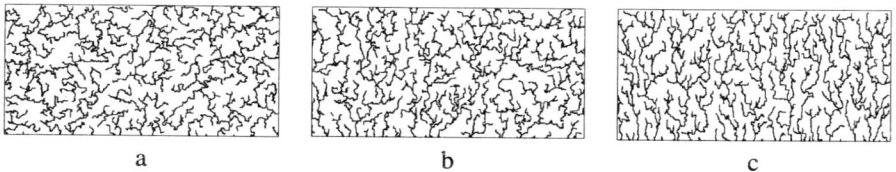

a   b   c

Figure 11.17. Examples of simulated drainage networks grown by the self-avoiding invasion percolation method. Numerical simulations were set up using a square site-bond lattice in which random bond strengths were normally distributed; lattice size is 512×256. Seeding was allowed from every other point along the bottom edge. In this case invasion of the lattice was allowed to continue until no further free sites were available. Strahler stream ordering was applied: streams of order four and above are shown. The substrate becomes steadily more susceptible to northward growth from (a) (isotropic lattice) to (c) (strongly anisotropic lattice). The fractal dimension D_{min} of the principal streams is invariant to this anisotropy and is on average 1.30 throughout. (From Stark, 1991.)

limited range, can provide a more realistic description of soil structures. This is probably what we need to do to understand growth processes through the full range of generations of vascular growth. The term that is used is multifractal. Prusinkiewicz and Hanan (1989) use Lindenmayer's recursive sets of rules to develop analogs to plants, and illustrate extensions of this to cell growth and division, and even to music. In any case, the applications to mammalian systems need to be begun.

11.8   Phylogeny versus Ontogeny

Presumably fractal growth processes do not apply to the growth of phylogenetically determined structures. While it is true that disturbing the environment for the growing structures disturbs the form of structures that have evolved through the phyla (Icardo, 1988), what is equally clear is that the body's major vessels, extending down to the form of the largest vessels feeding an organ, have very similar forms within a species and from species to species. Such uniformities cannot be attributed to opportunistic processes such as appear to work for capillary growth. But phylogenetically controlled processes cost genes, and given that we have so few, it makes sense that they be retained for the most important structural features only. Advanced animals have more complex structures. The capacity for complexity, measured in terms of the numbers of genes or the amount of DNA, seems to match the observed complexity, measured in terms of the numbers of cell types, as shown in Fig. 11.18 from Kauffman (1991).

11.9   Summary

The process of diffusion-limited aggregation (DLA) of particles diffusing freely in solution is an example of a natural self-organizing process that creates macroscopic structures from molecules. The form of these clusters depend on few physicochemical rules but provide an interesting range of shapes that are rather similar to natural and biological structures. Vascular growth patterns are often so similar to mathematical DLA patterns that one gains an impression that the driving forces for vascular growth may also be simple, if not in detail, then in overall control. Branching systems grown in accord to recursive rules give rise to correlation in regional vascular resistances and flows and the degrees of flow heterogeneity described in Chapter 10. Since it is practical and efficient for growth to occur via recursive rules, such as branch, grow, and repeat the branching and growing, it appears that fractals may be useful in understanding the ontological aspects of growth of tissues and organs, thereby minimizing the requirements for genetic material.

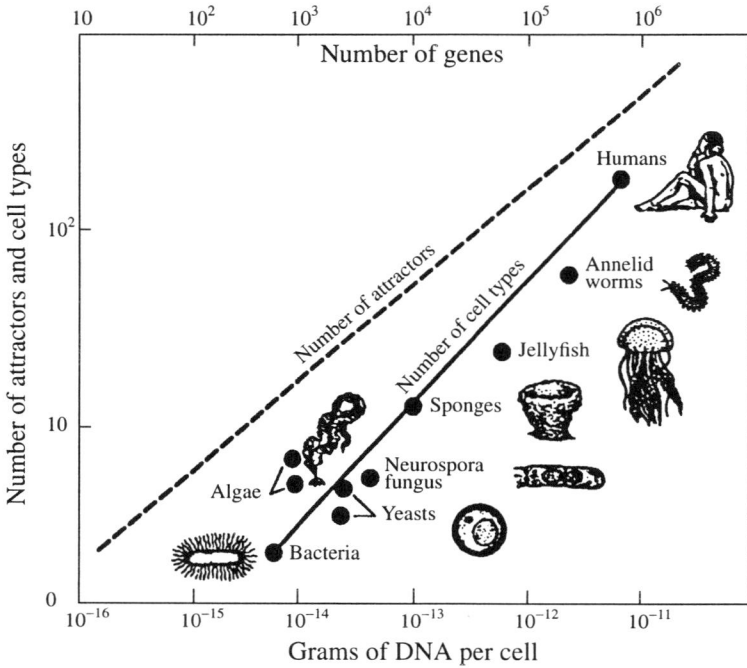

Figure 11.18. Number of cell types in organisms seems to be related mathematically to the number of genes in the organism. In this diagram the number of genes is assumed to be proportional to the amount of DNA in a cell. If the gene regulatory systems are "K = 2 networks," then the number of attractors in a system is the square root of the number of genes. The actual number of cell types in various organisms appears to rise accordingly as the amount of DNA increases. The "K = 2 networks" refers to a simple system wherein the connections between neighbors in a planar array control whether an element is active or not, a kind of binary level of control that is about the simplest conceivable for a multielement system. (Figure reproduced from Kauffman, 1991).

11.10   Some Fractal Growth Programs

Most algorithms are better for unconstrained branching than for constrained. Programs that run on home computers can illustrate the phenomena. Garcia (1991) reviews related ideas. Van Roy's "Designer Fractal" (1988) allows one to create complex patterns by repetitively replacing line segments. The iterated function systems of Barnsley (1988) are available on a Macintosh (Lee and Cohen, 1990). Lauwerier's (1991) book lists programs in Basic that make trees, etc. A set of programs on cellular automata and evolutionary processes has been collated by Prata (1993).

12

Mechanisms That Produce Fractals

> The Master of the oracle at Delphi does not say anything and does not conceal anything, he only hints.
>
> Heraclitus, fragment #93 (as cited in Diels, 1901)

12.1 Fractals Describe Phenomena and Give Hints about Their Causes

Fractals help us understand physiological systems in two different ways. First, they provide a *phenomenological* description. Since fractals have qualitatively different properties than nonfractals, just the realization that we are working with a fractal is important. This realization can actually be difficult, even if the pattern is "obvious." Thus, it has been a significant surprise each time yet another physiological system was found to be fractal, such as the branching of the airways in the lung, or the distribution of blood flow in the heart, or the kinetics of ion channels.

The realization that we are studying a fractal tells us what is meaningful to measure and what is not. It tells us how to correctly perform and interpret those measurements. For example, the *properties measured from a fractal are a function of the resolution used to make the measurement*. Thus, *a single measurement at a single resolution is not meaningful*. What is required is to determine how the values that are measured depend on the resolution used to make the measurement. If we are analyzing a fractal, we also know that measurements based on different measuring techniques, or performed at different laboratories, must be compared at the same resolution.

Correctly recognizing and analyzing phenomena, however, is only the first step. As suggested by the quotation above, we want to use the hints provided by the data to reveal the *mechanisms* that produced the data. Fractals are also helpful in this second step. Work in several different scientific fields has now shown that some generating mechanisms produce fractals of certain forms with certain dimensions. Thus, when we see one of these forms or dimensions in our data we should suspect that it may have been produced by one of these known mechanisms. Different mechanisms may produce the same form or dimensions. Thus, this procedure does not guarantee that we discover the correct generating mechanism. However, it may provide very helpful clues.

The list of mechanisms given here is incomplete and somewhat speculative, but serves as a starting point to uncover the appropriate mechanism when new fractal phenomena are recognized.

12.2   A Single Process or Many Processes?

The self-similarity of fractals means that structures or processes at one spatial or temporal scale are linked to those at other scales. This can happen in two ways.

1. *A single process may start at one scale and then extend to other scales.*

Many seemingly global phenomena have now been identified as due to a single process whose effects spread across many different scales. D'Arcy Thompson (1961) first showed that simple physical forces could produce complex patterns in the structures of cells, tissues, organs, and body parts. Alan Turing (1952) showed that complex patterns in the development of an organism could arise from the reaction and diffusion of chemical substances. Manfred Eigen (Eigen and Schuster, 1979; Eigen, 1992) described the likelihood of formation of molecules and of combinations of molecules in hierarchical "hypercycles" of increasing complexity. Molecular self-assembly makes sense, thermodynamically and kinetically. Prusinkiewicz et al. (1990) showed that iterated sets of simple instructions could produce the complex patterns of plants.

The process may start at small and extend to higher scales. For example, the growth of dendrites from neurons may depend on the release of growth factors. Faster-growing regions release more growth factors than do slower-growing regions and thus grow faster at their leading tips, until new tips branch off. Thus, a *global* branching pattern is formed from *local* interactions.

On the other hand, the process may start at large scales and extend to smaller scales. Large airways may grow into the developing lung and continue to bifurcate into smaller and smaller airways.

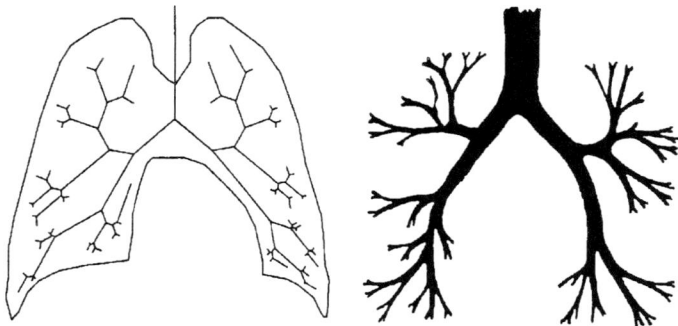

Figure 12.1. Tracheal branching: A local branching rule, repeated many times, can generate a global structure analogous to real structures. *Left:* A computer simulation based on repeated applications of a local branching rule of the airways of the lung. *Right:* Silhouette of the airways from a chest X ray. (From Nelson and Manchester, 1988, with permission.)

2. *Different processes that operate at different scales can somehow become linked together so that a single scaling relationship is produced over many scales.*

The kinetics of ion channels switching between closed and open states may involve many different physicochemical processes, each with its own characteristic time scale. Yet overall an approximate fractal scaling is often observed in the kinetics. The diffusion of substances throughout the body involves the passage through many different barriers, each having different spatial and temporal scales. Yet overall indicator dilution curves have an approximate fractal scaling. In such systems, in ways that are not well understood, different physical processes become linked together into a single fractal scaling.

Thus, our first task in analyzing a fractal phenomenon is to try to determine if it is due to a single cause that percolates through many scales, or many different causes that then become linked together. This is a very difficult task.

12.3   Single Causes That Spread Across Many Scales

Fractals due to a single cause that spreads across many scales can result from local interactions at each point, short-range interactions between nearby points, or long-range interactions between distant points.

Local Recursion: Repeat the Same Process Many Times

Self-similarity is produced if the same rule is applied over and over again. Such mechanisms are local because each part of the structure interacts repeatedly with itself but not with other parts of the structure. These iterations may be triggered by physical forces, chemical stimuli, or surface antigen markers in development.

For example, as shown in Fig. 12.1, the branching of the airways of the lung can be described as repeated applications of a branching rule that is modified to respond to the predetermined boundary of the lung. The distribution of blood flow in the vessels of the heart can be described by repeated applications of the rule that when the vessels branch a slightly larger fraction of the flow enters one branch and the remaining fraction enters the other branch (Bassingthwaighte and van Beek, 1988). This is undoubtedly a contributory factor in the tracer washout curve in Fig. 12.2 from Bassingthwaighte et al. (1994), which shows power law scaling of the tail of the curves for ^{15}O-labeled water washing out of the heart. Repeated branches or foldings can give rise to the structures seen in the bile duct, the urinary collecting tubes in the kidney, the convoluted surfaces of the brain, the lining of the bowel, the placenta, and the His-Purkinje system in the heart.

Recursive rules can also apply to functional networks as well as morphological structures. For example, the immune system can be analyzed as repeated applications of the rule: Create new antigens to all existing epitopes (Farmer, 1986, Crutchfield et al., 1986). Evolution can be described as repeated additions and deletions (Eigen and Schuster, 1979; Katz and Feigl, 1987). Learning and memory can be modeled as a neural network where associations are repeatedly strengthened between similar concepts (Waner and Hastings, 1988).

The repeated application of a rule can also be thought of as subsequent iterations in time. A system that evolves in time is called a dynamical system. This relationship is exemplified by the surprisingly complex and beautiful images of the Mandelbrot set (Peitgen and Richter, 1986). Each Julia set, such as in Chapter 5, corresponds to a dynamical system. The parameters of each dynamical system depend on the (x, y) coordinates of the initial point taken from the Mandelbrot set. To create color images, the points produced at each iteration are assigned colors representing how the value of the variable of that dynamical system evolves in time. Barnsley (1988) has shown how to determine the iterated function system that produces any given fractal pattern. Thus, any fractal pattern in space can be generated from a system that evolves in time.

Grow Where the Local Gradient Is Largest or Keep Walking until You Hit and Stick

In many cases, *the local structure itself precipitates its own further growth at that point*. Thus, at each point, the fractal structure and its environment interact locally. For example, in many cases the growth of a structure is proportional to a local gradient, which is greatest at the tips of a structure. Examples include the viscous fingering of low-density fluids injected into high-density fluids such as water into oil, the spread of an electric spark through a dielectric media as in lightning, the propagation of cracks in brittle materials, the deposition of material on electrodes, and the solidification of crystals (see, e.g., Stanley and Ostrowsky, 1986, 1988).

These same patterns can also be generated in another way (Stanley and Ostrowsky, 1986, 1988). One particle at a time is released outside the object. It continues to walk in a random way until it hits the object, and then it sticks where it landed. It can be shown that this procedure is mathematically equivalent to the prescription that the growth is proportional to a local gradient. Particles are more likely to be captured by the waiting tips of the exterior arms and are less likely to survive the perilous journey deep into the interior of the structure. Thus, the exterior arms grow near their tips, until they are long enough so that significant branches develop and grow, and then these new branches dominate the growth.

The patterns formed by these processes are called *diffusion-limited aggregation*, or DLA. When formed in a plane, these structures have a theoretical fractal dimension of approximately 1.7. DLA generated by using this algorithm is shown in Figs. 9.8 and 12.3. These structures grow fastest at the tips, where the gradient is greatest. However, when the tips have grown very far without branching, then there is a good chance that new tips will bud off from that long linear segment. This yields a fractal structure with larger and larger branches at larger and larger scales. Thus, the *local growth condition produces a global branched pattern*.

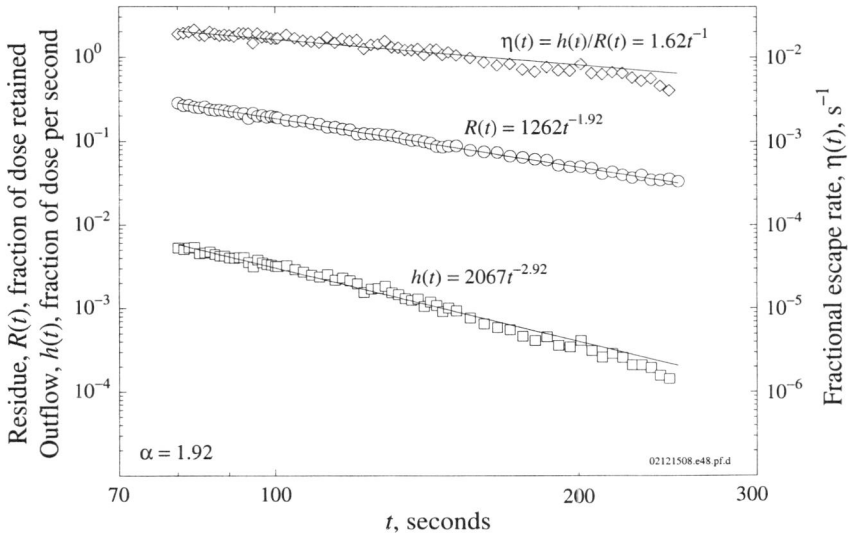

Figure 12.2. Washout of ^{15}O-water from an isolated blood-perfused rabbit heart after pulse injection into the coronary artery inflow. The residue signal $R(t)$ was recorded from the whole heart using coincidence detection from three-inch NaI crystals. The normalized outflow concentration time curves, $h(t)$, were recorded via a flow-through beta detector on the effluent. The fractional escape rate $\eta(t)$ varies with $1/t$. The tails of the curves follow power law relationships, $R(t) \sim t^{-\alpha}$ and $C(t) \sim t^{-\alpha-1}$, with $\alpha = 1.92$.

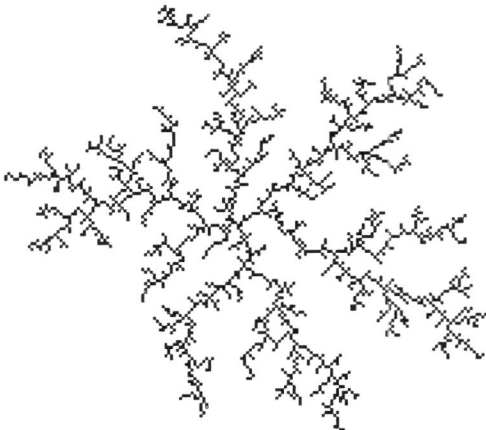

Figure 12.3. Simulation of a diffusion-limited aggregation (DLA). The particles are released far from the structure, randomly diffuse, and stick when they hit the structure. The fractal dimension determined by box counting is approximately 1.6. (This DLA was produced using a Macintosh 512K computer, taking three days; see also Fig. 9.8.)

Many physiological structures formed along planar surfaces have this type of branched pattern and a fractal dimension close to 1.7. Examples include the neurons in the retina shown in Fig. 12.4 (Caserta et al., 1990b), neurons grown in cell culture (Smith, Jr., et al., 1989), and the major blood vessels of the retina (Family et al., 1989; Mainster, 1990). Thus, these neurons and blood vessels may arise from local growth that is proportional to the diffusive gradient of a growth-inducing agent. The agents that stimulate growth could be chemical growth factors, oxygen, electrical fields, or hydrostatic pressure forces.

Fractals with different dimensions or asymmetrical forms can be formed by changing these rules or the boundary conditions. This is equivalent to changing the rule that the growth probability is proportional to the local gradient. For example, structures can be grown from a line or plane rather than free-standing. The probability that the particle sticks when it hits the structure can be varied between 0 and 1. Rather than random walks, the particles can travel in straight lines, producing a ballistic aggregation. The particles need not travel at all.

The probability of the appearance of new particles attached to the structure can be made proportional to the number of occupied nearest neighbors. The latter Eden model (Eden, 1961) was developed to simulate the growth of colonies of cells. Only a relatively small number of such rules have been studied by numerical simulations. It is not known how to predict if a newly devised rule will produce a fractal, or, if it does, how to predict its fractal dimension.

The original Eden model illustrated growth on a square lattice by random budding from an occupied site onto an adjacent occupied site. Wang et al. (1994) developed a "lattice-free" model for cluster growth, an analog to the growth of a solid tumor without an enclosing capsule. As seen in Fig. 12.5, the body of the tumor is not

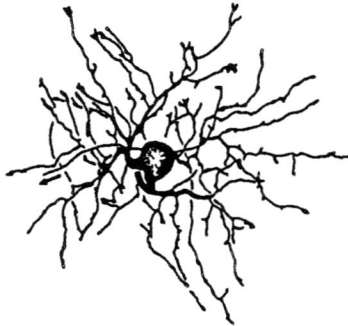

Figure 12.4. Digitized image of β ganglion cell in a cat retina (Maslim et al., 1986). Caserta et al. (1990b) measured the spatial correlation of pixels in this image and found that this neuron has a fractal dimension of approximately 1.6. The shape of this neuron has a form and a fractal dimension similar to the DLA in Fig. 12.3. This suggests that the pattern of these dendrites may arise from growth that is proportional to the gradient of a diffusible growth-inducing substance. (Image reproduced from Maslim et al, 1986.)

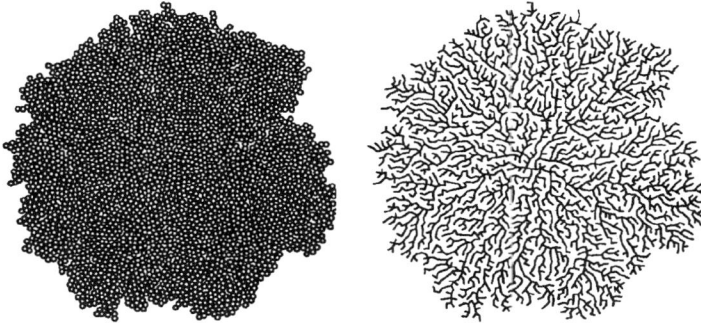

Figure 12.5. Lattice-free Eden-type cell-growth model in which new cells are added randomly to the periphery but only at empty locations adjacent to a cell. *Left panel:* Cell positions. The peripheral layer is a fractal. *Right panel:* Paths along which growth occurred. (From Wang et al., 1994.)

fractal in mass or cell density, but approaches a constant value of 0.65% of the space. The number of living cells, L, on the surface is proportional to the mean radius, R, of the boundary, $L = 3.58R$. This primitive model for tumor growth is simple enough to allow many variations—in cell sizes and growth rates of individual cells, for example. In reality, tumor cell growth and invasion of the tumor by blood vessels to provide nutrients and remove metabolites are approximately balanced. However, since nutrient supply to the periphery of a solid tumor is inevitably better than to its center, the peripheral growth is often followed by necrosis at the center.

Do What Your Nearest Neighbors Tell You to Do

In the previous subsection the fractal object was generated from the growth induced by local gradients between the fractal and its environment. In this section we see that *local interactions between nearest neighbors on the structure itself can generate a fractal in space or time.*

React with your neighbor. There are many important biochemical reactions where two types of molecules interact. Such reactions include the change in conformation of the sodium channel when bound by acetylcholine, the binding of oxygen to hemoglobin, the cleavage of large proteins by trypsin, or the trans-acting binding of a protein gene product to regulate the expression of another gene. In all these reactions one species, A, must encounter another species, B. Often, the rates of these reactions are limited by diffusion, that is, the rate at which the A's diffuse to encounter the B's, rather than the reaction time of the AB complex. In volumes with unequal initial amounts of A's and B's, when the reaction is complete, there will be leftover A's or B's. Thus, there will be regions of only A's or B's. The reaction will continue only on the boundary between these regions. This boundary has a fractal dimension less than 3. Thus, the reaction has created for itself a space of fractal dimension that is less than that of the full three-dimensional space. This is called self-organization. As time goes by the A-rich and B-rich regions grow in size. Thus, the *local* mechanism whereby A and B interact only when they are nearby leads to a *global* organization having a fractal spatial pattern with larger and larger areas of leftover A's and B's. As time passes the volume of the regions of leftover A's and B's grows relatively faster than the size of the boundaries between them. Thus, the relative rate of the reaction occurring in these boundaries decreases in time. The fractal scaling that develops in space thus leads to a reaction rate that has a fractal scaling in time similar to that of the simplest fractal scaling of the kinetic rate constant found for the kinetics of ion channels described in Chapter 8.

Diffusion-limited reactions, such as those described by Kopelman (1988), were studied on fractal networks such as are created by an aggregation process. Such a network is termed a "percolating network" if there is a path for a diffusing particle across the network (Orbach, 1986). The critical concentration of occupied sites on a square lattice that allows infinite percolation (where infinite means *any* finite size) across a connected set of sites is 59.28% of the total number of sites. (Connections are only vertical and horizontal, not diagonal.) A computer-generated percolation cluster is shown in Fig. 12.6. A plot of the instantaneous reaction rate coefficient in a mixture of reactants at the critical percolation concentration is shown in Fig. 12.7 (by the x's), illustrating the log-log diminution in the reaction rate as the reaction proceeds. This is in accord with theory where the rate coefficient, which isn't a constant in this situation, decreases:

$$k = k_1 t^{-h}, \tag{12.1}$$

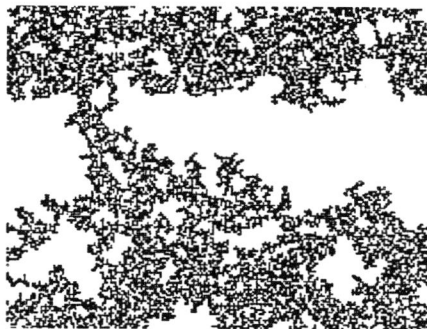

Figure 12.6. A percolation cluster. Here 59.3% of the points of a square lattice are "occupied" (the rest are empty). Only the points that belong to the connected ("percolating") giant cluster are shown. This aggregate is a "random fractal" with $D_f = 1.896 \ (= 91/48)$ and $D_s = 1.333 \ (= 4/3)$, where D_s is the spectral or fracton dimension. (From Kopelman, 1988, Fig. 2, with permission from AAAS.)

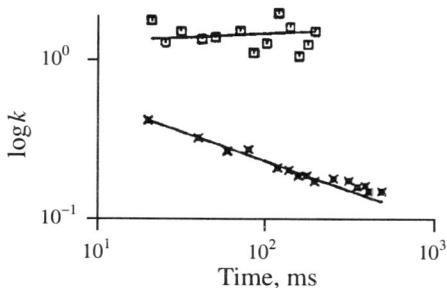

Figure 12.7. Log-log plot of instantaneous rate constant k versus time t, for the exciton fusion reaction in isotopically mixed naphthalene crystals at the critical percolation composition (bottom curve) and well above it (top curve). The analysis of this experiment gave $h = 0.32 \pm 0.03$ for the bottom curve (compared with the theoretical value of 0.33) and $h = 0.03$ for the top curve (theoretical value of 0). (From Kopelman, 1988, Fig. 1, with permission from AAAS.)

where h is the power law exponent, and which is calculated from the spectral dimension D_s:

$$h = 1 - D_s/2 . \tag{12.2}$$

Orbach (1986) gives a calculation and definition for the spectral or fracton dimension D_s:

$$D_s = \frac{2D_f}{2 + \theta}, \tag{12.3}$$

where θ takes values around 0.8 in two-dimensional space and 1.5 in three-dimensional space.

It is often assumed in chemical kinetic analysis that the reaction rate is a constant that is independent of time. The consequences of this assumption are rarely explicitly stated. This assumption is equivalent to assuming that the remaining A's and B's are continually mixed to homogeneity, which is rarely the case. It is certainly not the case when the reaction rate is limited by the speed of the diffusion of the reactive species. The self-organization property of the diffusion-limited reaction amplifies any original inhomogeneity. It is only a question of time until a diffusion-limited reaction reaches the regime of fractal organization. The more homogeneous the starting system, the longer it will take for the fractal pattern to appear.

Different reaction mechanisms generate different fractal patterns in space and time (Kang and Redner, 1985; Kopelman, 1988). It is normally assumed that if a reaction rate is proportional to the nth power of the concentration of a reactant, that n

of those reactants must be involved in the reaction at the same time. In biochemistry this n is called the Hill coefficient. However, it has been clearly shown that when fractal spatial patterns form in a diffusion-limited reaction, n can have any value greater than 2, even though only two reactants are involved at any one time in the reaction (Kopelman, 1988; West et al., 1989).

Walk onto your neighbor's space. A *random walk* where the direction of *each step depends only on the available adjacent locations* is another mechanism whereby local interactions between nearest neighbors result in global patterns. For example, the fractal scaling in time of the distributions of closed durations of an ion channel can be produced by a random walk in a one-dimensional configurational space consisting of one open state and many closed states, as shown in Fig. 12.8 (Millhauser et al., 1988b). In another type of random walk, shown in Fig. 12.9, *one piece at a time* of the channel protein *moves into an available adjacent space* (Läuger, 1988). This also produces an approximate fractal scaling in time of the closed durations. Ion channel models with multiple available spaces have also been analyzed (Condat and Jäckle, 1989). In physics, various models with multiple random walkers and multiple available spaces to trap those walkers have been used to describe the decay of polarization after an electric field is removed from a material called a spin glass (Shlesinger, 1984).

The random walks described above occur in spaces with integer dimension, but similar random walks in spaces of fractal dimension have also been analyzed (Pietronero and Tosatti, 1986; Stanley and Ostrowsky, 1988).

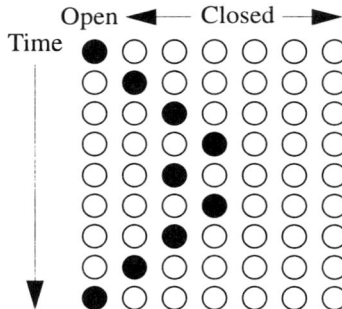

Figure 12.8. A cell membrane ion channel protein can have conformations open (left hand column of circles) to the passage of ions and conformations closed to such flows (right hand columns of circles). Successive rows illustrate the possible states at successive moments in time. This succession shows that the conformational state of the channel at a given time (black circle) can be described as a one-dimensional random walk among these states. The present state of the channel determines the probabilities to move to the left and the right on the next time step. In contrast to the random walk in Fig. 9.12, this random walk between adjacent sites produces distributions of durations of times spent in the closed state that have a power law fractal scaling. (From Millhauser et al., 1988b.)

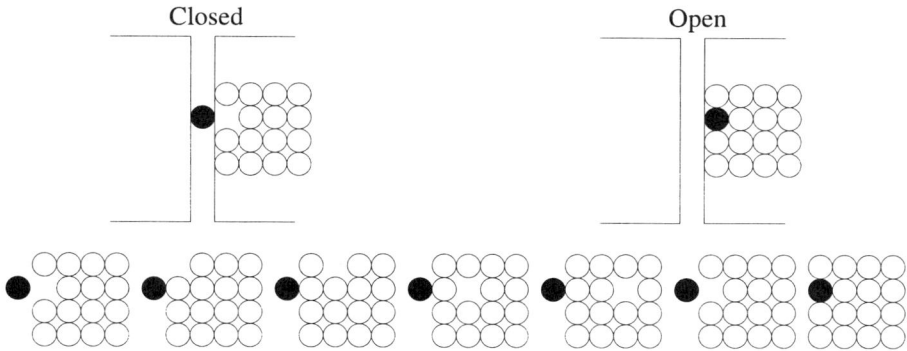

Figure 12.9. A cell membrane ion channel protein can be described as movable pieces (circles). One piece (black circle) moves, thus blocking the flow of ions through the channel. There is then an equal probability for each piece to move into the defect created by the movement of the first piece. Eventually, the original piece returns to its original position, opening the channel. Such a sequence is shown at the bottom row. These local interactions between adjacent pieces produce global spatial motions and distributions of the durations of times spent in the closed state that have an approximate fractal scaling (From Läuger, 1988).

The random walk description is sometimes equivalent to other mathematical descriptions. For example, as noted above, the pattern generated when one random walker at a time sticks to the growing object is equivalent to the mathematical description that the growth of the structure is proportional to the local gradient between the structure and its environment. However, the mathematical description corresponding to many other random walks is not known.

Feel your neighbor's force: Ising models, phase transitions, spin glasses, and self-organizing critical systems. Nearest-neighbor interactions in condensed matter physics are the most extensively studied (Ma, 1976; Peitgen and Richter, 1986). The archetypal example is a solid that has small magnetic domains whose magnetization can have only one of two distinct values. The energy state of the entire solid depends on the value of the magnetization in adjacent domains. This is called an *Ising model*. At low temperatures there is little random energy for the domains to change and so the domains can be frozen into a highly ordered pattern. At high temperatures each domain has enough energy to spontaneously change, so that the highly ordered pattern is destroyed. This change from order to disorder occurs sharply when the temperature reaches a critical value. Such a change is characteristic of a *phase transition*, such as the change in density when a liquid vaporizes into a gas. Near the critical temperature the system has fractal properties. There are spatial correlations at all length scales. One way to characterize such a system is by an *order parameter*, such as the total magnetization of a magnetic solid or the density of a fluid. The order parameter depends on the temperature offset from the critical value. It has a power law form that is characteristic of fractal scaling.

The properties of such systems can be predicted by the method of renormalization groups. The properties are recalculated by grouping together values of nearest neighbors into new regions, then regrouping the values of those larger regions and continuing this process. Thus, local interactions between neighboring domains result in global fractal self-similar spatial patterns and global fractal power law scalings. This is analogous to dispersional analysis (Chapter 4.)

Physiological systems often have many locally interacting pieces whose global characteristics undergo sharp changes with small changes in a parameter. Thus, such phase transitions may also be important in many physiological systems. For example, in an associative memory, the accuracy of identifying an input with the stored memory traces undergoes a phase transition from very good to very poor, as the number of stored memories increases (Sompolinsky, 1988). The activity of a neural network undergoes a phase transition from very little to very much activity as the connectivity of the network is increased (Shrager et al., 1987). In fact, the recent rapid developments in models of neural networks have grown out of the success of analyzing them by using the techniques developed to study the magnetic properties of spin glasses (Hopfield, 1982). The local interaction between tubulin molecules may result in the sudden phase change from tubules being unformed or formed in a cell to provide for locomotion or intracellular transport.

The fractal characteristics near a phase transition occur only when a parameter, such as the temperature, is near its critical value. However, in some systems, the value of that parameter is determined by the system itself. If the system itself causes the value of that parameter to be near its critical one, then the system will always be near a phase transition. A system with this property is called a self-organizing critical system (Bak et al., 1987). Since it is always near a phase transition it will have fractal properties. A self-organizing system results if stress is added from the environment and if the excessive stress at a point is relieved by being shared among its nearest neighboring points.

The prototypical example of a self-organizing critical system is a sandpile. Sand is dropped at random onto a sandpile. The sand will accumulate at each point until the local slope increases too much, at which time the sand will flow downhill, possibly triggering other points into an avalanche. These avalanches are fractal in both space and time. A single added grain of sand may cause no avalanche, or it may trigger an avalanche that involves the entire sandpile. The distribution of the amount of sand in the avalanches triggered by each added grain is a fractal power law. The power spectrum of the amount of sand moved downhill in time is also a fractal power law. The sandpile self-organizes into a phase transition state because the avalanches relieve only the stress above the maximum. Thus each point is always poised just below the value needed to be moved by the next added grain of sand. The sandpile is a system far from equilibrium. It is maximally unstable. Each point in the sandpile is poised at the transition between remaining static and moving downhill when the next grain of sand reaches it.

Self-organizing criticality has been used to explain the fractal patterns observed in the turbulence in fluids, fluctuations in the economy, and the spread of forest fires. However, not all nearest-neighbor stress relief rules result in fractal behavior. For example, the flow of water droplets on a pane of glass has some fractal and some nonfractal properties (János and Horváth, 1989). It is not known which rules result in fractal properties and which do not.

Thus, local nearest-neighbor interactions can cause a system to self-organize into a phase transition state that has fractal properties. Fractals in physiological systems caused by this mechanism may include the interaction between residues of an ion channel protein that cause the channel to open and close and the interactions between susceptible individuals in the spread of epidemics.

Do the Same (or the Opposite) of What You Used to Do

The mechanisms described above involve *short-range* interactions either at one point or between adjacent nearest neighbors. *Long-range interactions can also produce fractals.* The *fractal dimension of a time series is related to the strength of the memory* in the system. As shown in Fig. 12.10, the stronger the persistence of values from one time to another, the less jagged the time series, and thus the lower its fractal dimension. The values computed from random walks with memory have been used to describe many series of data that are fractal in space or time (Mandelbrot, 1983; Peitgen and Saupe, 1988; Feder, 1938), including the amount of rainfall in the watershed of the Nile river, the roughness of the edges of rocks, the

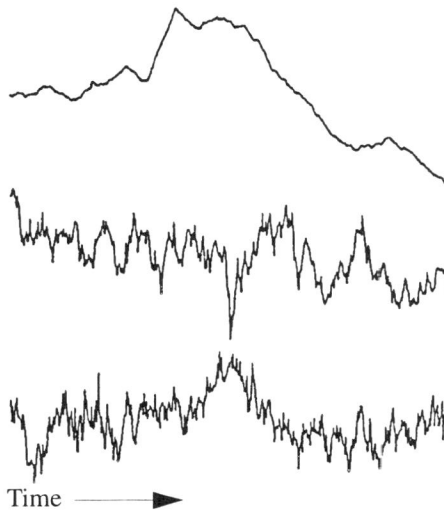

Figure 12.10. Random walks with memory yield fractal time series. *Top panel:* The value at each point is likely to be close to the value of the previous point. The fractal dimension D of the time series is 1.1 or $H = 0.9$. *Middle panel:* The value of the next point is not affected by the value of the previous point. $D = 1.5$; $H = 0.5$. *Bottom panel:* The value at each point is less likely to be close to the value at the previous point. $D = 1.9$; $H = 0.1$. (Reproduced from Feder, 1988.)

surface of the sea floor, the light output of quasars, the sound fluctuations of music (including medieval, Beethoven's Third Symphony, Debussy's piano works, R. Strauss's *ein Heldenlebe*, and the Beatles' *Sgt. Pepper*), and the prices paid for cotton.

These long-term correlations are also seen in many physiological processes. The tidal volume of each breath is more likely to be similar to that of the previous breath (Hoop et al., 1993). The time between action potentials recorded from primary auditory fibers is more likely to be similar to that of the time between the previous action potentials (Teich, 1989). This type of memory may occur in physiological processes having feedback loops that adjust their output depending on the previous input or output of the system.

The random walk can be literal, such as in the diffusion of small membrane molecules around the islands of integral proteins in the cell membrane. The random walk can also be a mathematical description of the changing value of a variable in a physiological system. For example, the electrical state of a neuron can be described by a random walk with memory, where the potential of the cell varies between the resting potential and the threshold of the action potential. Such a model yields fractal distributions of interspike intervals similar to that recorded from neurons (Gerstein and Mandelbrot, 1964).

Long-range interactions can also occur in space as well as time. In physics the magnetic properties of certain spin glasses can also have such long-range interactions (Fisher et al., 1988). These models may be applicable to neural networks with long-range correlations, for example, the lateral connections in the visual cortex.

12.4   Different Causes That Become Linked Across Many Scales

The mechanisms described above are all local interactions caused by a single mechanism whose cooperative effects result in a global scaling. However, there are many *fractals where the physical mechanisms must be different at different scales*. The power law scaling of the length of the west coast of Great Britain extends over many length scales, although different forces must sculpt it at scales of centimeters and kilometers. There is an approximate power law scaling of the dilution of concentration of tracers measured in the body over time that must represent many different transport processes (Wise and Borsboom, 1989). The power law scaling of the kinetics of ion channels extends over time scales representing different physicochemical processes (Liebovitch et al., 1987a). Yet in these and many other cases, the *fractal dimension remains approximately constant over many different scales*. How can that be? How can different physical processes acting at different scales self-organize into fractal patterns? The mechanisms that link processes at different scales are mysterious. We make only a few tentative and speculative remarks.

The first possibility is that *something is shared across scales that causes them to adjust together*. It must be something conserved (so that its balance is equalized across scales) and something minimized or maximized (so that it can be optimized at different scales). Thus, it must be something like free energy or entropy, and yet it is clearly not the free energy or entropy.

The second possibility is that *nothing is shared across scales*. That is, the *mechanisms at different scales are truly independent, and that true independence implies self-similarity*. For example, consider an ion channel that randomly switches between open and closed states. Following the adage that "what is not restricted is required" we ask, "Which are the least restrictive assumptions we need to make to describe the kinetics of switching from one state to another?" Until now, the answer to this question has been that there is equal probability per unit time of switching states. That is, the kinetics is a Markov process with no memory. However, such a process defines a time scale that is equal to the reciprocal of the time constant of the probability to change states. The probability distribution of the time spent in each state is a single exponential with this time constant. Thus, the channel knows what time it is. A less restrictive assumption is to assume that the kinetics of switching looks the same at all time scales. That is, it is self-similar. The probability distribution of the time spent in each state is a power law. At all time scales we see the channel switching between its two states. Thus, the channel doesn't know what time it is. In this sense, fractals arise because the processes are independent at different scales and the strongest way they can be independent is if the system has no preferred time scale, and this happens only when the system has a fractal form. As noted by Machlup (1981), "We do not need special mechanisms. We need simply a large ensemble of mechanisms with no prejudice about scale. The conjecture is that nature is sufficiently chaotic to possess this lack of prejudice."

12.5   Summary

We presented a partial list of mechanisms that produce fractals in space and time. This shopping list may be a useful starting point to think about appropriate mechanisms when a new fractal is discovered. Fractals may result from: 1) *a single process at one scale that extends to other scales*, or 2) *a set of different processes at different scales that become linked together*. We described three types of single process mechanisms: 1) *Local interactions*. These processes act independently at each point. They include recursive rules and growth proportional to a gradient between the structure and the environment. 2) *Short-range interactions*. Typically, these processes involve interactions between nearest neighbors. They can remove material, as in chemical reactions, or they can add material, as in growing structures. These processes also include random walks and systems that self-organize near a phase transition. 3) *Long-range interactions*. These processes generate long-term memory and thus long-term correlations in space or time.

It is surprising that local interactions from a single process can cause an overall global scaling. It is even more surprising that different processes over different spatial or temporal scales can organize into a structure with a consistent overall fractal dimension.

13

Chaos? in Physiological Systems

> In physics it is our habit to try to approach the explanation of a physical process by splitting this process into elements. We regard all complicated processes as combinations of simple elementary processes . . . that is, we think of the wholes before us as the sum of their parts. But this procedure presupposes that the splitting of the whole does not affect the character of the whole . . . Now, when we deal with the irreversible processes in this fashion the irreversibility is simply lost. One cannot understand such processes on the assumption that all properties of the whole may be approached by a study of its parts.
>
> Max Planck (1915)

13.1   Introduction

Over the last two decades chaotic dynamical systems analysis has become a standard tool for systems analysis in the hydrodynamics of turbulence, in mechanical systems, and in electrical signals. The label "chaotic" referes to the lack of predictability of the weather and other complex systems. We can use "chaos" theory to enhance our understanding of cardiovascular pressures, flows, heart rates, and tracer exchanges. "Chaos" is the descriptor applied to the state of a system that is unpredictable, even though bounded. Poincaré (1880) observed that some simple differential equations resulted in unpredictable behavior. But, as brought forcefully to our attention in modern times by Lorenz (1963), the unpredictability is not necessarily equivalent to randomicity. There is no reason to think that biological systems must be chaotic, and Glass and Malta (1990) argue that biochemical systems are in general not chaotic, so in this essay the focus is on the idea that consideration of chaotic and fractal aspects of systems augments our power to examine and interpret data, and perhaps to characterize the system.

Most data on biological systems are in the form of time series, the measures of one or sometimes more variables taken at even intervals over time. Inevitably, most of the system variables are unmeasured, but a theorem by Takens points out that the influences of all of the unmeasured variables that are a part of the "system" are reflected in the dynamics of the one or two measured variables. (See Section 7.4.) Such a statement applies to both continuous systems and discrete systems (Ruelle, 1989). This is the basis on which virtually all of the methods of analysis are based.

Studies of cardiovascular and neural systems have historically often been the first to be subjected to new methods of analysis, simply because the methods for recording quantitative information have been more readily available in these fields. Fractals and chaos stimulate new qualitative thinking, but the applications are quantitative; the power of the methods is in providing numbers that give clues as to mechanism and give descriptions of behavior. All of the methods and applications are recent, and proofs of the theory and statistical assessments of the analyses are often lacking. Nevertheless, the approaches have such a huge potential that exemplifying the possibilities seems to have much merit. It is in this spirit that these examples are provided.

The reasons for wanting to make the distinction between a *truly chaotic signal* and a merely *noisy signal* are both theoretical and practical: for description, for insight and prediction, or for providing a basis for intervention. If a signal is merely a constant summed with Gaussian noise, then the constant and the standard deviation of the noise constitute a full description. If it is chaotic, then the apparent dimension of the attractor provides a beginning of the description. A simple mathematical analog may augment this and provide some predictive capability that could not be reached for a noisy system. As we show below, even an incomplete and insecure level of description and understanding can be sufficient to allow intervention and control.

In the case of fatal cardiac arrhythmias, improving the capability for therapeutic intervention is a driving incentive for investigation, whether or not the signal is truly chaotic (Jalife, 1990). Also raised with respect to the cardiac rhythm is the question of the complexity of the variation in the signal: Goldberger makes the case that a healthy young person shows greater complexity in heart rate than an older person, and that great loss of complexity (reduction to an almost constant rhythm) may be premonitory to a fatal arrhythmia (Lipsitz et al., 1990; Kaplan et al., 1991).

Bio-oscillators. Nature abounds with rhythmic behavior that closely intertwines the physical and biological sciences. The diurnal variations in dark and light give rise to circadian physiological rhythms. Luce (1971) gives an incomplete list of such daily rhythms: the apparent frequency in fetal activity, variations in body and skin temperature, the relative number of red and white cells in the blood along with the rate at which blood will coagulate, the production and breakdown of ATP (adenosine triphosphate), cell division in various organs, insulin secretion in the pancreas, susceptibility to bacteria and infection, allergies, pain tolerance, and on and on. The shorter periods associated with the beating of the heart and breathing are also modulated by a circadian rhythm.

The tendency is to think of the rhythmic nature of many biological phenomena as arising from the dominance of one element of a biosystem over all other elements. A logical consequence of this tendency is the point of view that much of the biosystem is passive, taking information from the dominant element and passing it along through the system to the point of utilization. This perspective is being questioned by scientists who regard the rhythmic nature of biological processes to be a consequence of dynamic interactive nonlinear networks, e.g., nonlinear dynamics systems theory. The application of nonlinear equations to describe biorhythms dates back to the 1929 work of van der Pol and van der Mark on the *relaxation oscillator*. It is to these two that Glass et al. (1991) dedicate their book *Theory of Heart*, a valuable compendium linking mechanical and electrical aspects of cardiac function.

Oscillations in biological processes are commonly more complex than sinusoidal variation. The more usual situation is one in which the period of oscillation is dependent on a number of factors, some intrinsic to the system but others external to it, such as the amplitude of the oscillation, the period at which the biological unit may be externally driven, the internal dissipative properties, fluctuations, etc. In particular, since *all* biological systems are thermodynamically open to the environment they are dissipative, that is, they give up energy to their surroundings in the form of heat. Thus, for the oscillator to remain periodic, energy must be supplied to the system in such a way as to balance the continual loss of energy due to dissipation. If such a balance is maintained, then the phase space orbits becomes a stable *limit cycle*, that is, all orbits in the neighborhood of this orbit merge with it asymptotically. Such a system is called a bio-oscillator, which left to itself begins to oscillate without apparent external excitation. One observes that the self-generating or self-regulating characteristics of bio-oscillators depend on the intrinsic nonlinearity of the biological unit. Here we are concerned not with the continuous periodic behavior of bio-oscillators but rather with how perturbations of such systems can be used to explore their physiological properties.

13.2   Cardiovascular Chaos

Basic Mechanisms: Embryonic Heart Cells

Glass et al. (1984, 1986) and Glass and Winfree (1984) measured the transmembrane electrical potential of aggregates of embryonic chick heart cells that either were beating spontaneously or were periodically stimulated, as seen in Fig. 13.1. Embryonic chick heart cells have a different number of beats per stimulation when stimulated at different frequencies. They measured the phase, that is, the time between the stimulus and the next action potential. If we measure the durations x_n between successive events such as heartbeats, breaths, or ovulations, then we can think of the data as the time series of the values of those durations. The time lag can then be chosen as equal to one and the coordinates of the phase space are the successive durations $(x_n, x_{n+1}, x_{n+2}, \ldots)$. Or, if the process or the experimental protocol is periodic, then the time lag can be set equal to that period.

In Fig. 13.1 the phase in the bio-oscillator is the delay between the stimulus and the next beat as a fraction of the period of the stimulation, T_0, the reciprocal of the fundamental frequency. The *phase* of any given cycle at a time t, $0 < t < T_0$ is defined as $\phi = t/T_0$. In this nomenclature the phase is defined to lie in the interval $0 < \phi < 1$. An external perturbation (stimulus) is applied at a time t corresponding to a relative phase of t/T_0.

To construct the phase space set they used the period of the stimulation as a natural choice for the time delay, and constructed a two-dimensional phase space set by plotting the phase of the action potential after the $(i+1)^{th}$ stimulus against the phase of the action potential after the ith stimulus. When the response is periodic, the plot reveals nothing new, but when the response is irregular, the plot is most revealing. The result, shown in Fig. 13.2, showed that the phase of a beat could be accurately predicted from the phase of the preceding beat, although the relationship

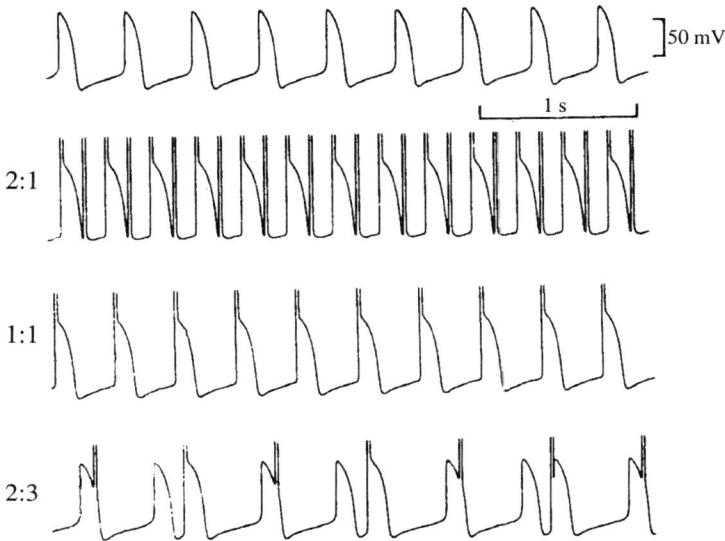

Figure 13.1. Establishing the deterministic nature of beating embryonic chick heart cells by the changes observed when a parameter is varied. *Top row:* Spontaneously beating cells with no external stimulation. *Bottom three rows:* The cells can also be stimulated (vertical spikes) to beat. When stimulated at high frequency the cells beat once for every two stimulations. As the stimulation frequency is lowered, the cells beat once for each stimulation. As the stimulation frequency is lowered again, the cells beat three times for every two stimulations. For some frequencies of stimulation, the cells beat in a seemingly irregular pattern. (From Glass et al., 1984, as adapted by Glass et al., 1986.)

was not directly evident from the recorded signal. The deterministic nature of this system is established by how the qualitative features of the system, that is, the variation of the number of beats per stimulation, changes as a function of the parameter, the stimulus frequency. Using a similar approach, Chialvo and Jalife (1987) and Chialvo et al. (1990a, 1990b) found that the number of beats per stimulation varied with the stimulus rate in nonpacemaker cells from the heart.

Glass et al. (1983) took the analysis to a deeper level, considering the system as a bio-oscillator. Their analysis was based on the three explicit assumptions:

> (i) A cardiac oscillator under normal conditions can be described by a system of ordinary differential equations with a single unstable steady state and displaying an asymptotically stable limit cycle oscillation that is globally attracting except for a set of singular points of measure zero.

> (ii) Following a short perturbation, the time course of the return to the limit cycle is much shorter than the spontaneous period of oscillation of the time between periodic pulses.

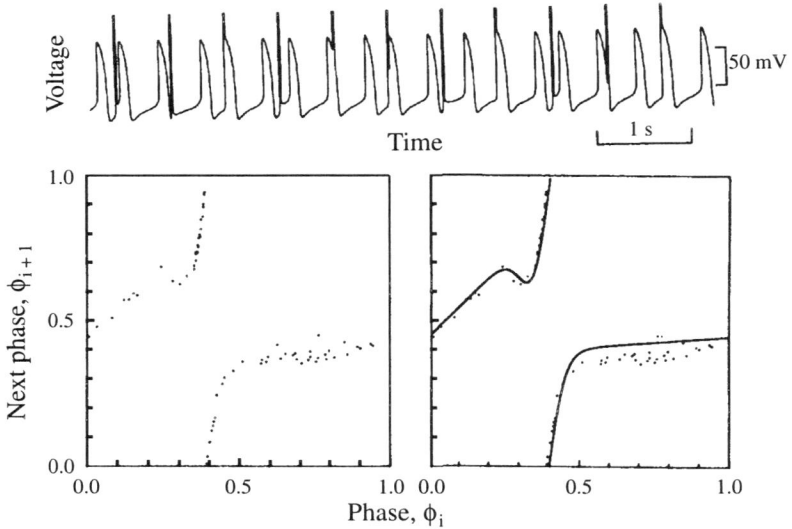

Figure 13.2. Transmembrane voltage in aggregates of embryonic chick heart cells stimulated periodically. *Top:* Voltage response of the cells as a function of time. The periodic stimulating pulses are indicated by vertical lines. *Bottom:* The phase space set does not fill the two-dimensional space, indicating that the variation in the intervals of the action potential is deterministic chaos rather than random noise. The deterministic relationship revealed in the data is modeled by the equation defined by the lines in the graph on the right. (Adapted from Glass et al., 1986, their Figs. 11.9 and 11.10.)

(iii) The topological characteristics of the phase transition curve change in stereotyped ways as the stimulus strength increases.

Continuous dynamical systems are modeled by a mapping that determines the effect of the perturbation on the cycle to which it is applied and to subsequent cycles. We denote the phase of the bio-oscillator immediately before the jth stimulus of a periodic external stimulation with a period t_s by ϕ_j. The recursion relation, or mapping, is

$$\phi_{j+1} = g(\phi_j) + t_s/T_0, \qquad (13.1)$$

where $g(\phi)$ is the experimentally determined phase transition curve for that stimulus strength, and T_0 is the period of the limit cycle. Eq. 13.1 measures the contraction of the aggregate as a function of the time of the perturbation. Using the phase-resetting data a Poincaré map was constructed to determine the phase transition function shown in Fig. 13.2. This function is obtained by plotting the new phase following a stimulation against the old phase. The theoretical Eq. 13.1 is now iterated, using the experimentally determined $g(\phi)$ to compute the response of the aggregate to periodic stimulation. The observed responses to such perturbation are phase locking, period doubling, and chaotic dynamics as the frequency of the periodic driver is increased. This is not unlike the sequence of the responses undergone by the logistic map as the control parameter is increased.

We start from an initial phase ϕ_0 and use Eq. 13.1 to generate $\phi_1 = f(\phi_0)$, where $f(\bullet)$ denotes the right-hand side of Eq. 13.1. By successive substitution of the phases into Eq. 13.1, we obtain $\phi_1 = f(\phi_0)$, $\phi_2 = f(\phi) = f^2(\phi_0)$, . . . , $\phi_N = f(\phi_{N-1}) = f^N(\phi_0)$. The initial state ϕ_0 is said to be a fixed point of period N if $\phi_N = \phi_0^*$ modulus 1, and $\phi_j \neq \phi_0$ for $1 \leq j < N$. The sequence $\phi_0, \phi_1, \ldots, \phi_N$ constitutes a cycle of period N. The cycle is stable, if the slope of the map generating this fixed point $\phi_0 = \phi_N$ is less than unity.

The dynamical responses to periodic stimulation are predicted by iterating the experimentally derived map and bear a close resemblance to that observed experimentally. Glass et al. (1983) point out that the experimentally observed dynamics show patterns similar to many commonly observed cardiac arrhythmias. Ikeda et al. (1981) use the properties of the phase response model to explain ventricular parasystoles. Guevara and Glass (1982) associate intermittent or variable AV block with complex irregular behavior characteristic of chaotic dynamics observed in the phase response model. The subject is nicely reviewed by Glass and Shrier (1991).

The Electrocardiogram

Glass and Hunter (1990), in their paper "There is a theory of heart," discussed the application of mathematical techniques to the analysis of mechanics and dynamics in the heart. Glass et al. (1991) consider the mechanics of the passive and active heart, ionic mechanisms of cardiac arrhythmias, and reentrant reexcitation and its connection with cardiac tachycardia and fibrillation.

Different areas of the mammalian heart are capable of spontaneous, rhythmic self-excitation, but under physiological conditions the normal pacemaker is the sino-atrial (SA) node, a small mass of pacemaker cells embedded in the right atrial wall near the superior vena cava. An impulse generated by the SA node spreads through the atrial muscle, giving rise to the P wave of the ECG and triggering atrial contraction. The depolarization wave then spreads through the atrioventricular (AV) node and down the His-Purkinjé system into the right and left ventricles, giving rise to the QRS complex and ventricular contraction. The third main component of the ECG, the T wave, represents the wave of recovery or repolarization of the ventricles. Traditional wisdom and everyday experience tells us that the ECG time series is periodic; however, quantitative analysis of the time series reveals quite a number of irregularities in the ECG record. In other words, regular sinus rhythm is not regular.

A model of the mammalian heart rate discussed by van der Pol and van der Mark (1929) gives a qualitative likeness to ECG time series, but does not account for the observed fluctuations in the data. The question arises as to whether these fluctuations are the result of the oscillations' being unpredictably perturbed by the cardiac environment, or are a consequence of cardiac dynamics being given by a chaotic attractor, or both. Measures must be developed to distinguish between these two types of time series. Spectral analysis, temporal autocorrelation function, and the phase space reconstruction method are qualitative measures, whereas the correlation dimension, Lyapunov exponents, and Kolmogorov entropy are potentially quantitative measures. However, we shall find that none of these are completely adequate for discriminating between noise and chaos.

The power spectrum of the QRS complex of a normal heart fits the hypothesis that the fractal structure of the His-Purkinjé network serves as a structural substrate for the observed broadband spectrum (Goldberger et al., 1985a). Babloyantz and Destexhe (1988) showed that the power spectrum of a four-minute record of ECG had a broad-band structure, something that can arise from stochastic or deterministic processes. Unlike the power-law spectra found for the single QRS complex, they find an exponential power spectrum (Fig. 13.3, left panel). The exponential form has been observed in a number of chaotic systems and has been used to characterize deterministic chaos by a number of authors (Greenside et al., 1982; Sigeti and Horsthemke, 1987). Frisch and Morf (1981) presented an intuitive argument that systems of coupled nonlinear rate equations that have strange attractor solutions give rise to chaotic time series whose power spectra decay more rapidly than algebraic, e.g., *exponential* in frequency at high frequencies, i.e., $P(f) \propto e^{-\lambda f}/f^{\alpha}$ where λ and α are positive constants. (This is hardly diagnostic of chaos, since any low-pass filter or merely capacitances in the signal-acquisition equipment will give the same result.) This argument was numerically verified by Greenside et al. (1982) using systems with known strange attractor solutions. Sigeti and Horsthemke (1987)

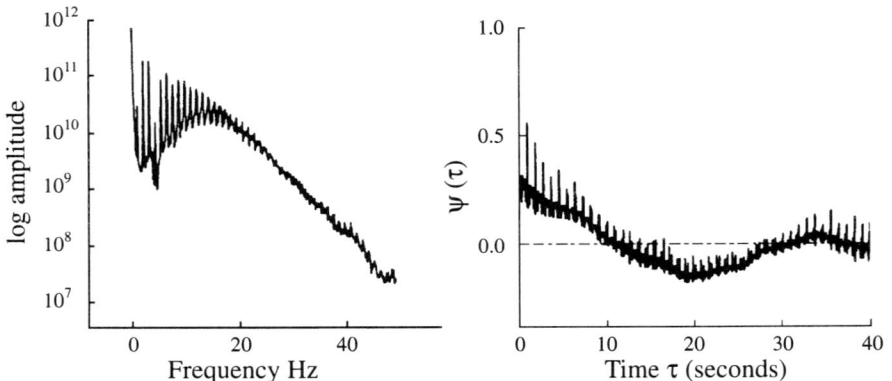

Figure 13.3. Analysis of a four-minute recording of the electro-cardiogram. *Left panel:* Semilogarithmic plot of the power spectrum showing exponential diminution at high frequencies. *Right panel:* Autocorrelation function versus time lag, showing a form characteristic of aperiodic dynamics. (From Babloyantz and Destexhe, 1988, with permission.)

give a general proof establishing the power law form of the spectrum for stochastic systems driven by additive or multiplicative noise. No such proof exists for the exponential decay of spectra for chaotic systems, although mathematical investigations have been initiated, see, e.g., Ruelle (1986), to establish such a proof. Contrary to this expectation, inverse power law spectra have been found for certain deterministic dynamical systems, e.g., the driven Duffing oscillator seen in Fig. 6.2 (Arecchi and Lisi, 1982) and the driven Helmholtz oscillator (Rubio et al., 1986) in which low frequency inverse power law spectra are observed in both analog simulations and numerical integration of the equations of motion. Inverse power law spectra also appear in discrete dynamical systems with intermittent chaos, i.e., systems that manifest randomly spaced bursts of activity that interrupt the normal regular behavior (Frijisaka and Yamada, 1987). *Therefore, an inverse power law spectrum is an ambiguous measure associated with both colored noise and chaos.*

Fig. 13.4 depicts a portrait of the ECG in three-dimensional phase space using two different delay times. The two-phase portraits look different, yet their topological properties are identical. The portraits seem to depict an attractor, unlike the closed curves of a limit cycle describing periodic dynamics. Further evidence for this is obtained by calculating the correlational dimension using the Grassberger-Procaccia correlation function; this dimension is found to range from 3.5 to 5.2, using four-minute segments of data or 6×10^4 data points sampled every 4 ms.

Figure 13.4. Phase portraits of human ECG reconstructed in three-dimensional space. A two-dimensional projection is displayed for two values of the delay, 12 ms (*left*) and 1200 ms (*right*). These portraits are quite unlike a single closed curve, which would describe periodic activity. (From Babloyantz and Destexhe, 1988, with permission; their Figs. 3a and 3b.)

The successive intersections of the trajectory with a plane located at Q in Fig. 13.4 constitutes a Poincaré surface of section; shown in Fig. 13.5 (left). The right panel is a return map between successive points of intersection, i.e., the set of points $P_0, \ldots, P_N$ are related by $P_n = f(P_{n-1})$, where f is the return map. This noninvertible functional relationship between the points indicates the presence of a deterministic chaotic dynamics (cf. Fig. 13.5, right panel). Babloyantz and Destexhe (1988) qualify this result by noting that because of the high dimensionality of the cardiac attractor, no coherent functional relationships between successive points are observed in other Poincaré surfaces of section, however, correlational dimensions were calculated for a total of 36 phase portraits and yielded the results quoted previously, that is, the correlation dimension spans the interval 3.6 to 5.2. The analysis, however, did not include the testing of surrogate data sets, as suggested in Chapter 8, which might have strengthened an argument against colored noise.

Another indicator that the normal sinus rhythm is not strictly periodic is the broadband $1/f$-like spectrum observed by analysis of interbeat interval variations in healthy subjects (Kobayashi and Musha, 1982; Goldberger et al., 1985). The heart rate is modulated by a complicated combination of respiratory, sympathetic, and parasympathetic regulators. Akselrod et al. (1981) showed that suppression of these effects considerably alters the interbeat (R-R) interval power spectrum in healthy individuals, but a broadband spectrum persists. Using the interbeat sequence as a discrete time series, Babloyantz and Destexhe evaluated the correlation dimension of R-R intervals to be 5.9 ± 0.4 with typically 1000 intervals in the series. This dimension is significantly higher than that of the overall ECG, but we do not as yet understand the relation in the dynamics of the two quantities. The apparently high dimension may be a consequence of insufficient data and the influence of noise.

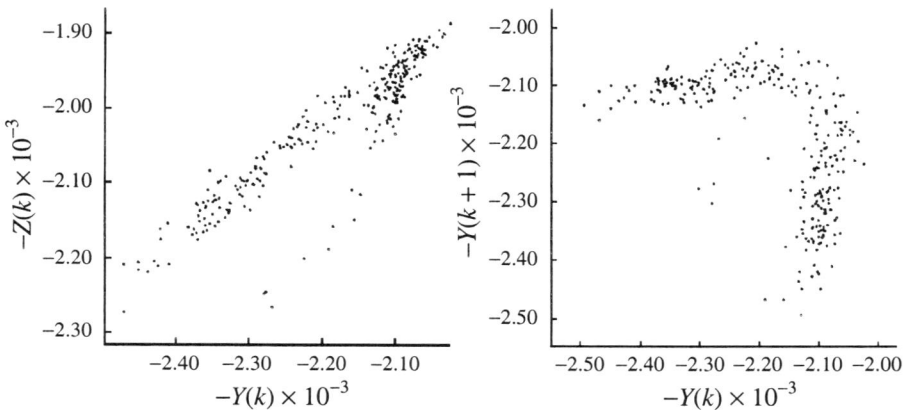

Figure 13.5. Analysis of three-dimensional phase plane trajectories. *Left:* The Poincaré map of normal heart activity: intersection of the phase plane trajectories with the Y-Z plane (X = constant) in the region Q of Fig. 13.4. *Right:* First return map constructed from the Y-coordinate of the previous intersection. (From Babloyantz and Destexhe, 1988, with permission; their Fig. 4.)

Babloyantz and Destexhe (1988) arrive at the conclusion reached earlier by Goldberger et al. (1985b), among others, that the normal human heart follows deterministic dynamics of a chaotic nature. The unexpected aspect of the present results are the high dimensions of the chaotic attractors. In any event, there is no way that the "conventional wisdom" of the ECG consisting of periodic oscillations can be maintained in light of these new results.

Complexity and Aging, or Loss of Complexity Presages Pathology

Spectral analysis of electrocardiographic data in clinical disorders has shown that there is a loss of the normal variability in heart rate in situations that proved to be unhealthy. The narrowing of the power spectrum toward uniform periodicity or the loss in richness of broadbands in the power spectrum is termed loss of spectral reserve or reduced complexity of the signal. Premature babies showed such reduction in complexity (Aarimaa et al., 1988), as did infants being monitored who subsequently succumbed to sudden infant death syndrome (Kluge et al., 1988). Goldberger and colleagues (Goldberger et al., 1985; Goldberger et al., 1988) and Kleiger et al. (1987) noted that heart-rate variability decreased in patients who were at higher risk of sudden cardiac death. They conjectured that the loss of complexity in the heart rhythm was useful as a warning of heightened risk of ventricular fibrillation. An example of the contrast between the ECGs of normal subjects and those at risk is shown in Fig. 13.6, taken from the work of Goldberger (1992).

Lipsitz and Goldberger (1992) gave an example of the difference in the form of the ECG at two different ages, shown in Fig. 13.7. The range of the variation in heart rate is not very different in the two cases, but the difference in degree of smoothness is striking. The mechanisms are not known, but they conjecture that reduced beta-adrenoreceptor responsiveness and reductions in parasympathetic tone are in line with this interpretation. They cite evidence on reduction in cerebral complexities in anatomic features and in dynamics with the aging process as a parallel situation. The spectral loss tended to be greater in the high-frequency ranges, as Figs. 13.6 and 13.7 show, and so a tentative quantitative measure was to use the slope of the power spectrum: the value of β in a fitting of $1/f^\beta$ would be increased in signals of reduced complexity. This doesn't work very well, because the power spectral density function often fails to show a simple $1/f^\beta$ structure and is more often like those shown in Fig. 13.6 (middle row).

Probably a more generally applicable quantitative measure to define changes in complexity is the approximate entropy, *ApEn*, proposed by Kaplan et al. (1991), Pincus (1991) and Pincus et al. (1991). The basic method is described in Chapter 7, section 7.6. Pincus recognized the need to define a statistic that is useful for short data sets and for abbreviated practical computation, and which could be used on correlated stochastic processes (autoregression models, coupled stochastic systems, Markov processes, etc.) for which the correlation dimension and Kolmogorov-Sinai entropy are infinite. Pincus devised this measure, which omits measurements of high-frequency short distances in the embedding space since they can be regarded as (and usually are) noise. It is a comparative statistic, to be applied in comparing members of a population by using ECG signals of the same duration (in terms of

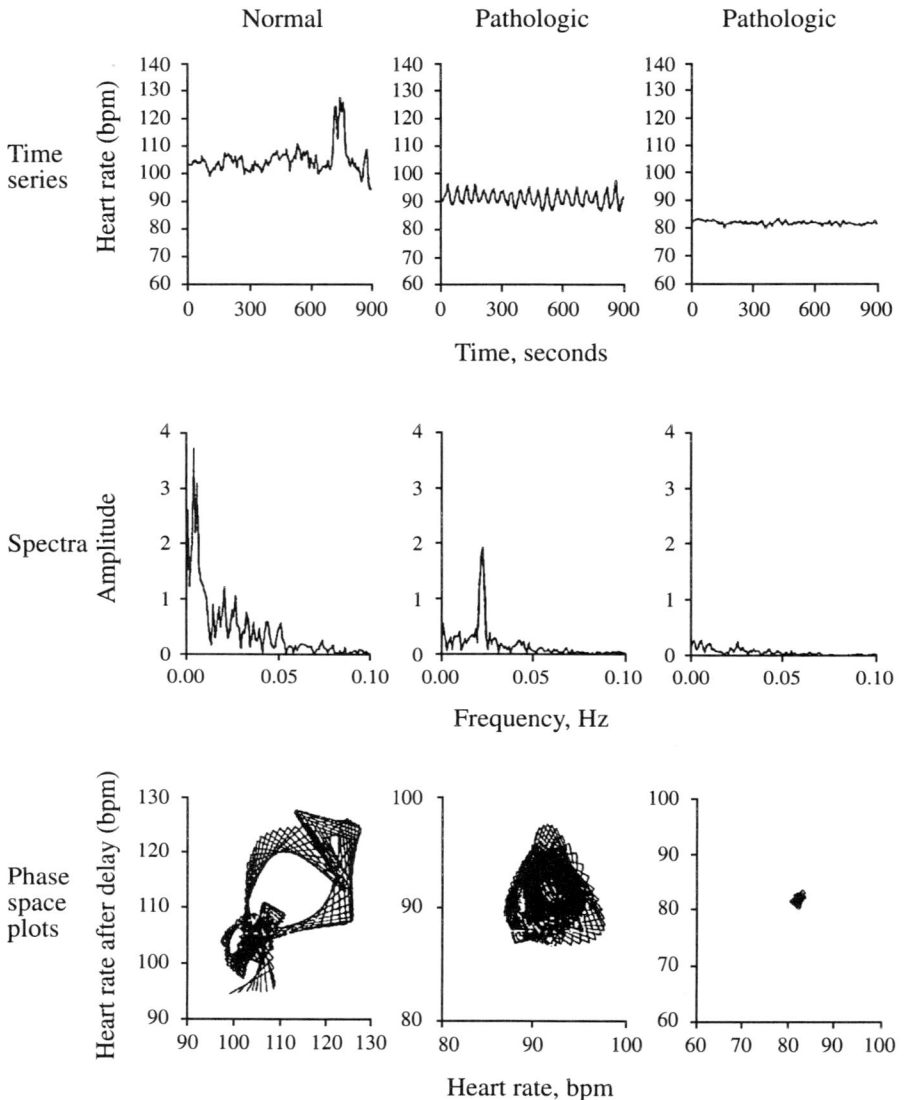

Figure 13.6. Normal sinus rhythm in healthy subjects (*left*) shows complex variability with a broad spectrum and a phase space plot consistent with a strange (chaotic) attractor. Patients with heart disease may present altered dynamics, sometimes with oscillatory sinus rhythm heart rate dynamics (*middle*) or an overall reduction of sinus variability. With the oscillatory pattern, the spectrum shows a sharp peak, and the phase space plot shows a more periodic attractor, with trajectories rotating about a central hum. With the flat pattern (*right*), the spectrum shows an overall loss of power, and the phase space plot is more reminiscent of a fixed-point attractor. (From Goldberger, 1992, Fig. 6, after adaptation from Goldberger et al., 1990.)

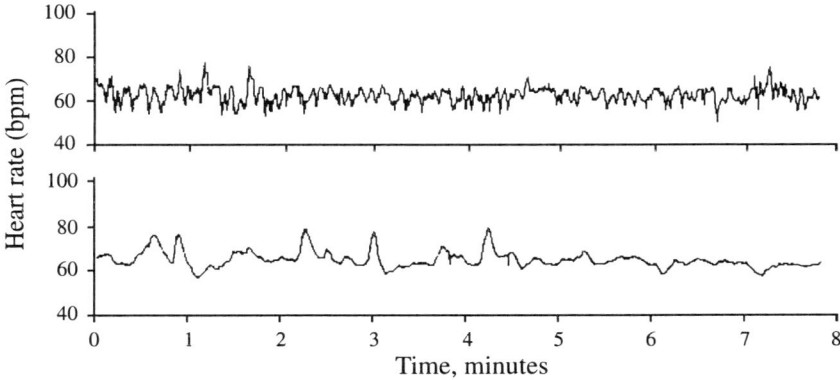

Figure 13.7. Heart rate time series for a 22-year-old female subject (*top panel*) and a 73-year-old male subject (*bottom panel*). The mean heart beats per minute for the young subject was 64.7; *SD*, 3.9; and approximate entropy, 1.09. The mean heart beats per minute for the old subject was 64.5; *SD*, 3.8; and approximate entropy, 0.48. Approximate entropy is a measure of "nonlinear complexity." Despite the nearly identical means and *SD*s of heart rate for the two time series, the "complexity" of the signal from the older subject is markedly reduced. (From Lipsitz and Goldberger, 1992, with permission.)

numbers of heartbeats), omitting the same fraction of the signal as noise in each case, and using the same level of embedding for all cases. A specific *ApEn* measure they have found useful uses 1000 beats, omits fluctuations less than 20% of the standard deviation of the beat length, and uses embedding dimensions of 2 and 3. The value of *ApEn* ($E_m = 2$, $r = 0.2$, $N = 1000$) is a measure of the increase in entropy that is revealed by going from an embedding dimension of 2 to that of 3. It is therefore a measure of the (logarithmic) probability that a greater distance between points will be found by an embedding in E_m of 3 compared to an E_m of 2. Otherwise stated, patterns of heartbeat lengths L that are close together when plotting L_{i+1} versus L_i (embedding dimension $E_m = 2$) are still close together in runs of three ($E_m + 1$) observations plotted in three-dimensional space as L_{i+2}, L_{i+1}, L_i. Fig. 13.8 shows the data, acquired in 74 subjects ranging in age from 20 to 90, for heart rate as a function of age. The results suggest the merit of the concept that complexity diminishes with age, even in apparently healthy subjects. The topic is one of continuing research.

Control of Chaos, and Perhaps a Lead Toward Treating Arrhythmias

A study by Garfinkel and colleagues (1992) illustrates the potential for incorporating the ideas of chaotic dynamics into therapeutic situations. The

Heart rate, quiet

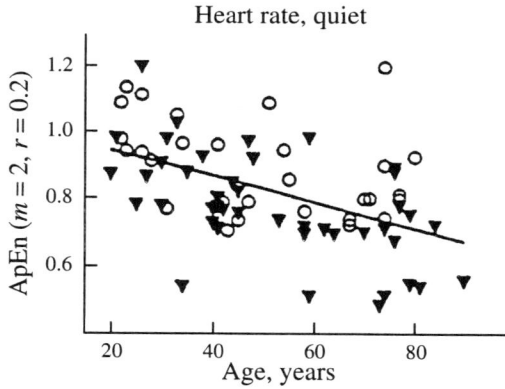

Figure 13.8. Approximate entropy for heart rate versus age in healthy subjects at rest. The line is the mean linear regression. Triangles indicate males; circles, females. (From Pincus and coworkers, unpublished, with permission.)

experiment was performed in an artificial situation, namely, in the isolated perfused rabbit interventricular septum, in which chaotic cycling of excitation was induced by high dosage of oubain, blocking the sodium pump. The situation was found to be describable and understandable by using a two-dimensional delay plot, plotting the duration of each interval against that of the preceding one. The trajectory of points revealed a saddle bifurcation: along one line in this plane the interval durations tended to go from alternating large or small values toward a middle point, moving closer to this point with successive iterations. This is like bouncing a ball in a bowl. In a second direction more or less at right angles to this line, however, the successive durations rapidly diverged, oscillating from below to above this same middle point. (The line along which convergence occurred is the stable manifold; the other direction is the unstable manifold.) The divergence does not continue and ergodically there are shifts from the unstable toward the stable manifold, a truly chaotic behavior.

Control was attempted by stimulating the preparation at appropriate times. One must find the right time for the stimulus, since the only intervention that can be accomplished is to shorten the interbeat interval. The trick is to put the trajectory onto the stable manifold, more or less, and this is done by predicting from the trajectory the next long beat which will occur along the unstable manifold and then giving a stimulus to create a shorter interval that is closer to the stable manifold. The stimulus thus shortens the next interval, though it is longer than the preceding interval. Bringing it closer to the stable manifold means that the interval will tend toward the intersection between the stable and unstable manifolds. Of course, the basic instability remains and without repeated intervention the rhythm would return to the chaotic mode. In Fig. 13.9 is shown one example from Garfinkel et al. (1992) in which the control effected a triple rhythm with a mean interbeat interval longer

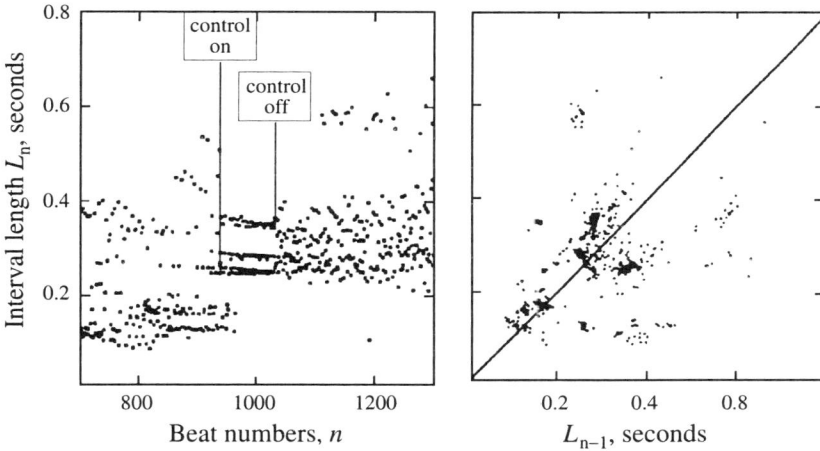

Figure 13.9. Controlling chaos in cardiac rhythm. Interbeat intervals in an oubain-treated rabbit septal preparation. *Left panel:* Interval length L versus beat number for a 600-beat sequence. Computer control of a stimulus was started at "control on" and after some further cycling in the "chaotic" mode resulted in a stable 3-cycle. *Right panel:* Lag or phase plane plot of the intervals in the left panel. The denser groups of dots represent data in the "controlled" period of data collection. The period of control resulted in a 3-cycle with the middle interval probably representing points on the stable manifold and the other two points probably lying on either side of the stable manifold and approximately in the line of the unstable manifold. (From Garfinkel et al., 1992, with permission.)

than that occurring in the chaotic state. We look on this as "effective treatment" in the sense that prolongation of the average interval length would give greater ventricular filling time. In this particular instance there also appeared to be some continuing benefit after the controlling stimulations were turned off, for the average interval was longer after the series of stimuli than it was before.

Vasomotion

Small arterioles of skeletal muscle exhibit pronounced vasomotion (Meyer, 1988; Tangelder et al., 1984, 1986), which appears to be altered by a reduction in perfusion pressure (oude Vrielink et al., 1990). However, the dynamics involved appear to be quite complex, since they do not conform to simple models such as the myogenic hypothesis of Folkow (1964). Oude Vrielink et al. (1990) found that vasomotion cycle length does lengthen following a reduction in perfusion pressure, which is in accord with the myogenic hypothesis, but vasomotion amplitude did not show a decrease due to pressure reduction as predicted by Folkow's hypothesis. In fact, changes in cycle length and amplitude showed no correlation. In an attempt to unravel some of these compdlexities, the dynamics of contraction and dilation were analyzed in this study using the techniques of nonlinear dynamics.

The main motivation behind applying nonlinear dynamics methods is in the possibility that rather than being nonspecific "noise," the observed variability is due to complex dynamics of simple sets of equations. This leads to the question of whether a set of observations corresponds to randomness or chaos. The typical observation set for vasomotion consists of diameter measurements at different times. Fig. 13.10 shows this as a plot of diameter fluctuations versus time in records on a transverse arteriole of rabbit tenuissimus muscle from Slaaf et al. (1994).

To apply the methods of nonlinear dynamics it is necessary to convert this time series to a geometric object. To do this we plot the original time series against a delayed version of the same series. This is shown in Fig. 13.11 for one complete cycle of dilation and constriction. The ordinate is the time series shifted by 25 time samples (0.5 seconds) and the abscissa is the original time series.

Visualizing vasomotion in this way calls one's attention to the asymmetry between trajectories above the diagonal (dilation) and below (contraction). Fig. 13.12 shows the phase plane plot for all the data in Fig. 13.10. Plateau formation is evident by the clustering of trajectories running vertically from the line of identity line near the origin. Fig. 13.12 is a two-dimensional representation of vasomotion that can be expanded to higher dimensions by further delays of the original time series. Given this geometric representation, the question of whether the observations are due to noise or chaos can be addressed. The answer comes from fractal analysis of the multidimensional representation or phase space.

To pick the time lag, Slaaf et al. (1994) found the dominant frequency by Fourier analysis. Linear trends present in the data were removed prior to calculating the Fourier spectrum. Two spectral peaks were usually present, one about 0.5 Hz and the other about 0.05 Hz. If these peaks were treated as sine waves, then the best choice of a lag for a phase space plot is 90 degrees or a quarter of a cycle. This value converts a sine to a cosine, which is equivalent to differentiation. If the diameter-time data were really differentiable, then the phase space plot could be made in terms of orthogonal variables, the first derivative, second derivative, and so on. In theory, the value chosen for the lag does not matter. However, when limited data points are involved, it seems more critical to use the "right" lag.

The data were analyzed using the box-counting method, the algorithm developed by Liebovitch and Tóth (1989), yielding estimates of the fractal dimension, D_{Box}, which increased with increasing embedding dimensions as seen in Fig. 13.13. For an embedding dimension of 2, the estimated fractal dimension for the data shown in Fig. 13.10 was 1.51 ± 0.04 (s.e.m.) using a lag of 0.5 seconds (2.0 seconds/4), corresponding to the higher-frequency peak, and 1.46 ± 0.04 (s.e.m.) using a lag of 5.0 seconds (20 seconds/4), which matched the lower-frequency peak. The choice of lag time did not appear to be critical.

Surrogate data sets prepared by shuffling the phases in the frequency spectrum have the same power spectrum as the original data set. The box-counting estimates of fractal dimension for the phase-randomized set were statistically distinguished from the values for the ordered set over the whole range of embedding dimensions.

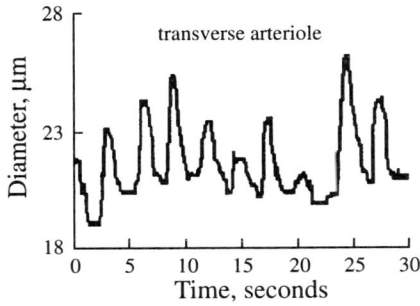

Figure 13.10. Transverse arteriole diameter fluctuations. Femoral artery pressure 79 mm Hg. (From Slaaf et al., 1994.)

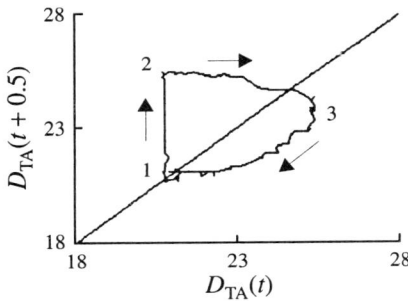

Figure 13.11. Phase plane plot of one complete cycle (dilation-constriction) from data shown in Fig. 13.10 using a lag of 0.5 s. Point 1 corresponds to a plateau occurring at $t = 5$ seconds; points 2 and 3 are peaks of $D_{TA}(t + 0.5)$ and $D_{TA}(t)$, respectively. Line of identity is also shown. (From Slaaf et al., 1994.)

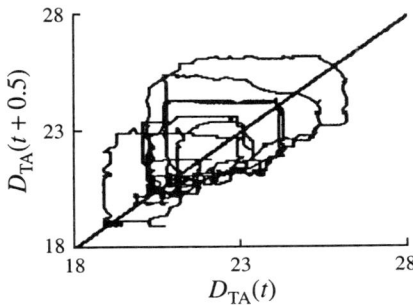

Figure 13.12. Vasomotion: Delay plot of the diameter of a transverse arteriole from the rabbit tenuissimus muscle under physiological conditions ($\Delta t = 0.5$ seconds). Phase plane plot of data in Fig. 13.10. Identity line is shown. (From studies of oude Vrielink et al. (1990) using the analysis of Slaaf et al., 1994.)

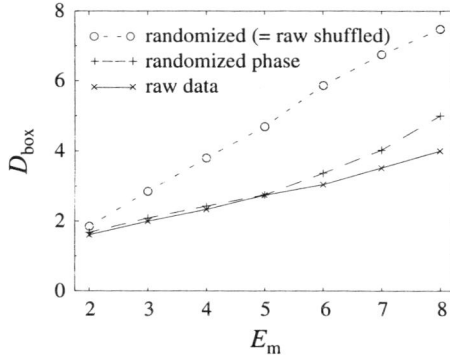

Figure 13.13. Estimates of D_{Box} versus embedding dimension E_m for a vascular diameter signal. The estimates of D_{Box} do not plateau as desired, necessitating comparison with a surrogate data set. The surrogate sets, obtained by data shuffling (randomization) and by phase randomization, showed higher values for D_{Box}, at each embedding. (From Slaaf et al., 1994.)

Reliable slopes could not be calculated from higher-dimensional embeddings because the data sets were too short (only 1500 points). For an embedding $E_m = 3$ the original diameter signals gave values for D_{Box} of 1.49 ± 0.10 ($N = 6$ studies), while the phase-shuffled surrogate data set gave $D_{Box} = 1.89 \pm 0.06$. Because the ordered and phase randomized data sets were statistically different for the six studies even with an embedding dimension of 3 or less using this strategy, the analysis denies the null hypothesis that the data were due to a linear system with noise. This same strategy motivates the use of measures of approximate entropy, $ApEn$. The inference that the signal is due to a nonlinear system of low dimension remains unproven. Nevertheless, the result stimulates and justifies application of further methods for nonlinear analysis, and for seeking explanations in terms of low-order systems that have the potential for chaotic behavior.

The physiological issue is how to use this information to learn about the system. Since one is always working on the "inverse problem," able to get data but never knowing exactly what underlies them, the question is not easily answered. A useful strategy is to try to develop a model whose behavior has similar characteristics; when this is a physiologically realistic model then one can use it as a working hypothesis and a vehicle for designing the next experiments. For example, can vasomotion be described by spontaneous fluctuations in smooth-muscle contraction given a specific set of time- and voltage-dependent conductances, ionic pumps, calcium release, and storage systems? Or are waves of propagated depolarization spreading along intercellular connections between smooth-muscle cells or endothelial cells required to give a spatial spreading component also required to produce the observations? Certainly the studies of Segal et al. (1989) showed that signals were propagated along an arteriole, taking several seconds. Temporal fluctuations in regional flows in heart muscle were observed by Yipintsoi et al. (1973), who labeled them "twinkling"; their observations were made in regions large compared to single arterioles, in tissue masses of about 150 mg, and the fluctuations were relatively smaller in larger regions. The probability is that there is a hierarchy of control levels, with extensive interaction between the levels mediated via such mechanisms as shear-dependent vasodilation, intercellular communications via endothelial and smooth-muscle cells, all organized along the branching structures of the vasculature. The general idea expressed in Fig. 13.14 is that the flow into a branching arteriolar network fed from a single parent vessels has a regulatory hierarchy. Large relative fluctuations in flow at the smallest levels may be partially reciprocated via common parent vessel, implying that the flow response at the next higher generation is to some extent smoothed out. Whether or not the system behaves locally as a chaotic system depends on the combination of the nonlinear processes of membrane transport, membrane potential changes, the calcium fluctuations, and the influences of the longer-range signals propagated by humoral or electrical means along the branching structures. While it is known that myogenic reflexes occur within vessel segments, it is not yet known how the sensitivity to stretch may propagate, or by what means it propagates. But the connectivity and the basic nonlinearities and delays in the system make it an ideal candidate for chaotic behavior.

13.3   Metabolism

The potential for chaotic behavior in larger-scale biochemical systems would seem to be immense. Not only are such systems nonlinear, but there are usually sequences of reactions lying between major control points in metabolic reaction sequences. If there are high gain controllers or significant delays, then chaos is likely. An apparently chaotic system is glycolysis, as illustrated by the work of Markus and colleagues and shown in Figs. 13.15 and 13.16.

Markus and Hess (1985), Markus et al. (1984), and Hess and Markus (1987) used this method to study the metabolic pathway that converts the free energy available in glucose into adenosine triphosphate (ATP), which is used as an energy source of the cell to drive many other biochemical reactions. As shown in the top of Fig. 13.15, the concentrations of the products in these reactions exert positive and negative feedback control on the rates of the enzymes. They solved the nonlinear ordinary differential equations for this reaction scheme to determine the time course of the concentrations of the products, such as adenosine diphosphate (ADP). When the input flux of sugar was constant, they predicted that the output concentrations would be either constant or periodic depending on the value of the input flux. When the input flux of sugar was sinusoidal, they predicted that the output concentrations would be either periodic or chaotic, depending on the amplitude and frequency of the input flux. These predictions were tested by injecting glucose into a cell-free extract of commercial baker's yeast and measuring the fluorescence of the reduced form of nicotinamide adenine dinucleotide (NADH). As predicted, the product concentrations varied periodically or chaotic, depending on the amplitude and frequency of the input sugar flux.

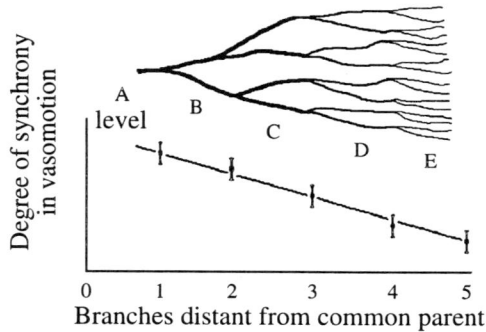

Figure 13.14. Levels of microvascular coordination. Communication up and down the fractal branching structures of the microcirculation allows the transmission of signals, with some delay (*upper panel*). Local dilation of a small branch would result in a "steal" of flow from a sibling branch, partially damping the degree of fluctuation at the levels of the parent branch and higher. Correlation between flows would be highest in vessels served by a common parent, less in branches more distantly connected (*lower panel*).

Their model predicted how the qualitative form of the variation in the product concentrations should change with the frequency of the input sugar flux. As shown in the top of Fig. 13.16, the qualitative feature they studied is the ratio of the period of the ADP fluctuations to the period of the input sugar flux. At low input sugar flux frequencies, the ADP concentration oscillates in step with the sugar flux and thus both have the same period. At higher values of the frequency, the period of the ADP concentration is a multiple of the input sugar frequency. For some values of the frequency, the ADP concentration seems to fluctuate randomly. These fluctuations are not noise but chaos. As shown in the bottom of Fig. 13.16, they measured the period of the fluctuations of the ADP concentration for different values of the frequency of the input sugar flux. The experimental results are in beautiful agreement with the predictions of the model. Thus, the deterministic nature of this system is established by how the qualitative features of the system, that is, the length of the ADP cycle, changes as a function of the parameter, the input sugar flux frequency.

This work also illustrates an important physiological point. In the chaotic regime, the product concentrations in glycolysis or the durations between beats of the heart cells, are changing with time in a seemingly random way. If you tried to publish such measurements you might, in an earlier era, have been heavily criticized. It might be said that these fluctuations were due to inability to conduct these experiments properly: it was too noisy, there were variations in the input, the measurements were scattered by experimental errors. That is, it has commonly been believed that if the experimental conditions are kept constant then the output of a system will stay approximately constant or periodic. This is *not* true for some

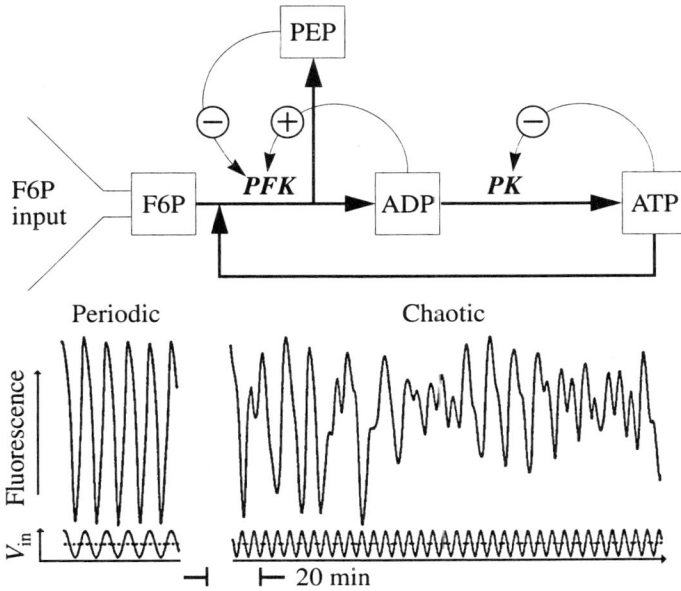

Figure 13.15. *Upper panel*: In the part of the metabolic pathway of glycolysis analyzed by Markus and Hess (1985), Markus et al. (1984), and Hess and Markus, (1987), the sugar fructose 6-phosphate (F6P) is converted into phosphoenolpyruvate (PEP), adenosine diphosphate (ADP), and adenosine triphosphate (ATP) by the enzymes phosphofructokinase (PFK) and pyruvate kinase (PK). The strong positive (+) and negative (−) effects of the products PEP, ADP, and ATP on the enzymes PFK and PK produce a system that is deterministic and highly nonlinear. They solved the differential equations of this system for the concentrations of the products. They found that the amplitude and period of the input flux of sugar determined whether the output concentrations of the products were periodic or chaotic. *Lower panel*: The reduced form of nicotinamide adenine dinucleotide (NADH), which is related to the ADP concentration, was measured in a cell-free extract from baker's yeast. The time course of the NADH concentration (large traces) is shown for two different input glucose fluxes (small traces). On the left the input glucose flux causes periodic fluctuations of the NADH concentration. On the right, an input glucose flux of higher frequency causes chaotic fluctuations of the NADH concentration. (Top: modified from Markus and Hess, 1985, their Fig. 1. Bottom: Hess and Markus, 1987, their Fig. 2. With permission.)

systems. Nonlinear systems in chaos can produce such seemingly random output that is *not* due to noise or experimental error. This does not mean that all experimental scatter is due to chaos. Indeed, the conditions in an experiment can be poorly controlled and the measurements done inaccurately. But it does mean that some systems that seem to be random may not be random at all, but deterministic.

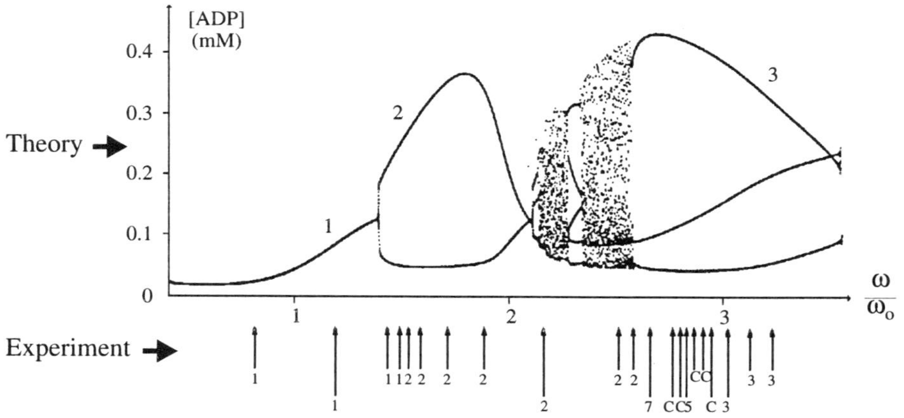

Figure 13.16. Establishing the deterministic nature of glycolysis by the changes observed when the frequency of a sinusoidal input of glucose is varied (Markus and Hess, 1985; Markus et al., 1984; Hess and Markus, 1987) *Upper*: Theoretical predictions from the differential equation model of glycolysis (Fig. 13.15). The values of ADP predicted once during each cycle of the oscillatory sugar input are shown for different sugar-input oscillation frequencies, ω. A single value (1) indicates the ADP concentration has the same period as the input sugar flux. Two values (2) indicate that the cycle of the ADP concentration is twice as long as that of the input sugar flux, which is called a period doubling. Chaotic (*C*) regimes appear as a broad smear of points. This is called a bifurcation diagram because it shows how a qualitative feature of the system, the cycle length of ADP concentration, varies with a parameter, ω/ω_o, which is the frequency of the input sugar flux. *Lower*: Experimental measurements of the cycle length of ADP concentration relative to the cycle of the glucose input. Although the experimental results are shifted in frequency, the bifurcation pattern beautifully matches the predictions, including such details as the sequence of periods: 2, 7, chaos, 5, chaos, 3 as the frequency increases. The deterministic nature of this system is established by how the qualitative features of the system, that is, the length of the ADP cycle, changes as a function of the parameter, the input sugar flux frequency. (From Markus et al., 1984, their Fig. 1.)

13.4   The Chaotic Brain

The electroencephalogram (EEG) measures the activity of the brain by recording the voltages from electrodes on the head. Considerable effort has now gone into analyzing the phase space sets of the EEG recordings to determine if they are due to low-dimensional systems, which would imply that brain activity can be reduced to a small set of simpler influences. As shown in Fig. 13.17, Rapp et al. (1985, 1988) found that the phase space set of the EEG is indeed low-dimensional, and was lower

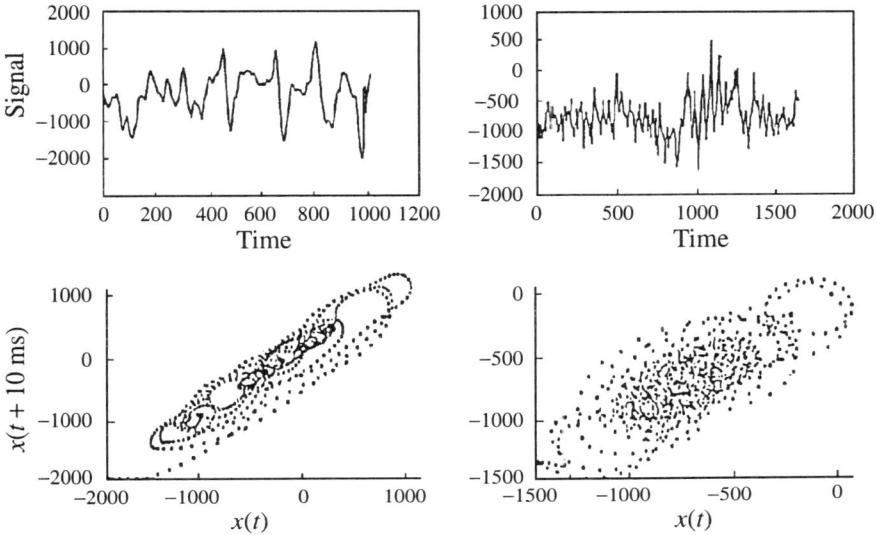

Figure 13.17. Rapp et al. (1985) recorded the electroencephalogram (EEG) when a person was resting quietly (*left*) and performing the mental task of counting backwards from 300 in steps of 7 (*right*). *Top:* The time series recorded by sampling the voltage of the EEG at intervals of 2 ms. *Bottom:* A two-dimensional phase space set was constructed from a time lag of 10 ms. The phase space set is more extended, that is, has higher fractal dimension during the mental task (*right*) than when the person was resting (*left*). Embeddings of the same time series in higher-dimensional spaces reveal that the fractal dimension of the phase space during the mental task is approximately 3.0 compared to 2.4 during rest. (From Rapp et al., 1985, Figs. 5A, 7A, 8A, and 8C.)

when the subject rested than when required to do mental arithmetic. Similar results were also reported by Babloyantz and Destexhe, (1986, 1988), Layne et al. (1986), Mayer-Kress and Layne (1987), Watt and Hameroff (1987), Skarda and Freeman (1987), and Xu and Xu (1988). For example, Fig. 13.18 shows the analysis by Mayer-Kress and Layne (1987), where it can be seen that the fractal dimension of the phase space set reaches a plateau, indicating that the fluctuations in the EEG may be a result of a deterministic process. However, they emphasize that because the EEG may be constantly changing and because of the large amount of data needed for this analysis, there is considerable uncertainty in determining the dimension of the phase space set, as evidenced by the size of the errors bars in Fig. 13.18. Thus, they suggest that their calculations of " 'dimension' have meaning only in a comparative sense" between different data sets (Layne et al., 1986).

The traditional methods of analyzing EEG time series rely on the paradigm that all temporal variations are decomposable into harmonic and periodic vibrations. Pseudo-phase plane reconstruction, however, reinterprets the time series as a

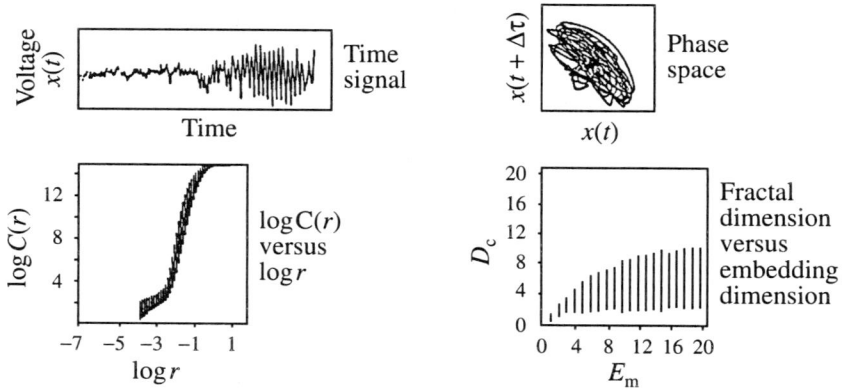

Figure 13.18. Mayer-Kress and Layne (1987) analyzed the time series of an EEG recorded from a person resting quietly. Shown are 1) the time series; 2) the two-dimensional phase space set constructed with a time lag $L\Delta t = 20$ ms; 3) the estimates of the correlation integral $C(r)$ from the slope of which they calculated the fractal dimension $D_c = d[\log C(r)]/d[\log(r)]$ of the two-dimensional phase space set determined by counting the number $C(r)$ of pairs of points of the phase space set within radius r; 4) the fractal dimension D of the phase space set as a function of the dimension E_m of the phase space. Since, the fractal dimension of the phase space set D *reaches a plateau*, this seemingly "noisy random-like" EEG is shown to be probably the result of a *deterministic process* based on only a few independent variables (like that shown in Fig. 7.10) rather than a truly random process (like that shown in Fig. 7.9). (From Mayer-Kress and Layne, 1987, Figs. 2, 6, and 7.)

multidimensional geometrical object generated by a deterministic dynamical process that is not necessarily a superposition of periodic oscillations. If the dynamics are reducible to deterministic laws, then the phase portraits of the system converge to a finite region of phase space. This invariant region is the attractor. In this way the phase space trajectories constructed from the data are confined to lie along such an attractor, if one exists. Fig. 13.19 is composed of a number of EEG time series under different physiological circumstances. There is a striking contrast between the EEG form during an epileptic seizure compared with that during normal brain activity. A reduction in the dimensionality of the time series is measured for a brain in seizure as compared with normal activity.

Gallez and Babloyantz (1991) point out that in the past few decades there has been great progress in the understanding of the molecular basis of activity in the cerebral cortex, consisting as it does of an interconnected network of 10^{10} neurons. Başar (1990) reviews how chaos theory has been applied to bridge the gap between neuronal phenomena and the oscillatory behavior observed in EEG data. Therein he emphasized: *"The EEG is not noise, but is a quasi-deterministic signal."*

The brain-wave activity of an individual during four stages of sleep was analyzed by Babloyantz (1986). In stage one, the individual drifts in and out of sleep. In stage two, the slightest noise will arouse the sleeper, whereas in stage three a loud noise is

Eyes open

Eyes closed

Sleep 2

Sleep 4

REM sleep

Epilepsy

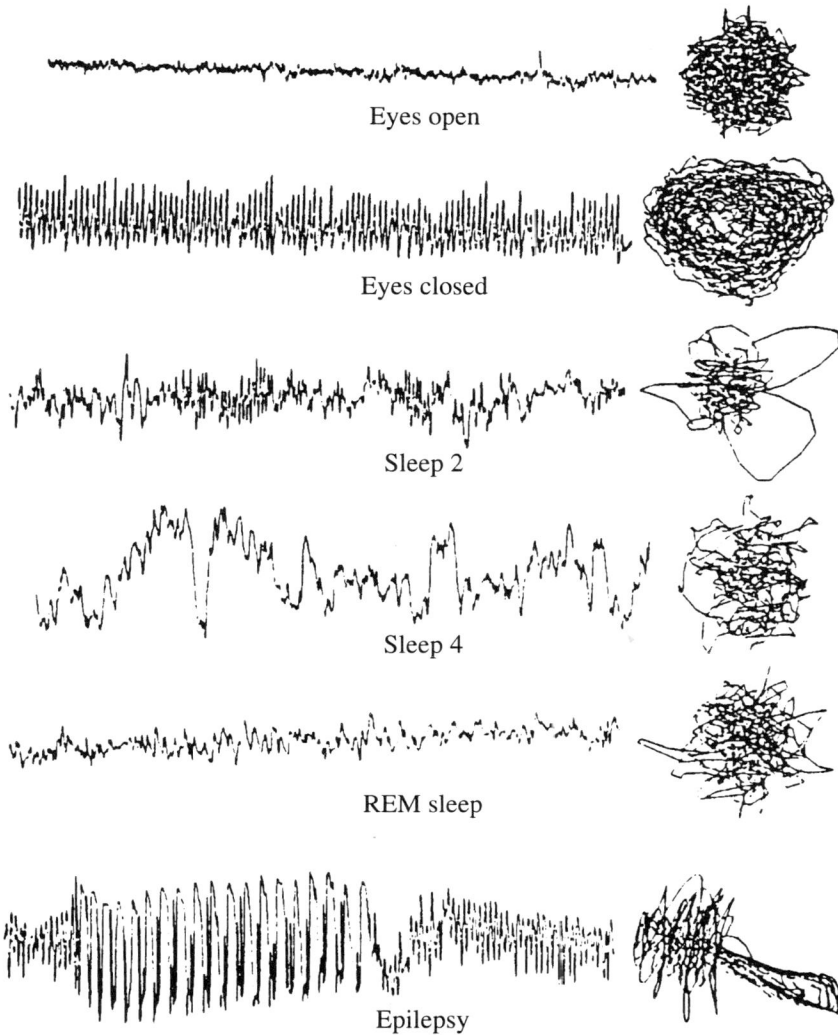

Figure 13.19. Electrical activity of the human brain. *Left panels:* EEG recordings. *Right panels:* Corresponding two-dimensional pseudo-phase plane trajectories. These are two-dimensional projections of three-dimensional constructions. The EEG was digitized at 4-ms intervals after filtering with a fourth-order 120-Hz low-pass filter. (From Babloyantz and Destexhe, 1987, with permission.)

required. The final stage is one of deep sleep. This is the normal first sequence of stages one goes through during a sleep cycle. Afterward the cycle is reversed back through stages three and two, at which time dreams set in and the sleeper manifests rapid eye movements (REM). The dream state is followed by stage two, after which the initial sequence begins again. An EEG phase portrait for each of those stages of sleep is depicted in the right column of Fig. 13.19. The form of the hypothesized attractor is not static, that is to say, it varies with the level of sleep. Correspondingly, the correlational dimension D_{corr} decreases in value as sleep deepens.

One of the more dramatic results that has been obtained in recent years has to do with the relative degree of order in the electrical activity of the human cortex in epileptic patients versus in normal persons engaged in various activities. The correlation dimensions obtained in a variety of situations have been summarized by Başar (1990) and in a comprehensive chapter by West (1990). We again note, as discussed in Section 7.7, how long the time series must be to obtain a high correlation dimension reliably. Since the *cognitive attractor* has the highest dimension for individuals that are awake and engaged in cognitive activity, it is the calculation of these dimensions that is the least reliable. In the same way, those dimensions obtained for sleeping individuals are the most stable.

In addition to the correlation dimension we can use the Lyapunov exponent as a measure of the phase space attractor. This exponent measures the rate at which two initially nearby trajectories exponentially separate from one another with increasing time. An attractor with at least one positive Lyapunov exponent is chaotic, with this exponent determining the time scale over which memory of the initial state of the system is lost. The method for calculating a Lyapunov exponent from a time series is given by Wolf et al. (1985), Parker and Chua (1989), and West (1990). The numerical values of the Lyapunov exponents are sensitive to the number of data points, the evolution time, and the embedding dimension.

Gallez and Babloyantz (1991) determined the spectrum of Lyapunov exponents for two individuals, and within the limit of errors inherent in the algorithm computing the exponents they state that it can be verified that:

(i) The number of positive, zero and negative exponents is the same, so that the overall characteristics of the Lyapunov spectrum are conserved in three different situations for the alpha stage of brain activity. Again, the *dynamical silhouette* of the attractor is unchanged.

(ii) The absolute values of the positive exponents may vary from one individual to another, but are similar for two different recordings from the same individual at two-year intervals.

(iii) The negative exponents are remarkably stable from one individual to another as well as for different recordings of the same individual.

They also take note of the data requirements specified by Eckmann and Ruelle (1985), which are that if N data points are required to provide a stable estimate of the correlation dimension, then N^2 data points are required to estimate the Lyapunov exponent for the same attractor. This means in general that the method is of limited value. However, Sano and Sawada (1985), Briggs (1990), and Kennel et al. (1992) have found methods for calculating the Lyapunov exponent using shorter data sets, giving useful approximations.

The general finding is therefore that the phase space set of the EEG has the lowest dimension when the brain has latched on to a highly repetitive state, such as that induced by disease or meditation (Babloyantz and Destexhe, 1988; Xu and Xu, 1988). The dimension increases as the brain confronts tasks of increasing difficulty. The dimension increases as the subjects shift from states of quiet sleep to quiet awake with eyes closed, to performing a mental task (Rapp et al., 1988; Babloyantz and Destexhe, 1988; Mayer-Kress and Layne, 1987). It appears that the phase space set of the EEG has the highest dimension while the brain is processing new information. This suggests an important new physiological result, namely, that the "noise" in brain activity is neither counterproductive nor accidental, but that it is there by design and that it may also be controllable. It has been proposed that this chaotic activity may provide a way to exercise neurons, to stay alert to many different possibilities covering many possible states, until a decision is made and then the brain settles into a more focused, lower-dimensional mode (Rapp et al., 1985; Skarda and Freeman, 1987; Pool, 1989).

These analyses of the EEG remain controversial for several reasons. The EEG recordings are not always a pure measure of a single mental state. For example, the amplitude of the voltage from the EEG is much smaller than that produced by muscles in the head, such as the twitch of an eyebrow, or an eyeblink. Extraneous stimuli, such as a distant noise, can provoke attention and thus a response in the EEG unrelated to the test state. It is not clear how many data are needed to accurately determine the dimension of a phase space set, but it appears that some of the EEG studies use time series that may not be large enough to accurately determine dimensions as large as those reported. Although different groups are (mostly) in qualitative agreement on the results, they have also reported different quantitative values for the fractal dimension of the phase space set recorded under similar experimental conditions (see, for example, Table 1 in Mayer-Kress and Layne, 1987, or Table 2 in Rapp et al., 1988).

13.5   Physiological Advantages of Chaos

The existence of physiological systems with a few independent influences that have complex chaotic behavior suggests that we ask if these types of systems have properties that are physiologically advantageous. It has been proposed that chaos may have several important physiological roles.

Sustained periodicities are often unhealthy; the highly rhythmic oscillations in the brain during epilepsy are an extreme example. To maintain health, the variables of a physiological system must be able to extend over a wide range to provide flexible adaptation. Healthy people have greater variability in their heart rates than those with heart disease. The *variables of a chaotic system explore a much wider range of values* than those of a system in rhythmic oscillations. Thus, chaos may provide the plasticity to cope with the exigencies of an unpredictable and changing environment (Goldberger and West, 1987c; Pool, 1989b; Goldberger et al., 1990).

Chaotic activity of the brain may be a way of exercising a wide range of different neural pathways. Thus, the activity of the brain would be constantly shifting between revisiting old memory patterns and exploring possible new memory patterns. When a new stimulus is presented, the brain could reinforce a new pathway to fix a new memory trace. When an old stimulus is presented, the chaotic switching among previously stored memory patterns could help to find the pattern associated with the stimulus (Rapp et al., 1985; Skarda and Freeman, 1987; Pool, 1989a).

Chaotic systems can be *controlled more finely* and *more quickly* than linear systems. In linear systems, the response of the output depends linearly on the input. Small changes in a parameter of a linear system produce only small changes in the output. However, chaotic systems can have bifurcations, qualitative changes in behavior, as a parameter is varied. Thus, small changes in a parameter of a chaotic system can produce very large changes in the output. Hence, a chaotic system is under much finer control. The variable controlling a chaotic physiological response may need to change only a small amount to induce the desired large change in the physiological state. Moreover, the changes of state in a linear system happen slowly. However, the sensitivity to initial conditions of a chaotic system means that very small changes in the values of the variables soon result in very different states of the system. Thus, a chaotic physiological system can switch very rapidly from one physiological state to another. These advantages of sensitivity and speed in the control of chaotic systems have only recently been derived and shown to work on mathematical and physical models, and soon may be used in a wide range of engineering applications (Ott et al., 1990; Shinbrot et al., 1990; Singer et al., 1991; Ditto et al., 1990; Azevedo and Rezende, 1991; Langreth, 1991).

13.6   Special Situations

Even low-order processes of generation and decay (chemical reaction networks, predator-prey relationships, etc.) can exhibit dependence on a rate coefficient. A cycling or periodic system in which an increase of this rate results in a bifurcation or period doubling is diagnosed as being potentially, if not actually, chaotic. A bifurcation is a doubling of the number of stable settings for the observed variable, which is equivalent to doubling the length of the period required to repeat the pattern. In an iterated system in which a rate increase causes a steady signal to take 2 alternating values, then 4, 8, and 16, and thereafter appears unpredictable at a higher value of the rate, the system is said to follow period-doubling bifurcations on the route to chaos. In continuously observable systems the same kind of thing can occur: a stable variable shifts to a single cycle (circular or elliptical in the phase plane), then to a two-cycle (two linked but distorted ellipses in the phase plane), and so on; at higher rates the signal may never be exactly repetitive and is called chaotic. Even though the trajectories are nicely bounded and smooth, they are not exactly predictable.

These kinds of shifting in periodicity can occur spontaneously in living systems or may sometimes be induced by temperature changes or another stimulus. Observations of jumps in behavior patterns should therefore raise one's suspicions that the system operates on a low-dimension attractor in either a periodic or chaotic mode. The attractor, the region of phase space covered by the trajectories, is commonly fractal; like taking a slice through Saturn's rings, the distances between neighboring slices tends to show fractal variation in the intervals, which means the spacings show correlation structure, near-neighbor spacings being more alike than spacings between N^{th} neighbors. Thus, the spacings are neither random nor uniform.

13.7   Summary

"Homeostasis," the stability of the "milieu interieure," has been regarded as the *modus operandi* of physiological systems since the time of Claude Bernard (1878). The principle was that the body is designed so that concentrations and rates of processes tended toward a stable state, through multiple feedback mechanisms. This was not a passing idea, for it had many affirmations and was firmly positioned in the medical literature by Cannon (1929), whose elegant experiments gave immense support. Now we question whether or not "homeostasis" ought to be replaced by the concept of "homeodynamics," allowing a more flexible view of how the systems work and making room for the concept of systems with complex responses, even to the point of inherent instability (see Goldberger et al., 1990).

This is an introductory overview of chaotic dynamics and of the potential utility of using dynamical approaches for understanding and even controlling cardiovascular processes. The applications to biology are indeed very new, and the analytical techniques are a long way from being nicely defined or even thoroughly proven. The mammalian organism is composed of multiple nested loops of nonlinear interacting systems, and so is of the sort susceptible to chaotic behavior. This is at the physiological level. How much greater are the possibilities for chaotic behavior at the psychological and societal levels!

Works Cited

We are each other's ladders and weights at once.

James Lawson (1994)

Aarimaa, T., R. Oja, K. Antila, and I. Valimaki. Interaction of heart rate and respiration in newborn babies. *Pediatr. Res.* 24:745–750, 1988.

Aarts, E., and J. Korst. *Simulated Annealing and Boltzmann Machines.* New York: John Wiley & Sons, 1989.

Abarbanel, H. D. I., R. Brown, and J. B. Kadtke. Prediction in chaotic nonlinear systems: Methods for time series with broadband Fourier spectra. *Phys. Rev. A* 41:1782–1807, 1989.

Abelson, H., and A. A. diSessa. *Turtle Geometry.* Cambridge, MA: MIT Press, 1982.

Acton, F. S. *Numerical Methods That Work.* New York: Harper, 1970, 541 pp.

Agmon, N., and J. J. Hopfield. Co binding to heme proteins: a model for height distributions and slow conformational changes. *J. Chem. Phys.* 79:2042–2053, 1983.

Akselrod, S., D. Gordon, F. A. Ubel, D. C. Shannon, A. C. Barger, and R. J. Cohen. Power spectrum analysis of heart rate fluctuation: a quantitative probe of beat-to-beat cardiovascular control. *Science* 213:220–222, 1981.

Albano, A. I., J. Muench, C. Schwartz, A. I. Mees, and P. E. Rapp. Singular-value decomposition and the Grassberger-Procaccia algorithm. *Phys. Rev. A* 38:3017–3026, 1988.

Alcala, J. R., E. Gratton, and F. G. Prendergast. Interpretation of fluorescence decays in proteins using continuous lifetime distributions. *Biophys. J.* 51:925–936, 1987.

Amato, I. DNA shows unexplained patterns writ large. *Science* 257:747, 1992.

Ansari, A., J. Berendzen, S. F. Bowne, H. Frauenfelder, I. E. Iben, T. B. Sauke, E. Shyamsunder, and R. D. Young. Protein states and proteinquakes. *Proc. Natl. Acad. Sci.* 82:5000–5004, 1985.

Antonini, E., and M. Brunori. *Hemoglobin and Myoglobin in their Reactions with Ligands.* Amsterdam: North Holland, 1971, 436 pp.

Archie, J. P., Jr., D. E. Fixler, D. J. Ullyot, J. I. E. Hoffman, J. R. Utley, and E. L. Carlson. Measurement of cardiac output with and organ trapping of radioactive microspheres. *J. Appl. Physiol.* 35:148–154, 1973.

Arecchi, F. T., and F. Lisi. Hopping mechanism generating $1/f$ noise in nonlinear systems. *Phys. Rev. Lett.* 49:94–98, 1982.

Arnold, V. I., and A. Avez. *Ergodic Problems in Classical Mechanics.* New York: Benjamin, 1968.

Austin, R. H., K. W. Beeson, L. Eisenstein, H. Frauenfelder, and I. C. Gunsalus. Dynamics of ligand binding to myoglobin. *Biochemistry* 14:5355–5375, 1975.

Avnir, D. *The Fractal Approach to Heterogeneous Chemistry: Surfaces, Colloids, Polymers.* New York: John Wiley and Sons, 1989.

Azevedo, A., and S. M. Rezende. Controlling chaos in spin-wave instabilities. *Phys. Rev. Lett.* 66:1342–1345, 1991.

Babloyantz, A., and A. Destexhe. Low-dimensional chaos in an instance of epilepsy. *Proc. Natl. Acad. Sci.* 83:3513–3517, 1986.

Babloyantz, A. Evidence of chaotic dynamics of brain activity during the sleep cycle. In: *Dimensions and Entropies in Chaotic Systems*, edited by G. Mayer-Kress. Berlin: Springer-Verlag, 1986, pp. 241–245.

Babloyantz, A., and A. Destexhe. Chaos in neural networks. In: *Proceedings of the International Conference on Neural Networks*, edited by M. Caudill and C. Butler. Piscataway, NJ: SOS Print, 1987.

Babloyantz, A., and A. Destexhe. Is the normal heart a periodic oscillator?. *Biol. Cybern.* 58:203–211, 1988.

Babloyantz, A. Some remarks on nonlinear data analysis. In: *Measures of Complexity and Chaos*, edited by N. B. Abraham, A. M. Albano, A. Passamante, and P. E. Rapp. New York: Plenum, 1989, pp. 51–62.

Bak, P., C. Tang, and K. Wiesenfeld. Self-organized criticality: an explanation of $1/f$ noise. *Phys. Rev. Lett.* 59:381–384, 1987.

Bak, P., and K. Chen. Self-Organized Criticality. *Sci. Am.* January:46–53, 1991.

Barcellos, A. The fractal geometry of Mandelbrot. *College Math. J.* 15:98–114, 1984.

Barnsley, M. F. *Fractals Everywhere.* Boston: Academic Press, Inc., 1988.

Barnsley, M. F. "Desktop Fractal Design Systems." Atlanta: Iterated Systems, 1990.

Barrow, G. M. *Physical Chemistry, 4th Edition.* New York: McGraw Hill, 1979, 832 pp.

Basar, E. Chaotic dynamics and resonances phenomena in brain function: progress, perspectives, and thoughts. In: *Chaos in Brain Function*, edited by E. Basar. Berlin: Springer-Verlag, 1990, pp. 1–30.

Bassingthwaighte, J. B., T. Yipintsoi, and R. B. Harvey. Microvasculature of the dog left ventricular myocardium. *Microvasc. Res.* 7:229–249, 1974.

Bassingthwaighte, J. B., and C. A. Goresky. Modeling in the analysis of solute and water exchange in the microvasculature. In: *Handbook of Physiology. Sect. 2, The Cardiovascular System. Vol IV, The Microcirculation,* edited by E. M. Renkin and C. C. Michel. Bethesda, MD: Am. Physiol. Soc., 1984, pp. 549–626.

Bassingthwaighte, J. B., M. A. Malone, T. C. Moffett, R. B. King, S. E. Little, J. M. Link, and K. A. Krohn. Validity of microsphere depositions for regional myocardial flows. *Am. J. Physiol.* 253 (Heart Circ. Physiol. 22):H184–H193, 1987a.

Bassingthwaighte, J. B., L. Noodleman, R. T. Eakin, and R. B. King. Integrated phenomenology of transport, permeation, and metabolic reactions in the heart. In: *Activation, Metabolism and Perfusion of the Heart. Simulation and Experimental Models*, edited by S. Sideman and R. Beyar. Dordrecht: Martinus Nijhoff, 1987b, pp. 621–642.

Bassingthwaighte, J. B. Physiological heterogeneity: Fractals link determinism and randomness in structures and functions. *News Physiol. Sci.* 3:5–10, 1988.

Bassingthwaighte, J. B., and J. H. G. M. van Beek. Lightning and the heart: Fractal behavior in cardiac function. *Proc. IEEE* 76:693–699, 1988.

Bassingthwaighte, J. B., I. S. Chan, A. A. Goldstein, and I. B. Russak. GGOPT—an unconstrained nonlinear optimizer. *Comput. Meth. Prog. Biomed.* 26:275–281, 1988a.

Bassingthwaighte, J. B., R. B. King, J. E. Sambrook, and B. H. Van Steenwyk. Fractal analysis of blood–tissue exchange kinetics. In: *Oxygen Transport to Tissue X, Adv. Exp. Med. Biol. 222*, edited by M. Mochizuki et al.. New York: Plenum Press, 1988b, pp. 15–23.

Bassingthwaighte, J. B., R. B. King, and S. A. Roger. Fractal nature of regional myocardial blood flow heterogeneity. *Circ. Res.* 65:578–590, 1989.

Bassingthwaighte, J. B., M. A. Malone, T. C. Moffett, R. B. King, I. S. Chan, J. M. Link, and K. A. Krohn. Molecular and particulate depositions for regional myocardial flows in sheep. *Circ. Res.* 66:1328–1344, 1990a.

Bassingthwaighte, J. B., J. H. G. M. van Beek, and R. B. King. Fractal branchings: The basis of myocardial flow heterogeneities?. In: *Mathematical Approaches to Cardiac Arrhythmias. Ann. N. Y. Acad. Sci. Vol.591*, edited by J. Jalife, 1990b, pp. 392–401.

Bassingthwaighte, J. B. The myocardial cell. In: *Cardiology: Fundamentals and Practice, 2nd Ed.*, edited by E. R. Giuliani, V. Fuster, B. J. Gersh, M. D. McGoon and D. C. McGoon. St. Louis, MO: Mosby-Year Book Inc., 1991a, pp. 113–149.

Bassingthwaighte, J. B. Chaos in the Fractal Arteriolar Network. In: *The Resistance Vasculature*, edited by J. A. Bevan, W. Halpern and M. J. Mulvany. Totowa, N.J.: Humana Press Inc., 1991b, pp. 431–449.

Bassingthwaighte, J. B., and R. P. Beyer. Fractal correlation in heterogeneous systems. *Physica D* 53:71–84, 1991.

Bassingthwaighte, J. B. Fractal vascular growth patterns. *Acta Stereol.* 11(Suppl.1):305–319, 1992.

Bassingthwaighte, J. B., G. M. Raymond, and R. B. King. Fractal dispersional analysis: An assessment on one-dimensional fractional Brownian signals. *Ann. Biomed. Eng.* 22, 1994 (in press).

Bassingthwaighte, J. B. Evaluating rescaled range analysis for time series. *Ann. Biomed. Eng.* 22, 1994 (in press).

Bassingthwaighte, J. B. Fractal ^{15}O-water washout from the heart. *Science*, 1994 (in review).

Batra, S., and K. Rakusan. Morphometric analysis of capillary nets in rat myocardium. *Adv. Exp. Med. Biol.* 277:377–385, 1990.

Bauer, R. J., B. F. Bowman, and J. L. Kenyon. Theory of the kinetic analysis of patch-clamp data. *Biophys. J.* 52:961–978, 1987.

Beck, C. Brownian motion from deterministic dynamics. *Physica A* 169:324–336, 1990.

Berkson, J. Are there two regressions? *J. Am. Stat. Assoc.* 45:164–180, 1950.

Bernard, C. *Leçons sur les phénomène de la vie communs aux animaux et aux végétaux.*. Paris: Baillière, 1878. (See also the translation: Bernard, C. *Lectures on the Phenomena of Life Common to Animals and Plants. Translation by Hebbel E. Hoff, Roger Guillemin and Lucienne Guillemin*. Paris: J. Libraire, 1878.)

Blake, W. *William Blake's Auguries of innocence, together with the Rearrangement by Dr. John Sampson and a comment by Geoffrey Keynes Kt.* Burford: Cygnet Press, 1975.

Blatz, A. L., and K. L. Magleby. Quantitative description of three modes of activity of fast chloride channels from rat skeletal muscle. *J. Physiol. (Lond.)* 378:141–174, 1986.

Block, A., W. von Bloh, and H. J. Schellnhuber. Efficient box-counting determination of generalized fractal dimensions. *Phys. Rev. A.* 42:1869–1874, 1990.

Bloom, W., and D. W. Fawcett. *A Textbook of Histology.* Philadelphia, PA: Saunders, 1975, 1033 pp.

Briggs, K. An improved method for estimating Liapunov exponents of chaotic time series. *Phys. Lett. A* 151:27–32, 1990.

Brown, S. R., and C. H. Scholz. Broad bandwidth study of the topography of natural rock surfaces. *J. Geophys. Res.* 90:12575–12582, 1985.

Buckberg, G. D., J. C. Luck, B. D. Payne, J. I. E. Hoffman, J. P. Archie, and D. E. Fixler. Some sources of error in measuring regional blood flow with radioactive microspheres. *J. Appl. Physiol.* 31:598–604, 1971.

Burrough, P. A. Multiscale sources of spatial variation in soil. I. The application of fractal concepts to nested levels of soil variation. *J. Soil Sci.* 34:577–597, 1983a.

Burrough, P. A. Multiscale sources of spatial variation in soil. II. A non-Brownian fractal model and its application in soil survey. *J. Soil Sci.* 34:599–620, 1983b.

Burrough, P. A. Fractals and Geochemistry. In: *The Fractal Approach to Heterogeneous Chemistry,* edited by D. Avnir. New York: John Wiley & Sons, 1989, pp. 383–406.

Cairns, J., J. Overbaugh, and S. Miller. The origin of mutants. *Nature* 335:142–145, 1988.

Calder III, W. A. *Size, Function, and Life History.* Cambridge: Harvard University Press, 1984.

Caldwell, J. H., G. V. Martin, G. M. Raymond, and J. B. Bassingthwaighte. Regional myocardial flow and capillary permeability-surface area products are proportional. *Am. J. Physiol.,* 1994 (in press).

Cambanis, S., and G. Miller. Some path properties of pth order and symmetric stable processes. *Ann. Prob.* 8:1148–1156, 1980.

Cambanis, S., and G. Miller. Linear problems in pth order and stable processes. *SIAM J. Appl. Math* 41:43–69, 1981.

Cannon, W. B. Organization for physiological homeostasis. *Physiol. Rev.* 9:399–431, 1929.

Careri, G., P. Fasella, and E. Gratton. Statistical time events in enzymes: A physical assessment. *Crit. Review. Biochem.* 3:141–164, 1975.

Cargill, E. B., H. H. Barrett, R. D. Fiete, M. Ker, D. D. Patton, and G. W. Seeley. Fractal physiology and nuclear medicine scans. *SPIE* 914:355–361, 1988.

Casdagli, M., S. Eubank, J. D. Farmer, and J. Gibson. State space reconstruction in the presence of noise. *Physica D* 51:52–98, 1991.

Casdagli, M., D. Des Jardins, S. Eubank, J. D. Farmer, J. Gibson, J. Theiler, and N. Hunter. Nonlinear modeling of chaotic time series: Theory and applications. In:

Applied Chaos, edited by J. H. Kim and J. Stringer. New York: John Wiley & Sons, 1992, pp. 335–380.

Caserta, F., H. E. Stanley, G. Daccord, R. E. Hausman, W. Eldred, and J. Nittmann. Photograph of a retinal neuron (nerve cell), the morphology of which can also be understood using the DLA archetype. *Phys. Rev. Lett.* 64:95, 1990.

Caserta, F., H. E. Stanley, W. D. Eldred, G. Daccord, R. E. Hausman, and J. Nittmann. Physical mechanisms underlying neurite outgrowth: a quantitative analysis of neuronal shape. *Phys. Rev. Lett.* 64:95–98, 1990.

Chialvo, D. R., and J. Jalife. Non-linear dynamics of cardiac excitation and impulse propagation. *Nature* 330:749–752, 1987.

Chialvo, D. R., R. F. Gilmour, Jr., and J. Jalife. Low dimensional chaos in cardiac tissue. *Nature* 343:653–657, 1990a.

Chialvo, D. R., D. C. Michaels, and J. Jalife. Supernormal excitability as a mechanism of chaotic dynamics of activation in cardiac Purkinje fibers. *Circ. Res.* 66:525–545, 1990b.

Churilla, A. M., W. A. Gottschalk, L. S. Liebovitch, L. Y. Selector, and S. Yeandle. Membrane potential fluctuations of human T-lymphocytes have the fractal characteristics of fractional Brownian motion. *Ann. Biomed. Eng.* (in press).

Coggins, D. L., A. E. Flynn, R. E. Austin, Jr., G. S. Aldea, D. Muehrcke, M. Goto, and J. I. E. Hoffman. Nonuniform loss of regional flow reserve during myocardial ischemia in dogs. *Circ. Res.* 67:253–264, 1990.

Cohn, D. L. Optimal systems, Part I. *Bull. Math. Biophys.* 16:59–74, 1954.

Cohn, D. L. Optimal systems, Part II. *Bull. Math. Biophys.* 17:219–227, 1955.

Colquhoun, D., and F. J. Sigworth. Statistical analysis of records. In: *Single-channel Recording*, edited by B. Sakmann and E. Neher. New York: Plenum, 1983, pp. 191–263.

Colquhoun, D., and A. G. Hawkes. The principles of the stochastic interpretation of ion-channel mechanisms. In: *Single-channel Recording*, edited by B. Sakmann and E. Neher. New York: Plenum, 1983, pp. 135–175.

Condat, C. A., and J. Jäckle. Closed-time distribution of ionic channels: Analytical solution to a one-dimensional defect-diffusion model. *Biophys. J.* 55:915–925, 1989.

Conrad, M. What is the use of chaos? In: *Chaos*, edited by A. V. Holden. Princeton, N.J.: Princeton Univ. Press, 1986, pp. 3–14.

Courtemanche, M., and A. T. Winfree. Re-entrant rotating waves in a Beeler-Reuter based model of two-dimensional cardiac electrical activity. *Int. J. Bifurcation Chaos* 1:431–444, 1991.

Cox, D. R. *Renewal Theory.* New York: Wiley, 1962, 142 pp.

Cramér, H. *Mathematical Methods of Statistics.* Princeton: Princeton University Press, 1945.

Crile, G., and D. P. Quiring. A record of the body weight and certain organ and gland weights of 3690 animals. *Ohio J. Sci.* 40:219–259, 1940.

Croxton, T. L. A model of the gating of ion channels. *Biochim. Biophys. Acta* 946:19–24, 1988.

Crutchfield, J. P., J. D. Farmer, N. H. Packard, and R. S. Shaw. Chaos. *Sci. Am.* 255:46–57, 1986.

Davis, B. D. Transcriptional bias: a non-Lamarckian mechanism for substrate-induced mutations. *Proc. Natl. Acad. Sci.* 86:5005–5009, 1989.

Dawant, B., M. Levin, and A. S. Popel. Effect of dispersion of vessel diameters and lengths in stochastic networks. I. Modeling of microcirculatory flow. *Microvasc. Res.* 31:203–222, 1986.

Deering, W., and B. J. West. Fractal physiology. *IEEE Eng. Med. Biol.* 11:40–46, 1992.

DeFelice, L. J. *Introduction to Membrane Noise.* New York: Plenum, 1981.

Deussen, A., and J. B. Bassingthwaighte. Blood–tissue oxygen exchange and metabolism using axially distributed convection-permeation-diffusion models (personal communication), 1994.

Deutch, J. M., and P. Meakin. *J. Chem. Phys.* 78:2093, 1983.

Dewey, T. G., and D. B. Spencer. Are protein dynamics fractal? Comments. *Mol. Cell. Biophys.* 7:155–171, 1991.

Dewey, T. G., and J. G. Bann. Protein dynamics and 1/f noise. *Biophys. J.* 63:594–598, 1992.

Diels, H. *Heraclitus of Ephesus.* Berlin: Weidmannsche Buchhandlung, 1901.

Ditto, W. L., S. N. Rauseo, and M. L. Spano. Experimental control of chaos. *Phys. Rev. Lett.* 65:3211–3214, 1990.

Dutka, J. On the St. Petersburg paradox. *Arch. History Exact. Sci.* 39:13–39, 1988.

Eckmann, J. P., and D. Ruelle. Ergodic theory of chaos and strange attractors. *Rev. Mod. Phys.* 57:617–656, 1985.

Edelstein-Keshet, L. *Mathematical Models in Biology.* New York: Random House, 1988, 586 pp.

Eden, M. *Proc. 4th Berkeley Symp. Math. Stat. Prob.* 4:233, 1961.

Edgar, G. A. *Measure, Topology, and Fractal Geometry.* New York: Springer-Verlag, 1990, 230 pp.

Edwards, W. D. Applied anatomy of the heart. In: *Cardiology: Fundamentals and Practice,* edited by R. O. Brandenburg, V. Fuster, E. R. Guiliani and E. R. McGoon. Chicago, London: Year Book Medical Publishers, 1987, pp. 47–112.

Eigen, M., and P. Schuster. *The Hypercycle, A Principle of Natural Self-Organization.* New York: Springer-Verlag, 1979.

Eigen, M. *Steps Towards Life.* Oxford: Oxford University Press, 1992, 173 pp.

Ekeland, I. *Mathematics and the Unexpected.* Chicago, IL: Univ. of Chicago, 1988.

Elber, R., and M. Karplus. Low-frequency modes in proteins: use of the effective-medium approximation to interpret the fractal dimension observed in electron-spin relaxation measurements. *Phys. Rev. Lett.* 56:394–397, 1986.

Eng, C., S. Cho, S. M. Factor, E. H. Sonnenblick, and E. S. Kirk. Myocardial micronecrosis produced by microsphere embolization: Role of an α-adrenergic tonic influence on the coronary microcirculation. *Circ. Res.* 54:74–82, 1984.

Falconer, K. J. *The Geometry of Fractal Sets.* Cambridge: Cambridge University Press, 1985.

Falconer, K. J. *Fractal Geometry: Mathematical Foundations and Applications.* Chichester, New York: Wiley, 1990, 288 pp.

Family, F., B. R. Masters, and D. E. Platt. Fractal pattern formation in human retinal vessels. *Physica D* 38:98–103, 1989.

Farmer, J. D. Scaling in fat fractals. In: *Dimensions and Entropies in Chaotic Systems*, edited by G. Mayer-Kress. Berlin, Heidelberg: Springer-Verlag, 1986, pp. 54–60.

Feder, J. *Fractals*. New York: Plenum Press, 1988.

Feller, W. The asymptotic distribution of the range of sums of independent random variables. *Ann. Math. Stat.* 22:427–432, 1951.

Feller, W. *An Introduction to Probability Theory and Its Applications*. New York: John Wiley & Sons, Inc., 1968.

Feller, W. *An Introduction to Probability Theory and Its Applications, Volume 2, Second Edition*. New York: John Wiley & Sons, 1971.

Fenton, B. M., and B. W. Zweifach. Microcirculatory model relating geometrical variation to changes in pressure and flow rate. *Ann. Biomed. Eng.* 9:303–321, 1981.

Ferris, T. *Coming of Age in the Milky Way*. New York: W. Morrow, 1988, 496 pp.

Fisher, D. S., G. M. Grinstein, and A. Khurana. Theory of random magnetics. *Phys. Today* December: 56–67, 1988.

Fitzhugh, R. Impulses and physiological states in theoretical models of nerve membrane. *Biophys. J.* 1:445–466, 1961.

Foerster, P., S. C. Müller, and B. Hess. *Science* 241:685, 1988.

Foerster, P., S. C. Müller, and B. Hess. *Proc. Natl. Acad. Sci. USA* 86:6831, 1989.

Folkman, J., and R. Cotran. Relation of vascular proliferation to tumor growth. *Int. Rev. Exp. Pathol.* 16:207–248, 1976.

Folkow, B. Description of the myogenic hypothesis. *Circ. Res.* 14215 (Suppl.):279–287, 1964.

French, A. S., and L. L. Stockbridge. Fractal and Markov behavior in ion channel kinetics. *Can. J. Physiol. Pharmacol.* 66:967–970, 1988.

Frijisaka, H., and T. Yamada. A new intermittency in coupled dynamical systems. *Prog. Theor. Phys.* 74:918, 1987.

Frisch, U., and R. Morf. Intermittency in nonlinear dynamics and singularities at complex times. *Phys. Rev. A* 23:2673–2705, 1981.

Gallez, D., and A. Babloyantz. Predictability of human EEG: A dynamical approach. *Biol. Cybern.* 64:381–391, 1991.

Galvani, L. *Commentary on the effect of electricity on muscular motion; a translation of Luigi Galvani's De viribus electricitatis in motu musculari commentarius*. Cambridge, MA: E. Licht, 1953, 97 pp.

Garcia, L. *The Fractal Explorer*. Santa Cruz, CA: Dynamic Press, 1991.

Gardner, M. The fantastic combinations of John Conway's new solitaire game, Life. *Sci. Am.* 223:120–123, 1970.

Garfinkel, A., M. L. Spano, W. L. Ditto, and J. N. Weiss. Controlling cardiac chaos. *Science* 257:1230–1235, 1992.

Gerhardt. M., H. Schuster, and J. J. Tyson. A cellular automaton model of excitable media including curvature and dispersion. *Science* 247:1563–1566, 1990.

Gershenfeld, N. An experimentalist's introduction to the observation of dynamical systems. In: *Directions in Chaos Vol. II*, edited by H. Bai-Lin. Teaneck, New Jersey: World Scientific, 1988, pp. 310–384.

Gerstein, G. L., and B. Mandelbrot. Random walk models for the spike activity of a single neuron. *Biophys. J.* 4:41–68, 1964.

Geweke, J., and S. Porter-Hudak. The estimation and application of long memory time series models. *J. Time Series Anal.* 4:221–238, 1983.

Ghyka, M. *The Geometry of Art and Life*. New York: Dover, 1977, 174 pp.

Gibson, J. F., J. D. Farmer, M. Casdaglia, and S. Eubank. An analytic approach to practical state space reconstruction. *Sante Fe Institute Rep* 92-04-021:1–47, 1992.

Glass, L., M. R. Guevara, A. Shrier, and R. Perez. Bifurcation and chaos in a periodically stimulated cardiac oscillator. *Physica* 7D:89–101, 1983.

Glass, L., and A. T. Winfree. Discontinuities in phase-resetting experiments. *Am. J. Physiol.* 246:R251–R258, 1984.

Glass, L., M. R. Guevara, J. Bélair, and A. Shrier. Global bifurcations of a periodically forced biological oscillator. *Phys. Rev. A* 29:1348–1357, 1984.

Glass, L., A. Shrier, and J. Bélair. Chaotic cardiac rhythms. In: *Chaos*, edited by A. V. Holden. Princeton: Princeton University Press, 1986, pp. 237–256.

Glass, L., and M. C. Mackey. *From Clocks to Chaos. The Rhythms of Life*. Princeton: Princeton University Press, 1988.

Glass, L., and C. P. Malta. Chaos in multi-looped negative feedback systems. *J. Theor. Biol.* 145:217–223, 1990.

Glass, L., and P. Hunter. There is a theory of heart. *Physica D* 43:1–16, 1990.

Glass, L., P. Hunter, and A. McCulloch. *Theory of Heart: Biomechanics, Biophysics, and Nonlinear Dynamics of Cardiac Function*. New York: Springer-Verlag, 1991, 611 pp.

Glass, L., and A. Shrier. Low-dimensional dynamics in the heart. In: *Theory of Heart: Biomechanics, Biophysics, and Nonlinear Dynamics of Cardiac Function*, edited by L. Glass, P. Hunter and A. McCullough. New York: Springer-Verlag, 1991, pp. 289–312.

Gleick, J. *Chaos: Making a New Science*. New York: Viking Penguin, Inc., 1987.

Glenny, R. W., and H. T. Robertson. Fractal properties of pulmonary blood flow: characterization of spatial heterogeneity. *J. Appl. Physiol.* 69:532–545, 1990.

Glenny, R. W., and H. T. Robertson. Fractal modeling of pulmonary blood flow heterogeneity. *J. Appl. Physiol.* 70:1024–1030, 1991.

Glenny, R., H. T. Robertson, S. Yamashiro, and J. B. Bassingthwaighte. Applications of fractal analysis to physiology. *J. Appl. Physiol.* 70:2351–2367, 1991a.

Glenny, R. W., W. J. E. Lamm, R. K. Albert, and H. T. Robertson. Gravity is a minor determinant of pulmonary blood flow distribution. *J. Appl. Physiol.* 71:620–629, 1991b.

Glenny, R. W., L. Polissar, and H. T. Robertson. Relative contribution of gravity to pulmonary perfusion heterogeneity. *J. Appl. Physiol.* 71:2449–2452, 1991c.

Goetze, T., and J. Brickmann. Self similarity of protein surfaces. *Biophys. J.* 61(1):109–118, 1992.

Goldberger, A. L., V. Bhargava, B. J. West, and A. J. Mandell. On a mechanism of cardiac electrical stability: The fractal hypothesis. *Biophys. J.* 48:525–528, 1985a.

Goldberger, A. L., B. J. West, and V. Bhargava. Nonlinear mechanisms in physiology and pathophysiology: Toward a dynamical theory of health and disease. Oslo: Proceedings International Association for Mathematics and Computers in Simulation, 1985b, pp. 1–3.

Goldberger, A. L., D. Goldwater, and V. Bhargava. Atropine unmasks bed-rest effect: a spectral analysis of cardiac interbeat intervals. *J. Appl. Physiol.* 61:1843–1848, 1986.

Goldberger, A. L., and B. J. West. Fractals in physiology and medicine. *Yale J. Biol. Med.* 60:421–435, 1987a.

Goldberger, A. L., and B. J. West. Applications of nonlinear dynamics to clinical cardiology. *Ann. NY Acad. Sci.* 504:195–213, 1987b.

Goldberger, A. L., and B. J. West. Chaos in physiology: Health or disease?. In: *Chaos in Biological Systems*, edited by H. Degn, A. V. Holden and L. F. Olsen. New York, NY: Plenum, 1987c.

Goldberger, A. L., D. R. Rigney, J. Mietus, E. M. Antman, and S. Greenwald. Nonlinear dynamics in sudden death syndrome: Heartrate oscillations and bifurcations. *Experientia* 44:983–987, 1988.

Goldberger, A. L., D. R. Rigney, and B. J. West. Chaos and fractals in human physiology. *Sci. Am.* 262:42–49, 1990.

Goldberger, A. L. Fractal mechanisms in the electrophysiology of the heart. *IEEE Eng. Med. Biol* 11:47–52, 1992.

Gonzalez-Fernandez, J. M. Theory of the measurement of the dispersion of an indicator in indicator-dilution studies. *Circ. Res.* 10:409–428, 1962.

Gottlieb, M. E. The VT model: A deterministic model of angiogenesis and biofractals based on physiological rules. In: *Proceedings of the 1991 IEEE Seventeenth Annual Northeast Bioengineering Conference*, edited by M. D. Fox, M. A. F. Epstein, R. B. Davis and T. M. Alward. New York: IEEE, 1991, pp. 38–39.

Grassberger, P., and I. Procaccia. Measuring the strangeness of strange attractors. *Physica* 9D:189–208, 1983.

Grassberger, P. Estimating the fractal dimensions and entropies of strange attractors. In: *Chaos*, edited by A. V. Holden. Princeton, N.J.: Princeton Univ. Press, 1986, pp. 291–311.

Grassberger, P., T. Schreiber, and C. Schaffrath. Nonlinear time sequence analysis. *Int. J. Bifurc. Chaos* 1:521–547, 1991.

Greenside, H. S., A. Wolf, J. Swift, and T. Pignarro. Impracticality of a box-counting algorithm for calculating the dimensionality of strange attractors. *Phys. Rev. A* 25:3453–3456, 1982.

Grünbaum, B., and G. C. Shephard. *Tilings and Patterns*. New York: W. H. Freeman and Company, 1987.

Guckenheimer, J. *Nonlinear Oscillations, Dynamical Systems, and Bifurcations of Vector Fields*. New York: Springer-Verlag, 1983, 453 pp.

Guevara, M. R., and L. Glass. Phase locking, period doubling bifurcations and chaos in a mathematical model of a periodically driven oscillator: a theory for the entrainment of biological oscillators and the generation of cardiac dysrhythmias. *J. Math. Biol.* 14:1–23, 1982.

Guyon, E., and H. E. Stanley. *Fractal Forms*. Haarlem, The Netherlands: Elsevier/North-Holland, 1991.

Halmos, P. R. *Measure Theory*. New York: Van Nostrand, 1950, 304 pp.

Ham, A. W. *Histology*. Philadelphia: Lippincott, 1957, 894 pp.

Hampel, F. R., E. M. Ronchetti, P. J. Rousseeuw, and W. A. Stahel. *Robust Statistics. The Approach Based on Influence Functions*. New York: J. Wiley & Sons, 1986.

Hao, B. L. *Chaos*. Singapore: World Scientific, 1984, 576 pp.

Hao, B. L. *Directions in Chaos Vol. II*. New Jersey: World Scientific, 1988.

Harvey, W. *Anatomical Lectures: Prelectiones Anatomie Universalis, De Musculis*. Edinburgh, Published for the Royal College of Physicians, London: E. & S. Livingstone, 1964, 503 pp.

Haslett, J., and A. E. Raftery. Space-time modelling with long-memory dependence: Assessing Ireland's wind power resource (with discussion). *Appl. Statist.* 36:1–50, 1989.

Havlin, S. Molecular diffusion and reactions. In: *The Fractal Approach to Heterogeneous Chemistry: Surfaces, Colloids, Polymers*, edited by D. Avnir. New York: John Wiley and Sons, 1990, pp. 251–269.

Hentschel, H. G. E., and I. Procaccia. The infinite number of generalized dimensions of fractals and strange attractors. *Physica D* 8:435–444, 1983.

Hess, B., and M. Markus. Order and chaos in biochemistry. *Trends Biochem. Sci.* 12:45–48, 1987.

Heymann, M. A., B. D. Payne, J. I. E. Hoffman, and A. M. Rudolph. Blood flow measurements with radionuclide-labeled particles. *Prog. Cardiovasc. Dis.* 20:55–79, 1977.

Hille, B. *Ionic Channels of Excitable Membranes*. Sunderland, Mass.: Sinauer Associates, 1984, 426 pp.

Hille, B., and D. M. Fambrough. *Proteins of Excitable Membranes*. New York: Wiley-Interscience, 1987, 331 pp.

Hodgkin, A. L., and A. F. Huxley. A quantitative description of membrane current and its application to conduction and excitation in nerve. *J. Physiol.* 117:500–544, 1952.

Hollander, M., and D. A. Wolfe. *Nonparametric Statistical Methods*. New York: Wiley, 1973, 503 pp.

Holt, D. R., and E. L. Crow. Tables and graphs of the stable density functions. *J. Res. NBS* 77B:143–198, 1973.

Holzfuss, J., and G. Mayer-Kress. An approach to error-estimation in the application of dimension algorithms. In: *Dimensions and Entropies in Chaotic Systems*, G. Mayer-Kress, ed. New York: Springer-Verlag, 1986, pp. 114–122.

Hoop, B., H. Kazemi, and L. Liebovitch. Rescaled range analysis of resting respiration. *Chaos* 3:27–29, 1993.

Hopfield, J. J. Neural networks and physical systems with emergent collective computational abilities. *Proc. Natl. Acad. Sci.* 79:2554–2558, 1982.

Horn, R. Statistical methods for model discrimination: Applications to gating kinetics and permeation of the acetylcholine receptor channel. *Biophys. J.* 51:255–263, 1987.

Horn, R., and S. J. Korn. Model selection: reliability and bias. *Biophys. J.* 55:379–381, 1989.

Horsfield, K. Morphometry of the small pulmonary arteries in man. *Circ. Res.* 42:593–597, 1978.

Horsfield, K., and M. J. Woldenberg. Diameters and cross-sectional areas of branches in the human pulmonary arterial tree. *Anat. Rec.* 223:245–251, 1989.

Hou, X. J., R. Gilmore, G. B. Mindlin, and H. G. Solari. An efficient algorithm for fast O(N*1n(N)) box counting. *Phys. Lett.* A151:43–46, 1990.

Hudetz, A. G., K. A. Conger, J. H. Halsey, M. Pal, O. Dohan, and A. G. B. Kovach. Pressure distribution in the pial arterial system of rats based on morphometric data and mathematical models. *J. Cereb. Blood Flow Metab.* 7:342–355, 1987.

Hudlická, O. Development of microcirculation: capillary growth and adaptation. In: *Handbook of Physiology. Section 2: The Cardiovascular System Volume IV*, edited by E. M. Renkin and C. C. Michel. Bethesda, Maryland: American Physiological Society, 1984, pp. 165–216.

Hudlická, O., and K. R. Tyler. *Angiogenesis. The Growth of the Vascular System.* London: Academic Press, 1986.

Hurst, H. E. Long-term storage capacity of reservoirs. *Trans. Amer. Soc. Civ. Engrs.* 116:770–808, 1951.

Hurst, H. E., R. P. Black, and Y. M. Simaiki. *Long-term Storage: An Experimental Study.* London: Constable, 1965.

Icardo, J. M. Cardiac morphogenesis and development: a symposium. *Experientia* 44:909–1032, 1988.

Ikeda, N., H. Tsuruta, and T. Sato. Difference equation model of the entrainment of myocardial pacemaker cells based on the phase response curve. *Biol. Cybern.* 42:117–128, 1981.

Jalife, J. *Mathematical Approaches to Cardiac Arrhythmias.* New York, NY: New York Acad. Sci., 1990, 417 pp.

János, I. M., and V. K. Horváth. Dynamics of water droplets on a window pane. *Phys. Rev. A* 40:5232–5237, 1989.

Jalife, J., J. M. Davidenko, and D. C. Michaels. A new perspective on the mechanisms of arrhythmias and sudden cardiac death: Spiral waves of excitation in heart muscle. *J. Cardiovasc. Electrophysiol.* 2:S133–S152, 1991.

Jensen, M. H. Multifractals in convection and aggregation. In: *Random Fluctuations and Pattern Growth: Experiments and Models*, edited by H. E. Stanley and N. Ostrowsky. Boston: Kluwer Academic, 1988, pp. 292–309.

Journel, A. G., and C. J. Huijbregts. *Mining Geostatistics.* London: Academic Press, 1978.

Kandel, E. R. Small systems of neurons. *Sci. Am.* 241:66–76, 1979.

Kaneko, N. The basic structure of intramyocardial microcirculation system in the normal human heart. In: *Microcirculation—An Update, Vol. 2*, edited by J. Tsuchiya, M. Asano, Y. Mishima and M. Oda. Amsterdam: Elsevier Science Publishers, 1987, pp. 159–160.

Kang, K., and S. Redner. Fluctuation dominated kinetics in diffusion-controlled reactions. *Phys. Rev. A* 32:435–447, 1985.

Kaplan, D. T. *The Dynamics of Cardiac Electrical Instability.* Harvard University: Ph.D. Thesis, 1989.

Kaplan, D. T., M. I. Furman, and S. M. Pincus. Techniques for analyzing complexity in heart rate and beat-to-beat blood pressure signals. *Computers in Cardiology*:243–246, 1990.

Kaplan, D. T., M. I. Furman, S. M. Pincus, S. M. Ryan, L. A. Lipsitz, and A. L. Goldberger. Aging and the complexity of cardiovascular dynamics. *Biophys. J.* 59:945–949, 1991.

Kaplan, J. L., and J. A. Yorke. Chaotic behavior of multidimensional difference equations. In: *Functional Differential Equations and the Approximation of Fixed Points*, edited by H. O. Peitgen and H. O. Walther. New York: Springer-Verlag, 1979, pp. 228–237.

Kariniemi, V., and P. Ämmälä. Short-term variability of fetal heart rate during pregnancies with normal and insufficient placental function. *Am. J. Obstet. Gynecol.* 139:33–37, 1981.

Karplus, M., and J. A. McCammon. The internal dynamics of globular proteins. *CRC Crit. Rev. Biochem.* 9:293–349, 1981.

Kassab, G. S., C. A. Rider, N. J. Tang, and Y. B. Fung. Morphometry of pig coronary arterial trees. *Am. J. Physiol. (Heart Circ. Physiol.)* 265:H350–H365, 1993.

Katz, S. A., and E. O. Feigl. Little carbon dioxide diffusional shunting in coronary circulation. *Am. J. Physiol.* 253 (Heart Circ Physiol. 22):H614–H625, 1987.

Kauffman, S. A. Antichaos and adaptation. *Sci. Am.* 265:78–84, 1991.

Kaye, B. H. *A Random Walk Through Fractal Dimensions.* New York, NY: VCH, 1989, 421 pp.

Keirsted, W. P., and B. A. Huberman. Dynamical singularities in ultradiffusion. *Phys. Rev. A* 36:5392–5400, 1987.

Kennel, M. B., R. Brown, and H. D. I. Abarbanel. Determining embedding dimension for phase-space reconstruction using a geometrical construction. *Phys. Rev. A* 45:3403–3411, 1992.

Kiani, M. F., and A. G. Hudetz. Computer simulation of growth of anastomosing microvascular networks. *J. Theor. Biol.* 150:547–560, 1991.

Kienker, P. Equivalence of aggregated Markov models of ion-channel gating. *Proc. R. Soc. Lond. (Biol.)* 236:269–309, 1989.

Kilgren, L. M. Issues involved in the extrapolation of information from single-channel currents of cultured myotubes: a kinetic and stochastic analysis. Philadelphia, PA: Ph.D. Thesis, Department of Biophysics, University of Pennsylvania, 1989.

King, R. B., J. B. Bassingthwaighte, J. R. S. Hales, and L. B. Rowell. Stability of heterogeneity of myocardial blood flow in normal awake baboons. *Circ. Res.* 57:285–295, 1985.

King, R. B., and J. B. Bassingthwaighte. Temporal fluctuations in regional myocardial flows. *Pflügers Arch.(Eur.J.Physiol.)* 413/4:336–342, 1989.

King, R. B., L. J. Weissman, and J. B. Bassingthwaighte. Fractal descriptions for spatial statistics. *Ann. Biomed. Eng.* 18:111–121, 1990.

Kirby, M. L. Role of extracardiac factors in heart development. *Experientia* 44:944–951, 1988.

Klafter, J., and M. F. Shlesinger. On the relationship among three theories of relaxation in disordered systems. *Proc. Natl.. Acad. Sci. USA* 83:848–851, 1986.

Kleiger, R. E., J. P. Miller, J. T. Bigger, Jr., and A. J. Moss. Decreased heart rate variability and its association with increased mortality after acute myocardial infarction. *Am. J. Cardiol.* 59:256–262, 1987.

Kluge, K. A., R. M. Harper, V. L. Schechtman, A. J. Wilson, H. J. Hoffman, and D. P. Southall. Spectral analysis assessment of respiratory sinus arrhythmia in normal infants and infants who subsequently died of sudden infant death syndrome. *Pediatr. Res.* 24:677–682, 1988.

Kobayashi, M., and T. Musha. $1/f$ fluctuations of heartbeat period. *IEEE Trans. Biomed. Eng.* 29:456–457, 1982.

Kopelman, R. Fractal reaction kinetics. *Science* 241:1620–1626, 1988.

Korn, S. J., and R. Horn. Statistical discrimination of fractal and Markov models of single-channel gating. *Biophys. J.* 54:871–877, 1988.

Krenz, G. S., J. H. Linehan, and C. A. Dawson. A fractal continuum model of the pulmonary arterial tree. *J. Appl. Physiol.* 72:2225–2237, 1992.

Krige, D. G. A statistical approach to some basic mine valuation problems in the Witwatersrand. *Chem. Metall. Mining Soc. South Africa Jour.* 52:119–139, 1951.

Künsch, H. Discrimination between monotonic trends and long range dependence. *J. Appl. Prob.* 23:1025–1030, 1986.

Langille, B. L., M. P. Bendeck, and F. W. Keeley. Adaptations of carotid arteries of young and mature rabbits to reduced carotid blood flow. *Am. J. Physiol. (Heart Circ. Physiol. 25)* 256:H931–H939, 1989.

Langreth, R. Engineering dogma gives way to chaos. *Science* 252:776–778, 1991.

Laplace, P. S. *A Philosphical Essay on Probabilities.* New York: Dover Publications, 1951, 999 pp.

Läuger, P. Internal motions in proteins and gating kinetics of ionic channels. *Biophys. J.* 53:877–884, 1988.

Lauwerier, H. *Fractals.* Princeton, N.J.: Princeton Univ. Press, 1991, 209 pp.

Layne, S. P., G. Mayer-Kress, and J. Holzfuss. Problems associated with dimensional analysis of electroencephalogram data. In: *Dimensions and Entropies in Chaotic Systems,* edited by G. Mayer-Kress. Berlin: Springer-Verlag, 1986, pp. 246–256.

Lea, D. E., and C. A. Coulson. The distribution of the number of mutants in bacterial populations. *J. Genetics* 49:264–285, 1949.

Lee, K., and Y. Cohen. "Fractal Attraction" St. Paul: Sandpiper Software, 1990.

Lefèvre, J. Teleonomical optimization of a fractal model of the pulmonary arterial bed. *J. Theor. Biol.* 102:225–248, 1983.

Levin, M., B. Dawant, and A. S. Popel. Effect of dispersion of vessel diameters and lengths in stochastic networks. II. Modelling of microvascular hematocrit distribution. *Microvasc. Res.* 31:223–234, 1986.

Levitt, D. G., B. Sircar, N. Lifson, and E. J. Lender. Model for mucosal circulation of rabbit small intestine. *Am. J. Physiol.* 237 (Endocrinol. Metab. Gastrointest. Physiol. 6): E373–E382, 1979.

Levitt, D. G. Continuum model of voltage-dependent gating. *Biophys. J.* 55:489–498, 1989.

Li, H. Q., S. H. Chen, and H. M. Zhao. Fractal mechanisms for the allosteric effects of proteins and enzymes. *Biophys. J.* 58:1313–1320, 1990.

Liebovitch, L. S., J. Fischbarg, and J. P. Koniarek. Optical correlation functions applied to the random telegraph signal: How to analyze patch clamp data *without* measuring the open and closed times. *Math. Biosci.* 78:203–215, 1986.

Liebovitch, L. S., J. Fischbarg, and J. P. Koniarek. Ion channel kinetics: a model based on fractal scaling rather than multistate Markov processes. *Math. Biosci.* 84:37–68, 1987a.

Liebovitch, L. S., J. Fischbarg, J. P. Koniarek, I. Todorova, and M. Wang. Fractal model of ion-channel kinetics. *Biochim. Biophys. Acta* 896:173–180, 1987b.

Liebovitch, L. S., and J. M. Sullivan. Fractal analysis of a voltage-dependent potassium channel from cultured mouse hippocampal neurons. *Biophys. J.* 52:979–988, 1987.

Liebovitch, L. S. Analysis of fractal ion channel gating kinetics: kinetic rates, energy levels, and activation energies. *Math. Biosci.* 93:97–115, 1989a.

Liebovitch, L. S. Testing fractal and Markov models of ion channel kinetics. *Biophys. J.* 55:373–377, 1989b.

Liebovitch, L. S., and T. Tóth. A fast algorithm to determine fractal dimensions by box counting. *Phys. Lett. A* 141:386–390, 1989.

Liebovitch, L. S., and T. I. Tóth. Fractal activity in cell membrane ion channels. *Ann. NY Acad. Sci.* 591:375–391, 1990a.

Liebovitch, L. S., and T. I. Tóth. Using fractals to understand the opening and closing of ion channels. *Ann. Biomed. Eng.* 18:177–194, 1990b.

Liebovitch, L. S., and T. I. Tóth. The Akaike information criterion (AIC) is not a sufficient condition to determine the number of ion channel states from single channel recordings. *Synapse* 5:134–138, 1990c.

Liebovitch, L. S., and T. I. Tóth. A model of ion channel kinetics using deterministic chaotic rather than stochastic processes. *J. Theor. Biol.* 148:243–267, 1991.

Liebovitch, L. S., and F. P. Czegledy. Fractal, chaotic and self-organizing critical system descriptions of the kinetics of cell membrane ion channels. In: *Complexity, Chaos, and Biological Evolution*, edited by E. Moskilde and L. Moskilde. New York: Plenum, 1991, pp. 145–153.

Liebovitch, L. S., and F. P. Czegledy. A model of ion channel kinetics based on deterministic, chaotic motion in a potential with two local minima. *Ann. Biomed. Eng.* 20:517–531, 1992.

Lindenmayer, A. Mathematical models for cellular interactions in development. I. Filaments with one-sided inputs. *J. Theoret. Biol.* 18:280–299, 1968a.

Lindenmayer, A. Mathematical models for cellular interactions in development. II. Simple and branching filaments with two-sided inputs. *J. Theoret. Biol.* 18:300–315, 1968b.

Lipsitz, L. A., J. Mietus, G. B. Moody, and A. L. Goldberger. Spectral characteristics of heart rate variability before and during postural tilt: Relations to aging and risk of syncope. *Circulation* 81:1803–1810, 1990.

Lipsitz, L. A., and A. L. Goldberger. Loss of 'complexity' and aging: Potential applications of fractals and chaos theory to senescence. *JAMA* 267:1806–1809, 1992.

Little, S. E., and J. B. Bassingthwaighte. Plasma-soluble marker for intraorgan regional flows. *Am. J. Physiol.* 245 (*Heart Circ. Physiol.* 14):H707–H712, 1983.

Liu, S. H. Fractal model for the ac response of a rough interface. *Phys. Rev. Lett.* 55:529–532, 1985.

Lorenz, E. N. Deterministic nonperiodic flows. *J. Atmos. Sci.* 20:130–141, 1963.

Lowen, S. B., and M. C. Teich. Doubly stochastic Poisson point process driven by fractal shot noise. *Phys. Rev. A* 43:4192–4215, 1991.

Luce, G. G. *Biological Rhythms in Human and Animal Physiology.* New York: Dover, 1971, 183 pp.

Luria, S. E., and M. Delbruck. Mutations of bacteria from virus sensitivity of virus resistance. *Genetics* 28:491–511, 1943.

Ma, S. K. *Modern Theory of Critical Phenomena.* Reading, Mass.: W. A. Benjamin, 1976, 561 pp.

McCammon, J. A., and S. C. Harvey. *Dynamics of Proteins and Nucleic Acids.* New York: Cambridge University Press, 1987.

MacDonald, N. *Trees and Networks in Biological Models.* New York: John Wiley & Sons, 1983.

McGee, R., Jr., M. S. Sansom, and P. N. Usherwood. Characterization of a delayed rectifier K^+ channel in NG108-15 neuroblastoma X glioma cells: gating kinetics and the effects of enrichment of membrane phospholipids with arachidonic acid. *J. Membr. Biol.* 102:21–34, 1988.

Machacek, M. Copernicus, Ptolemy and particle theory. *Phys. Today* November:13–15, 1989.

Machlup, S. Earthquakes, thunderstorms, and other 1/f noises. In: *Sixth International Conference on Noise in Physical Systems,* edited by P. H. E. Meijer, R. D. Mountain and R. J. Soulen. Washington, D.C.: National Bureau of Standards, 1981, pp. 157–160.

McMahon, T. A., and J. T. Bonner. *On Size and Life.* New York: Scientific American Books, Inc., 1983.

McManus, O. B., A. L. Blatz, and K. L. Magleby. Sampling, log binning, fitting, and plotting durations of open and shut intervals from single channels and the effects of noise. *Pflügers Arch.* 410:530–553, 1987.

McManus, O. B., D. S. Weiss, C. E. Spivak, A. L. Blatz, and K. L. Magleby. Fractal models are inadequate for the kinetics of four different ion channels. *Biophys. J.* 54:859–870, 1988.

McManus, O. B., C. E. Spivak, A. L. Blatz, D. S. Weiss, and K. L. Magleby. Fractal models, Markov models, and channel kinetics. *Biophys. J.* 55:383–385, 1989.

McNamee, J. E. Distribution of transvascular pathway sizes through the pulmonary microvascular barrier. *Ann. Biomed. Eng.* 15:139–155, 1987.

McNamee, J. E. Fractal character of pulmonary microvascular permeability. *Ann. Biomed. Eng.* 18:123–133, 1990.

Mainster, M. A. The fractal properties of retinal vessels: embryological and clinical implications. *Eye* 4:235–241, 1990.

Majid, N. M., G. J. Martin, and R. F. Kehoe. Diminished heart rate variability in sudden cardiac death. *Circulation* 72:III–240, 1985.

Mandelbrot, B. Calcul des probabilités et climatologie statistique—Une classe de processus stochastiques homothétiques à soi; application à la loi climatologique de H. E. Hurst. *C. R. Acad. Sc. Paris* 260:3274–3277, 1965.

Mandelbrot, B. B., and J. R. Wallis. Noah, Joseph, and operational hydrology. *Water Resour. Res.* 4:909–918, 1968.

Mandelbrot, B. B., and J. W. Van Ness. Fractional brownian motions, fractional noises and applications. *SIAM Rev.* 10:422–437, 1968.

Mandelbrot, B. B., and J. R. Wallis. Computer experiments with fractional Gaussian noises. Part 1, averages and variances. *Water Resour. Res.* 5:228–241, 1969a.

Mandelbrot, B. B., and J. R. Wallis. Computer experiments with fractional Gaussian noises. Part 2, rescaled ranges and spectra. *Water Resour. Res.* 5:242–259, 1969b.

Mandelbrot, B. B., and J. R. Wallis. Computer experiments with fractional Gaussian noises. Part 3, mathematical appendix. *Water Resour. Res.* 5:260–267, 1969c.

Mandelbrot, B. B., and J. R. Wallis. Some long-run properties of geophysical records. *Water Resour. Res.* 5:321–340, 1969d.

Mandelbrot, B. B., and J. R. Wallis. Robustness of the rescaled range R/S in the measurement of noncyclic long run statistical dependence. *Water Resour. Res.* 5:967–988, 1969e.

Mandelbrot, B. B. Intermittent turbulence in self-similar cascades: divergence of high moments and dimension of the carrier. *J. Fluid Mech.* 62:331–358, 1974.

Mandelbrot, B. *Fractals: Form, Chance and Dimension.* San Francisco: W.H. Freeman and Co., 1977.

Mandelbrot, B. B. Fractal aspects of the iteration of $z \mapsto \lambda z(1 - z)$ for complex λ and z. *Annals NY Acad. Sciences* 357:249–259, 1980.

Mandelbrot, B. B. *The Fractal Geometry of Nature.* San Francisco: W.H. Freeman and Co., 1983.

Mandelbrot, B. B., and C. J. G. Evertsz. (Cover). *Nature* 348, 1990.

Markus, M., D. Kuschmitz, and B. Hess. Chaotic dynamics in yeast glycolysis under periodic substrate input flux. *FEBS Lett.* 172:235–238, 1984.

Markus, M., and B. Hess. Input-response relationships in the dynamics of glycolysis. *Arch. Biol. Med. Exp.* 18:261–271, 1985.

Maslim, J., M. Webster, and J. Stone. Stages in the structural differentiation of retinal ganglion cells. *J. Comp. Neurol.* 254:382–402, 1986.

Maslow, A. H. *The Psychology of Science: A Reconnaissance.* New York: Harper and Row, 1966, 168 pp.

Mathieu-Costello, O. Capillary tortuosity and degree of contraction or extension of skeletal muscles. *Microvasc. Res.* 33:98–117, 1987.

Mausner, J. S., and A. K. Bahn. *Epidemiology: An Introductory Text.* Philadelphia: Saunders, 1974, 377 pp.

May, R. M., and G. F. Oster. Bifurcations and dynamic complexity in simple ecological models. *Amer. Naturalist* 110:573–599, 1976.

344   Works Cited

May, R. M. Simple mathematical models with very complicated dynamics. *Nature* 261:459–467, 1976.

Mayer-Kress, G., and S. P. Layne.   Dimensionality of the human electroencephalogram. *Ann. N.Y. Acad. Sci.* 504:62–87, 1987.

Meakin, P. Diffusion-controlled cluster formation in 2-6-dimensional space. *Phys. Rev. A.* 27:1495–1507, 1983.

Meakin, P. A new model for biological pattern formation. *J. Theor. Biol.* 118:101–113, 1986a.

Meakin, P. Computer simulation of growth and aggregation processes. In: *On Growth and Form: Fractal and Non-fractal Patterns in Physics*, edited by H. E. Stanley and N. Ostrowsky. Boston: M. Nijhoff, 1986b, pp. 111–135.

Meakin, P. Simulations of aggregation processes. In: *The Fractal Approach to Heterogeneous Chemistry*, edited by D. Avnir. New York: John Wiley & Sons Ltd., 1989, pp. 131–160.

Meier, P., and K. L. Zierler.   On the theory of the indicator-dilution method for measurement of blood flow and volume. *J. Appl. Physiol.* 6:731–744, 1954.

Meinhardt, H. *Models of Biological Pattern Formation*. New York: Academic Press, 1982.

Meyer, R. A. A linear model of muscle respiration explains monoexponential phosphocreatine changes. *Am. J. Physiol. (Cell Physiol. 23)* 254:548–553, 1988.

Millhauser, G. L., E. E. Salpeter, and R. E. Oswald.   Rate-amplitude correlation from single-channel records: A hidden structure in ion channel gating kinetics?. *Biophys. J.* 54:1165–1168, 1988a.

Millhauser, G. L., E. E. Salpeter, and R. E. Oswald.   Diffusion models of ion-channel gating and the origin of power-law distributions from single-channel recording. *Proc. Natl. Acad. Sci.* 85:1503–1507, 1988b.

Millhauser, G. L. Reptation theory of ion channel gating. *Biophys. J.* 57:857–864, 1990.

Montroll, E. W., and B. J. West.   On enriched collection of stochastic processes. In: *Fluctuation Phenomena, First Edition*, edited by E. W. Montroll and J. L. Lebowitz. Amsterdam: North-Holland Personal Library, 1979, pp. 61–173.

Montroll, E. W., and M. F. Shlesinger.   On the wonderful world of random walks. In: *Nonequilibrium Phenomena II, From Stochastics to Hydrodynamics*, edited by E. W. Montroll. Amsterdam: North-Holland, 1984, pp. 288–293.

Montroll, E. W., and B. J. West.   On enriched collection of stochastic processes. In: *Fluctuation Phenomena, Second Edition*, edited by E. W. Montroll and J. L. Lebowitz. Amsterdam: North-Holland Personal Library, 1987, pp. 61–206.

Moon, F. C. *Chaotic Vibrations: An Introduction for Applied Scientists and Engineers*. New York: John Wiley & Sons, Inc., 1987.

Moon, F. C. *Chaotic and Fractal Dynamics: An Introduction for Applied Scientists and Engineers*. New York: John Wiley & Sons, 1992, 508 pp.

Musha, T., Y. Kosugi, G. Matsumoto, and M. Suzuki.   Modulation of the time relaxation of action potential impulses propagating along an axon. *IEEE Trans. Biomed. Eng.* 28:616–623, 1981.

Nagumo, J., S. Arimoto, and S. Yoshizawa.   An active pulse transmission line simulating nerve axon. *Proc. I.R.E.* 50:2061–2070, 1962.

Neale, E. A., L. M. Bowers, and T. G. Smith, Jr.. Early dendrite development in spinal cord cell cultures: a quantitative study. *J. Neurosci. Res.* 34:54–66, 1993.

Nelson, A. D., R. F. Muzic, F. Miraldi, G. J. Muswick, G. P. Leisure, and W. Voelker. Continuous arterial positron monitor for quantitation in PET imaging. *Am. J. Physiol. Imaging* 5:84–88, 1990.

Nelson, T. R., and D. K. Manchester. Modeling of lung morphogenesis using fractal geometries. *IEEE Trans. Med. Imaging* 7:321–327, 1988.

Nerenberg, M. A. H., and C. Essex. Correlation dimension and systematic geometric effects. *Phys. Rev. A* 42:7065–7074, 1990.

Neubauer, N., and H. J. G. Gundersen. Analysis of heart rate variations in patients with multiple sclerosis. *J. Neurol. Neurosurg. Psychiatry* 41:417–419, 1978.

Nose, Y., T. Nakamura, and M. Nakamura. The microsphere method facilitates statistical assessment of regional blood flow. *Basic Res. Cardiol.* 80:417–429, 1985.

Obert, M., P. Pfeifer, and M. Sernetz. Microbial growth patterns described by fractal geometry. *J. Bacteriol.* 172:1180–1185, 1990.

Orbach, R. Dynamics of fractal networks. *Science* 231:814–819, 1986.

Osborne, A. R., and A. Provenzale. Finite correlation dimension for stochastic systems with power-law spectra. *Physica D* 35:357–381, 1989.

Oswald, R. E., G. L. Millhauser, and A. A. Carter. Diffusion model in ion channel gating: Extension to agonist-activated ion channels. *Biophys. J.* 59:1136–1142, 1991.

Ott, E., C. Grebogi, and J. A. Yorke. Controlling chaos. *Phys. Rev. Lett.* 64:1196–1199, 1990.

Oude Vrielink, H. H. E., D. W. Slaaf, G. J. Tangelder, S. Weijmer-Van Velzen, and R. S. Reneman. Analysis of vasomotion waveform changes during pressure reduction and adenosine application. *Am. J. Physiol.* 258 (*Heart Circ. Physiol.* 27):H29–H37, 1990.

Packard, N. H., J. P. Crutchfield, J. D. Farmer, and R. S. Shaw. Geometry from a time series. *Phys. Rev. Lett.* 45 (9):712–716, 1980.

Papoulis, A. *Probability, Random Variables, and Stochastic Processes.* New York: McGraw-Hill, 1984, 576 pp.

Parker, T. S., and L. O. Chua. *Practical Numerical Algorithms for Chaotic Systems.* New York: Springer-Verlag, 1989.

Paumgartner, D., G. Losa, and E. R. Weibel. Resolution effect on the stereological estimation of surface and volume and its interpretation in terms of fractal dimensions. *J. Microsc.* 121:51–63, 1981.

Peitgen, H. O., and P. H. Richter. *The Beauty of Fractals: Images of Complex Dynamical Systems.* Berlin/Heidelberg: Springer-Verlag, 1986.

Peitgen, H. O., and D. Saupe. *The Science of Fractal Images.* New York: Springer-Verlag, 1988.

Peitgen, H. O., H. Jürgens, and D. Saupe. *Fractals for the Classroom: Part One, Introduction to Fractals and Chaos.* New York: Springer-Verlag, 1992a, 450 pp.

Peitgen, H. O., H. Jürgens, and D. Saupe. *Fractals for the Classroom: Part Two, Complex Systems and the Mandelbrot Set.* New York: Springer-Verlag, 1992b, 500 pp.

Peitgen, H. O., H. Jürgens, and D. Saupe. *Fractals for the Classroom: Strategic Activities, Volume Two*. New York: Springer-Verlag, 1992c, 187 pp.

Peitgen, H. O., H. Jurgens, and D. Saupe. *Chaos and Fractals: New Frontiers of Science*. New York: Springer-Verlag, 1992d, 984 pp.

Pelosi, G., G. Saviozzi, M. G. Trivella, and A. L'Abbate. Small artery occlusion: A theoretical approach to the definition of coronary architecture and resistance by a branching tree model. *Microvasc. Res.* 34:318–335, 1987.

Penck, A. *Morphologie der Erdoberfläche*. Stuttgart, 1894.

Peng, C. K., S. V. Buldyrev, A. L. Goldberger, S. Havlin, F. Sciortino, M. Simons, and H. E. Stanley. Long-range correlations in nucleotide sequences. *Nature* 356:168–170, 1992.

Perkal, J. *On the length of empirical curves* Ann Arbor: Discussion Paper No. 10, Department of Geography, University of Michigan, 1966, pp. 1–34.

Perrin, J. L'agitation moléculaire et le mouvement brownien. *Compt. Rend. Acad. Sci. Paris* 146:967–970, 1908.

Pfeifer, P., and M. Obert. Fractals: basic concepts and terminology. In: *The Fractal Approach to Heterogeneous Chemistry*, edited by D. Avnir. New York: John Wiley & Sons, 1989, pp. 11–43.

Pietronero, L., and E. Tosatti. *Fractals in Physics: Proceedings of the Sixth Trieste International Symposium on Fractals in Physics*. New York: North-Holland, 1986, 476 pp.

Pincus, S. M. Approximate entropy as a measure of system complexity. *Proc. Natl. Acad. Sci.* 88:2297–2301, 1991.

Pincus, S. M., I. M. Gladstone, and R. A. Ehrenkranz. A regularity statistic for medical data analysis. *J. Clin. Monit.* 7:335–345, 1991.

Planck, M. *Eight Lectures on Theoretical Physics*. New York: Columbia Univ., 1915, 130 pp.

Plato *The Republic. Translated, with notes and an interpretive essay, by Howard Bloom*. New York: New York, Basic Books, 1968, 487 pp.

Platt, J. R. Strong inference. *Science* 146:347–353, 1964.

Poincaré, H. *Sur Les Proprietes Des Fonctions Definies Par Les Equations Aux Differences Partielles*. Paris: Gauthier-Villars, 1879.

Poincaré, H. *Mémoire sur les courbes définies par les equations différentielles, I–IV, Oevre I.*. Paris: Gauthier-Villars, 1880.

Pontryagin, L. S. *Foundations of Combinational Topology*. Rochester, N.Y.: Graylock Press, 1952.

Pool, R. Is it chaos, or is it just noise?. *Science* 243:25–28, 1989a.

Pool, R. Is it healthy to be chaotic?. *Science* 243:604–607, 1989b.

Prata, S. *Artificial Life Playhouse: Evolution at Your Fingertips*. Corte Madera, Calif.: Waite Group Press, 1993, 179 pp.

Press, W. H., B. P. Flannery, S. A. Teukolsky, and W. T. Vetterling. *Numerical Recipes: The Art of Scientific Computing*. Cambridge: Cambridge University Press, 1986, 818 pp.

Prusinkiewicz, P., and J. Hanan. *Lecture Notes in Biomathematics: Lindenmayer Systems, Fractals, and Plants*. New York: Springer-Verlag, 1989, 120 pp.

Prusinkiewicz, P., A. Lindenmayer, J. S. Hanan, F. D. Fracchia, D. R. Fowler, M. J. M. de Boer, and L. Mercer. *The Algorithmic Beauty of Plants.* New York: Springer-Verlag, 1990.

Pulskamp, R. J. Constructing a map from a table of intercity distances. *College Math J.* 19:154–163, 1988.

Raabe, D. S., Jr., J. C. Fischer, and R. L. Brandt. Cavernous hemangioma of the right atrium: presumptive diagnosis by coronary angiography. *Cathet. Cardiovasc. Diagn.* 2:389–395, 1976.

Rapp, P. E., I. D. Zimmerman, A. M. Albano, G. C. de Guzman, and N. N. Greenbaun. Dynamics of spontaneous neural activity in the simian motor cortex: The dimension of chaotic neurons. *Phys. Lett.* 110A:335–338, 1985.

Rapp, P. E., A. M. Albano, and A. I. Mees. Calculation of correlation dimensions from experimental data: Progress and problems. In: *Dynamic patterns in complex systems,* edited by J. S. A. Kelso, A. J. Mandell and M. F. Schlessinger. Singapore: World Scientific Publishers, 1988.

Rhodin, J. A. G., and H. Fujita. Capillary growth in the mesentery of normal young rats. Intravital video and electron microscope analyses. *J. Submicrosc. Pathol.* 21:1–34, 1989.

Richardson, L. F. The problem of contiguity: an appendix to *Statistics of Deadly Quarrels. Gen. Sys.* 6:139–187, 1961.

Ripley, B. D. Modelling spatial patterns. *J. Royal Statist. Society B* 39:172–212, 1977.

Ripley, B. D. Tests of 'randomness' for spatial point patterns. *J. R. Statist. Soc. B* 41 (3):368–374, 1979.

Roger, S. A., J. H. G. M. van Beek, R. B. King, and J. B. Bassingthwaighte. Microvascular unit sizes govern fractal myocardial blood flow distributions. Personal communication, 1993.

Rogers, A. *Statistical Analysis of Spatial Dispersion.* London: Pion, 1974.

Rössler, O. E. An equation for continuous chaos. *Phys. Lett. A* 57:397, 1976.

Rubinson, K. A. The effects of n-pentane on voltage-clamped squid nerve sodium currents: A reinterpretation using kinetics of ordered. *Biophys. Chem.* 25:43–55, 1986.

Rubio, M. A., M. de la Torre, and J. C. Antoranz. Intermittencies and power-law low-frequency divergencies in a nonlinear oscillator. *Physica D* 36:92–108, 1989.

Ruelle, D. Resonances of chaotic dynamical systems. *Phys. Rev. Lett.* 56:405–407, 1986.

Ruelle, D. *Chaotic evolution and strange attractors.* New York: Cambridge University Press, 1989.

Sakmann, B., and E. Neher. *Single-Channel Recording.* New York: Plenum Press, 1983, 503 pp.

Sano, M., and Y. Sawada. Measurement of the Lyapunov spectrum from a chaotic time series. *Phys. Rev. Lett.* 55:1082–1085, 1985.

Sansom, M. S. P., F. G. Ball, C. J. Kerry, R. McGee, R. L. Ramsey, and P. N. R. Usherwood. Markov, fractal, diffusion, and related models of ion channel gating. *Biophys. J.* 56:1229–1243, 1989.

Saupe, D. Algorithms for random fractals. In: *The Science of Fractal Images*, edited by H. O. Peitgen and D. Saupe. New York: Springer-Verlag, 1988, pp. 71–136.

Schaffer, W. M., and M. Kot. Differential systems in ecology and epidemiology. In: *Chaos*, edited by A. V. Holden. Princeton: Princeton University Press, 1986, pp. 158–178.

Schepers, H. E., J. H. G. M. van Beek, and J. B. Bassingthwaighte. Comparison of four methods to estimate the fractal dimension from self-affine signals. *IEEE Eng. Med. Biol.* 11:57–64x71, 1992.

Scher, H., M. F. Shlesinger, and J. T. Bendler. Time-scale invariance in transport and relaxation. *Phys. Today* 44:26–34, 1991.

Schierwagen, A. K. Scale-invariant diffusive growth: A dissipative principle relating neuronal form to function. In: *Organizational constraints on the dynamics of evolution*, edited by J. M. Smith and G. Vida. Manchester, UK: Manchester University Press, 1990, pp. 167–189.

Schmidt, P. W. Use of scattering to determine the fractal dimension. In: *The Fractal Approach to Heterogeneous Chemistry*, edited by D. Avnir. New York: John Wiley & Sons, 1989, pp. 67–79.

Schmidt-Nielsen, K. *Scaling; Why Is Animal Size So Important?* New York: Cambridge University Press, 1984.

Schroeder, M. R. Auditory paradox based on fractal waveform. *J. Acoust. Soc. Am.* 79:186–189, 1986.

Schuster, H. G. *Deterministic Chaos.* New York: VCH Publishers, Inc., 1988.

Segal, S. S., D. N. Damon, and B. R. Duling. Propagation of vasomotor responses coordinates arteriolar resistances. *Am. J. Physiol. (Heart Circ. Physiol. 25)* 256:H832–H837, 1989.

Sernetz, M., B. Gelléri, and J. Hofmann. The organism as bioreactor. Interpretation of reduction law of metabolism in terms of heterogeneous catalysis and fractal structure. *J. Theor. Biol.* 117:209–230, 1985.

Shannon, G. E. *Bell Systems Tech. J.* 27:379, 1948.

Shaw, G., and D. Wheeler. *Statistical techniques in geographical analysis.* New York: J. Wiley & Sons, 1985.

Shinbrot, T., E. Ott, C. Grebogi, and J. A. Yorke. Using chaos to direct trajectories to targets. *Phys. Rev. Lett.* 65:3215–3218, 1990.

Shlesinger, M. Williams-Watts dielectric relaxation: a fractal time stochastic process. *J. Stat. Phys.* 36:639–648, 1984.

Shlesinger, M. F. Fractal time and $1/f$ noise in complex systems. *Ann. NY Acad. Sci.* 504:214–228, 1987.

Shlesinger, M. F., and B. J. West. Complex fractal dimension of the bronchial tree. *Phys. Rev. Lett.* 67:2106–2109, 1991.

Shrager, J., T. Hogg, and B. A. Huberman. Observation of phase transitions in spreading activation networks. *Science* 236:1092–1094, 1987.

Shu, N. H., X. Wu, N. Chung, and E. L. Ritman. Fractal analysis of heterogeneity of myocardial perfusion estimated by fast CT. *FASEB J.* 3:A690, 1989.

Sigeti, D., and W. Horsthemke. High-frequency power spectra for systems subject to noise. *Phys. Rev. A* 35:2276–2282, 1987.

Sigworth, F. J., and S. M. Sine. Data transformations for improved display and fitting of single-channel dwell time histograms. *Biophys. J.* 52:1047–1054, 1987.

Singer, J., Y. Z. Wang, and H. H. Bau. Controlling a chaotic system. *Phys. Rev. Lett.* 66:1123–1125, 1991.

Singhal, S., R. Henderson, K. Horsfield, K. Harding, and G. Cumming. Morphometry of the human pulmonary arterial tree. *Circ. Res.* 33:190–197, 1973.

Skarda, C. A., and W. J. Freeman. How brains make chaos in order to make sense out of the world. *Behav. Brain Sci.* 10:161–195, 1987.

Slaaf, D. W., S. M. Yamashiro, G. J. Tangelder, R. S. Reneman, and J. B. Bassingthwaighte. Nonlinear dynamics of vasomotion. In: *Vasomotion and Flowmotion. Prog. Appl. Microcirc.Vol. 20:*, edited by C. Allegra, M. Intaglietta and K. Messmer, 1993, pp. 67–80.

Slaaf, D. W., S. M. Yamashiro, G. J. Tangelder, R. S. Reneman, and J. B. Bassingthwaighte. Nonlinear dynamics of vasomotion. Personal communication, 1994.

Smith, A. R. *Graftal formalism notes* San Rafael, CA: Technical Memo 114, Lucasfilm Computer Division, 1984.

Smith, L. A., J. D. Fournier, and E. A. Spiegel. Lacunarity and intermittency in fluid dynamics. *Phys. Lett.* A 114:465–468, 1986.

Smith, T. G., Jr., W. B. Marks, G. D. Lange, W. H. Sheriff, Jr., and E. A. Neale. Edge detection in images using Marr-Hildreth filtering techniques. *J. Neurosci. Methods* 26:75–81, 1988.

Smith, T. G., Jr., W. B. Marks, G. D. Lange, W. H. Sheriff Jr., and E. A. Neale. A fractal analysis of cell images. *J. Neurosci. Methods* 27:173–180, 1989.

Smith, T. G., Jr., and E. A. Neale. A fractal analysis of morphological differentiation of spinal cord neurons in cell culture. In: *Fractals in Biology and Medicine*, edited by T. F. Nonnenmacher, G. A. Losa and E. R. Weibel. Basel: Birkhäuser Verlag, 1994, pp. 210–220.

Snedecor, G. W., and W. G. Cochran. *Statistical Methods*. Ames, IA: Iowa State University Press, 1980.

Sompolinsky, H. Statistical mechanics of neural networks. *Phys. Today* December:70–80, 1988.

Stanley, H. E., and N. Ostrowsky. *On Growth and Form: Fractal and Non-fractal Patterns in Physics*. Boston: M. Nijhoff, 1986, 308 pp.

Stanley, H. E., and N. Ostrowsky. *Random Fluctuations and Pattern Growth: Experiments and Models*. Boston: Kluwer Academic, 1988, 355 pp.

Stapleton, D. D., J. H. G. M. van Beek, S. A. Roger, D. G. Baskin, and J. B. Bassingthwaighte. Regional myocardial flow heterogeneity assessed with 2-iododesmethylimipramine. *Circulation* 78:II-405, 1988.

Stapleton, H. J., J. P. Allen, C. P. Flynn, D. G. Stinson, and S. R. Kurtz. Fractal forms of proteins. *Phys. Rev. Lett.* 45:1456–1459, 1980.

Starace, D. A three state fractal model of ion channel kinetics. Syracuse, NY: Syracuse University, Department of Physics, 1991.

Stark, C. P. An invasion percolation model of drainage network evolution. *Nature* 352:423–425, 1991.

Stephenson, J. L. Theory of the measurement of blood flow by the dilution of an indicator. *Bull. Math. Biophys.* 10:117–121, 1948.

Stewart, F. M., D. M. Gordon, and B. R. Levin. Fluctuation analysis: the probability distribution of the number of mutants under different conditions. *Genetics* 124:175–185, 1990.

Stockbridge, L. L., and A. S. French. Characterization of a calcium-activated potassium channel in human fibroblasts. *Can. J. Physiol. Pharmacol.* 67:1300–1307, 1989.

Strahler, A. N. Quantitative analysis of watershed geomorphology. *Trans. Am. Geophys. Union* 38:913–920, 1957.

Stroud, R. M. Topological mapping and the ionic channel in acetylcholine receptor. In: *Proteins of Excitable Membranes*, edited by B. Hille and D. M. Fambrough. New York: Wiley-Interscience, 1987, pp. 67–75.

Suwa, N., T. Niwa, H. Fukasawa, and Y. Sasaki. Estimation of intravascular blood pressure gradient by mathematical analysis of arterial casts. *Tohoku J. Exp. Med.* 79:168–198, 1963.

Suwa, N., and T. Takahashi. *Morphological and Morphometrical Analysis of Circulation in Hypertension and Ischemic Kidney.* Munich: Urban & Schwarzenberg, 1971.

Swift, J. *The Poems of Jonathan Swift.* London: Oxford, 1937.

Szilard, A. L., and R. E. Quinton. An interpretation for DOL systems by computer graphics. *Sci. Terrapin* 4:8–13, 1979.

Takens, F. Detecting strange attractors in turbulence. In: *Dynamical systems and turbulence, Warwick 1980*, edited by D. A. Rand and L. S. Young. New York: Springer-Verlag, 1981.

Tangelder, G. J., D. W. Slaaf, and R. S. Reneman. Skeletal muscle microcirculation and changes in transmural and perfusion pressure. *Prog. Appl. Microcirc.* 5:93–108, 1984.

Tangelder, G. J., D. W. Slaaf, A. M. M. Muijtjens, T. Arts, M. G. A. oude Egbrink, and R. S. Reneman. Velocity profiles of blood platelets and red blood cells flowing in arterioles of the rabbit mesentery. *Circ. Res.* 59:505–514, 1986.

Taylor, C. R., G. M. Maloiy, E. R. Weibel, V. A. Langman, J. M. Kamau, H. J. Seeherman, and N. C. Heglund. Design of the mammalian respiratory system. III. Scaling maximum aerobic capacity to body mass: wild and domestic mammals. *Respir. Physiol.* 44:25–37, 1981.

Teich, M. C., and S. M. Khanna. Pulse-number distribution for the neural spike train in the cat's auditory nerve. *J. Acoust. Soc. Am.* 77:1110–1128, 1985.

Teich, M. C. Fractal character of the auditory neural spike train. *IEEE Trans. Biomed. Eng.* 36:150–160, 1989.

Teich, M. C., R. G. Turcott, and S. B. Lowen. The fractal doubly stochastic Poisson point process as a model for the cochlear neural spike train. In: *The Mechanics and Biophysics of Hearing: Lecture Notes in Biomathematics*, edited by P. Dallos, C. D. Geisler, J. W. Matthews, M. A. Ruggero and C. R. Steele. New York: Springer Verlag, 1990, pp. 354–361.

Teich, M. C. Fractal neuronal firing patterns. In: *Single Neuron Computation*, edited by T. McKenna, J. Davis and S. F. Zornetzer. Boston: Academic Press, 1992, pp. 589–625.

Theiler, J. Estimating fractal dimension. *J. Opt. Soc. Am.* A7:1055, 1990.

Theiler, J., S. Eubank, A. Longtin, B. Galdrikian, and J. D. Farmer. Testing for nonlinearity in time series: the method of surrogate data. *Physica D* 58:77–94, 1992.

Thompson, D. A. W. *On Growth and Form, Second Edition.* Cambridge: Cambridge Univ. Press, 1942, 1116 pp.

Thompson, D. A. W. *On Growth and Form.* Cambridge: Cambridge University Press, 1961, 346 pp.

Thompson, J. M. T., and H. B. Stewart. *Nonlinear Dynamics and Chaos.* New York: Wiley, 1986.

Toffoli, T., and N. Margolus. *Cellular Automata Machines: A New Environment for Modeling.* Cambridge, MA: MIT Press, 1987.

Tolstoy, L. *What Is Art? and Essays on Art.* London: Oxford University Press, 1930, 331 pp.

Trefán, G., P. Grigolini, and B. J. West. Deterministic Brownian motion. *Phys. Rev. A* 45:1249–1252, 1992.

Troy, W. C. Mathematical modeling of excitable media in neurobiology and chemistry. In: *Theoretical Chemistry, Volume 4,* edited by H. Eyring and D. Henderson. New York: Academic Press, 1978, pp. 133–157.

Turing, A. M. The chemical basis of morphogenesis. *Philos. Trans. R. Soc. London* B237:37–72, 1952.

Tyler, B., T. Wegner, M. Peterson, and P. Branderhorst. *Fractint* wegner@mwunix.mitre.org, 1991.

Upsher, M. E., and H. R. Weiss. Relationship between beta-adrenoceptors and coronary blood flow heterogeneity. *Life Sci.* 44:1173–1184, 1989.

Utley, J., E. L. Carlson, J. I. E. Hoffman, H. M. Martinez, and G. D. Buckberg. Total and regional myocardial blood flow measurements with 25 μ, 15 μ, 9 μ, and filtered 1-10 μ diameter microspheres and antipyrine in dogs and sheep. *Circ. Res.* 34:391–405, 1974.

van Bavel, E. Metabolic and myogenic control of blood flow studied on isolated small arteries. University of Amsterdam, The Netherlands: Ph.D. Thesis, Physiology, 1989.

van Beek, J. H. G. M., S. A. Roger, and J. B. Bassingthwaighte. Regional myocardial flow heterogeneity explained with fractal networks. *Am. J. Physiol.* 257(Heart.Circ.Physiol.26):H1670–H1680, 1989.

van Damme, H. Flow and interfacial instabilities in Newtonian and colloidal fluids. In: *The Fractal Approach to Heterogeneous Chemistry,* edited by D. Avnir. New York: John Wiley & Sons Ltd., 1989, pp. 199–225.

van den Berg, R. J., J. de Goede, and A. A. Verveen. Conductance functions in Ranvier nodes. *Pflügers Arch. Em. J. Physiol.* 360:657–658, 1975.

van der Pol, B., and J. van der Mark. De l'homme et des animaux. *Extr. Arch. Neerl. Physiol.* 14:418, 1929.

van Roy, P., L. Garcia, and B. Wahl. "Designer Fractal. Mathematics for the 21st Century." Santa Cruz, California: Dynamic Software, 1988.

Vicsek, T. *Fractal Growth Phenomena.* Singapore: World Scientific, 1989, 355 pp.

von Koch, H. Une méthode géométrique élémentaire pour l'étude de certaines questions de la théorie des courbes planes. *Acta Mathematica* 30:145–174, 1905.

Voss, R. F. Fractals in nature: From characterization to simulation. In: *The Science of Fractal Images*, edited by H. O. Peitgen and D. Saupe. New York: Springer-Verlag, 1988, pp. 21–70.

Voss, R. F. Evolution of long-range fractal correlations and 1/f noise in DNA base sequences. *Phys. Rev. Lett.* 68:3805–3808, 1992.

Waddington, J. L., M. J. MacCulloch, and J. E. Sanbrooks. Resting heart rate variability in men declines with age. *Experientia* 35:1197–1198, 1979.

Waner, S., and H. M. Hastings. History dependent stochastic automata: a formal view of evolutionary learning. *Int. J. Intelligent Sys.* 3:19–34, 1988.

Wang, C. Y., and J. B. Bassingthwaighte. Area-filling distributive network model. *Mathl. Comput. Modelling* 13:27–33, 1990.

Wang, C. Y., L. J. Weissman, and J. B. Bassingthwaighte. Bifurcating distributive system using Monte Carlo method. *Mathl. Comput. Modelling* 16:91–98, 1992.

Wang, C. Y., P. L. Liu, and J. B. Bassingthwaighte. A lattice-independent growth model. Personal communication, 1994.

Watt, R. C., and S. R. Hameroff. Phase space analysis of human EEG during general anesthesia. In: *Perspectives in Biological Dynamics and Theoretical Medicine*, edited by S. H. Koslow, A. J. Mandell and M. F. Shlesinger. New York, NY: N. Y. Acad. Sci., 1987, pp. 286–288.

Weibel, E. R., and D. M. Gomez. Architecture of the human lung. *Science* 137:577–585, 1962.

Weibel, E. R. *Morphometry of the Human Lung*. New York: Academic Press, 1963.

Weitz, D. A., and J. S. Huang. Self-similar structures and the kinetics of aggregation of gold colloids. In: *Aggregation Gelation*, edited by F. Family and D. P. Landau. Amsterdam: North-Holland, 1984, pp. 19–28.

Weitz, D. A., M. Y. Lin, J. S. Huang, T. A. Witten, S. K. Sinha, J. S. Gertner, and C. Ball. Scaling in colloid aggregation. In: *Scaling Phenomena in Disordered Systems*, edited by R. Pynn and A. Skjeltorp. New York: Plenum Press, 1985, pp. 171–188.

Welch, G. R. *The Fluctuating Enzyme*. New York: John Wiley & Sons, 1986.

West, B. J. *An Essay on the Importance of Being Nonlinear*. New York: Springer-Verlag, 1985, 204 pp.

West, B. J., V. Bhargava, and A. L. Goldberger. Beyond the principle of similitude: renormalization in the bronchial tree. *J. Appl. Physiol.* 60:1089–1097, 1986.

West, B. J., and A. L. Goldberger. Physiology in fractal dimensions. *Am. Scient.* 75:354–365, 1987.

West, B. J., and J. Salk. Complexity, organization and uncertainty. *Eur. J. Op. Res.* 30:117, 1987.

West, B. J., and M. F. Shlesinger. On the ubiquity of 1/f noise. *Int. J. Mod. Phys. B* 3:795–819, 1989.

West, B. J., R. Kopelman, and K. Lindenberg. Pattern formation in diffusion-limited reactions. *J. Stat. Phys.* 54:1429–1439, 1989.

West, B. J., and M. Shlesinger. The noise in natural phenomena. *Am. Sci.* 78:40–45, 1990.

West, B. J. *Fractal Physiology and Chaos in Medicine*. Singapore: World Scientific, 1990a, 278 pp.

West, B. J. The Disproportionate Response. In: *Mathematics and Science*, edited by R. E. Mickens. Singapore: World Scientific, 1990b.

Wiest, G., H. Gharehbaghi, K. Amann, T. Simon, T. Mattfeldt, and G. Mall. Physiological growth of arteries in the rat heart parallels the growth of capillaries, but not of myocytes. *J. Mol. Cell Cardiol.* 24:1423–1431, 1992.

Willis, J. C. *Age and Area; A Study in Geographical Distribution and Origin of Species.* Cambridge: Cambridge University Press, 1922, 259 pp.

Winfree, A. T. Spiral waves of chemical activity. *Science* 175:634–636, 1972.

Winfree, A. T. Wavelike activity in biological and chemical media. In: *Lecture Notes in Biomathematics 2*, edited by P. van der Driesche, 1974, pp. 243–260.

Winfree, A. T. *The Timing of Biological Clocks.* New York: Scientific American Library, 1987, 199 pp.

Winfree, A. T. Vortex action potentials in normal ventricular muscle. *Ann. NY Acad. Sci.* 591:190–207, 1990a.

Winfree, A. T. The electrical thresholds of ventricular myocardium. *J. Cardiovasc. Electrophysiology* 1:393–410, 1990b.

Winfree, A. T. *The Geometry of Biological Time.* Berlin: Springer-Verlag, 1990c.

Winfree, A. T. Rotors in normal ventricular myocardium. *Proc. Kon. Ned. Akad. v. Wetensch.* 94:257–280, 1991a.

Winfree, A. T. Varieties of spiral wave behavior: An experimentalist's approach to the theory of excitable media. *Chaos* 1:303–334, 1991b.

Wise, M. E. Spike interval distributions for neurons and random walks with drift to a fluctuating threshold. In: *Statistical Distributions in Scientific Work, Volume 6*, edited by C. E. A. Taillie. Boston: D. Reidel, 1981, pp. 211–231.

Wise, M. E., and G. J. J. M. Borsboom. Two exceptional sets of physiological clearance curves and their mathematical form: test case?. *Bull. Math. Biol.* 51:579–596, 1989.

Witten, T. A., and Sander L.M.. Diffusion-limited aggregation. *Phys. Rev. B* 27:5686–5697, 1983.

Woldenberg, M. J. A structural taxonomy of spatial hierarchies. In: *Regional Forecasting*, edited by M. Chisholm, A. E. Frey, and P. Haggett. London: Butterworths, 1970, pp. 147–175.

Woldenberg, M. J., and K. Horsfield. Relation of branching angles to optimality for four cost principles. *J. Theor. Biol.* 122:187–204, 1986.

Wolf, A., J. B. Swift, H. L. Swinney, and J. A. Vastano. Determining Lyapunov exponents from a time series. *Physica* 16D:285–317, 1985.

Wolfram, S. Preface to Cellular Automata: Proceedings of an Interdisciplinary Workshop, Los Alamos, NM, March 7–11, 1983. *Physica D* 10:vii–xii, 1984.

Xu, N., and J. Xu. The fractal dimension of EEG as a physical measure of conscious human brain activities. *Bull. Math. Biol.* 50:559–565, 1988.

Yamashiro, S. M., D. W. Slaaf, R. S. Reneman, G. J. Tangelder, and J. B. Bassingthwaighte. Fractal analysis of vasomotion. In: *Mathematical Approaches to Cardiac Arrhythmias. Ann. N. Y. Acad. Sci. Vol.591*, edited by J. Jalife, 1990, pp. 410–416.

Yipintsoi, T., W. A. Dobbs, Jr., P. D. Scanlon, T. J. Knopp, and J. B. Bassingthwaighte. Regional distribution of diffusible tracers and carbonized microspheres in the left ventricle of isolated dog hearts. *Circ. Res.* 33:573–587, 1973.

Andrei Andreevich Markov Youschkevitch. In: *Dictionary of Scientific Biography, Volume 4*, edited by C. C. Gillispie. New York: Charles Scribner's Sons, 1974, pp. 124–130.

Zaikin, A. N., and A. M. Zhabotinsky. Concentraion wave propagation in two-dimensional liquid-phase self-organization systems. *Nature* 225:535–537, 1970.

Zamir, M., and H. Chee. Segment analysis of human coronary arteries. *Blood Vessels* 24:76–84, 1987.

Zierler, K. L. Equations for measuring blood flow by external monitoring of radioisotopes. *Circ. Res.* 16:309–321, 1965.

Zipf, G. K. *Human Behavior and the Principle of Least Effort; An Introduction to Human Ecology.* Cambridge, Mass.: Addison-Wesley, 1949, 573 pp.

Index

Of the making of many books there is no end.

Eccl. 12:12

355